SOLAR ENERGY: NEW RESEARCH

SOLAR ENERGY: NEW RESEARCH

TOM P. HOUGH
EDITOR

Nova Science Publishers, Inc.
New York

For permission to use material from this book please contact us:
Telephone 631-231-7269; Fax 631-231-8175
Web Site: http://www.novapublishers.com

LIBRARY OF CONGRESS CATALOGING-IN-PUBLICATION DATA

Solar energy : new research / Tom P. Hough (editor).
p. cm.
Includes index.
ISBN 1-59454-630-4
1. Solar energy. I. Hough, Tom P.
TJ810.S4895 2006
621.47--dc22 2005031863

Published by Nova Science Publishers, Inc. ✢ New York

CONTENTS

PREFACE

Solar energy is derived ultimately from the sun. It can be divided into direct and indirect categories. Most energy sources on Earth are forms of indirect solar energy, although we usually don't think of them in that way. Coal, oil and natural gas derive from ancient biological material which took its energy from the sun (via plant photosynthesis) millions of years ago. All the energy in wood and foodstuffs also comes from the sun. Movement of the wind (which causes waves at sea), and the evaporation of water to form rainfall which accumulates in rivers and lakes, are also powered by the sun. Therefore, hydroelectric power and wind and wave power are forms of indirect solar energy. Direct solar energy is what we usually mean when we speak of solar power – it is the use of sunlight for heating or generating electricity. Solar energy research and applications have been receiving increasing attention throughout the world as solar energy must play a much greater role in the energy mix in upcoming years. This new book examines new research in this frontier field.

The low-temperature air conditioning technology employing the latent heat energy storage system provides an attractive alternative to air conditioners and process cooling, with the possibility of deloading peak energy consumption and making the environment pollution free. The thermal performance of the binary mixture of decanoic acid (capric acid, 65 mol%) and dodecanoic acid (lauric acid, 35 mol%) mixture, herein referred to as the C-L acid, was closely examined for thermal energy storage for cooling application in chapter one.

In chapter two, we present a review of our results on thermal energy storage in passive solar buildings. Thermophysical properties of a large number of phase change materials were studied, particularly those undergoing solid/liquid transition near ambient temperature. The materials studied can be applied in two typical, so called "intelligent" solar objects: a) in transparent walls of agricultural greenhouses and b) incorporated with building materials of wallboards in residential buildings.

On an international market where industries of traditional production are successfully developing, our purpose in chapter three is to present a solar flat-plate thermal collector integrating a production technology which is different from the industrials processes. The technological benefit from this new sensor resides in dividing by four the number of components, by means of the reduction of assembling steps, the thermoformed fabrication of two pieces of the totally symmetric absorber (key component, patent protected). There is no more box. For equivalent performances, the weight and the cost are strongly reduced. Thus this is not an incremental improvement of a product, but a new technology which allows industrial production. Industrials from the solar sector are not able to lower their production costs since they have reached a limit of productivity gain for their technologies.

Heating and cooling of buildings require a large amount of energy. For countries with good solar potential, solar energy can be considered a viable option for these applications. Traditionally solar energy is used for hot water production. With the increase in the price of fuels solar space heating became a viable option, however, the potential problem of such a system is that a lot of solar energy collectors need to be disconnected or shaded during summertime. Solar cooling offers possibilities of using the same solar collectors for space cooling as well through an absorption chiller. To be viable solar space heating systems need to utilize a low temperature supply system such as an air system. Absorption units on the other hand require temperature of the order of 90°C but the system during summertime is more effective and the available solar radiation is much more than in the wintertime. Additionally any effort to use solar energy for space heating and cooling must be done after the building is insulated well in order to reduce the thermal losses. A system suitable for a typical 196 m^2 house is modelled and simulated with TRNSYS program for two locations at different latitudes; Nicosia, Cyprus (35°) and Madison, Wisconsin (43°). These two locations offer two different climates; i.e. mild winters and severe summers for Nicosia and severe winters and mild summers for Madison. Three different types of solar collectors are considered; a flat plate, advance flat plate and evacuated tube collectors. The economic analysis performed in chapter four showed that both systems are viable as positive life cycle savings are obtained. Therefore it can be concluded that solar energy should be used whenever possible.

A "dynamic" solar power plant (which consists of a solar collector - thermal engine combination) is proposed as an alternative for the more usual photovoltaic cells. Upper bounds for the efficiency of solar thermal power plants operating in the Martian environment are first evaluated. A general thermodynamic approach, first presented here in chapter five, clearly shows which of the three theories usually quoted in literature gives the exergy of thermal radiation. Recent works reporting accurate upper bounds for the efficiency of thermal radiation energy conversion into work are subsequently used in this chapter. The results refer to thermal engines powered by direct or diffuse solar radiation on Mars.

Thermosyphons are active heat transfer devices that utilize natural convection in a closed or open loop for effective heat transfer. The thermosyphon idea has been commonly used in domestic solar water heaters as well as other solar energy systems and heat transfer processes. Based on the working principle of thermosyphon, several thermodiode designs which allow heat to be transferred in the desired direction but block the heat flow in the opposite direction were developed and tested for solar heating applications. More recently thermodiodes with easily alterable direction of heat flow, so-called bi-directional thermodiodes, were invented and their performances tested and optimized for both summer-cooling and winter-heating of buildings and shelters. Presented in chapter six are a review of single-phase thermosyphons and thermodiodes applied to solar energy utilization and energy-efficient buildings, and the authors' and co-workers' research results in this area.

The authors' work includes heat transfer analyses and optimizations of natural convection loops of different configurations and operating conditions, different designs of bi-directional thermodiodes, and lab test results as well as year-round test data collected from experiments on energy–efficient construction modules under actual weather conditions in the United

States and Korea. The performance and efficiencies of different thermodiode designs are compared and discussed.

Solar boosted heat pumps (SBHPs) can capture energy from low temperature ambient air and solar energy, then upgrading to a higher temperature level for useful applications. For water heating the highest delivered water temperature can be much higher than that provided by regular thermosyphon solar water heaters. Solar boosted heat pump water heaters do not require auxiliary back up heating, they could provide hot water even when the solar energy is not available. There is negligible collector heat loss compared to regular solar collectors since the collector temperature is lower, or about the same as the surrounding ambient air temperature. The collector of a SBHP water heater does not require a cover glazing; this makes the system cost competitive with conventional solar water heaters. A brief history of the development of SBHPs and fundamental principles are presented in chapter seven. The current and potential applications, and recent research directions are also discussed in this chapter.

One major source of energy consumption at homes and industry is water heating. Solar water heaters are important because they provide cheapest and clean energy to homeowners especially when cost of energy from fossil fuels is on the rise. In chapter eight we study types and components of solar water heating systems especially those based on thermosyphonic flow. Performance of a solar water heater depends on its orientation toward the equator with respect to the local latitude. Optimization of tilt angles and orientations can significantly improve solar water heater performance. We show how optimum collector slope with the horizon can be obtained based on a set of experimental data obtained using this type of collector in the Middle East.

Design and operational aspects of a thermosyphonic solar water heater made of polypropylene tubes is discussed. Also the rate of heat transfer to the water is measured and correlated through Nusselt, Reynolds and Prandtl numbers. It is also shown how one can estimate amount of increase in water temperature from the collector specifications, water flow rate and sun radiation. Finally a recently proposed method for calculation of thermal efficiency of thermosyphonic solar water heaters is presented.

To electrify remote areas, the use of solar energy is the best economical and technological solution. The choice of the sites for the installation of photovoltaic systems and analysis of their performance requires the knowledge of the solar irradiation source. The evolution of the solar irradiation obeys to seasonal variations and random fluctuations. These latter considerably affects the behavior of the PV systems, therefore it is necessary to quantify them. The concept of fractals may then be used for modeling the solar irradiances fluctuations. To measure the fractal dimension of time series, several methods and algorithms based on various coverings has been elaborated in chapter nine. In order to improve the complexity and the precision of the fractal dimension estimation of the discreet time series, the authors developed a simple method based on a multi-scale covering using the rectangle as a structuring element of covering. An optimization technique has been associated to this method in order to determine the optimal time interval $\Delta\tau_{max}$ through which the line log-log is fitted whose slope represents the fractal dimension. This optimization technique is permitted to improve the precision and the time computing of the method.

A PV systems is by nature a non linear power sources which delivers the maximum power output with the highest efficiency only for a particular load current and voltage termed the maximum power point (MPP). This point depends on the PV array's current-voltage (I-V) characteristics that in its turn varies instant to instant mainly due to random changes in insolation (incoming solar radiation). Moreover, the load connected to the PV array also changes as per consumers' needs and does not correspond to the MPP. It appears from a review of the literature that mainly different variants of artificial neural network (ANN) have been applied to find the MPP of a PV array. However, the methods suffer from one or more of the drawbacks such as requiring sophistication in the choice of various parameters which influence the success of ANN, use of difficult-to-measure input variables, comparison of the output with those obtained through conventional modeling based calculations but not with the measured values, and prediction of MPP on hourly basis whereas solar insolation changes every few minutes. In chapter ten, an algorithm based on autoregressive (AR) third order time series has been proposed to forecast a photovoltaic (PV) array's short circuit current and open circuit voltage at a given instant from directly and easily measurable load currents and voltages in shortly spaced three preceding intervals of time. The maximum power point (MPP) of the PV array's current-voltage characteristic curve corresponding to that instant can then be obtained from the forecasted short circuit current and open circuit voltage. This chapter has emphasized on laying the theoretical foundation, achieving preliminary results, their comparison with measured values and identifying further investigations needed for the proposed real-time application of time series which has so far been used mainly for forecasting long term events. The methodology presented in this chapter does not lack in generality and can easily be extended to include refinements. It is applicable for a PV array installed at any geographical location either in stand-alone or in grid connected mode.

The surface photovoltage method is described in chapter eleven and its use for evaluation of some parameters of semiconductor materials and diodes is shown. Theory of the photovoltaic effect is presented first in a commonly used form suitable for thick samples. At present, advanced technology produces thin films and thin wafers, which is the reason for our generalization of the theory for arbitrary thick samples. Relations for the photocurrent and the open - circuit voltage are derived including the effect of reflections from the sample surfaces, surface recombination and thickness of the samples. The surface photovoltage technique is modified for contactless measurements as well as for measurements with an auxiliary contact. The advantage of the latter method of measurement is that only a small number of parameters influence the photovoltage spectra. This allows reliable extraction of parameters by comparison of the experimental data with the theory. The main purpose is evaluation of diffusion length of minority carriers and surface recombination velocity in materials designated for solar cells and in the bulk of the absorber in the photocells.

In: Solar Energy: New Research
Editor: Tom P. Hough, pp.1-42
ISBN 1-59454-630-4

Chapter 1

THERMAL PERFORMANCE CHARACTERIZATION OF CAPRIC ACID AND LAURIC ACID MIXTURE FOR LOW TEMPERATURE ENERGY STORAGE

Takayuki Watanabe[1]* and Maria Natalia R. Dimaano[2]
[1] Department of Environmental Chemistry and Engineering,
Tokyo Institute of Technology, Nagatsuta, Yokohama, Japan
[2] Research Center for the Natural Sciences, University of Santo Tomas,
Manila, Philippines

ABSTRACT

The low-temperature air conditioning technology employing the latent heat energy storage system provides an attractive alternative to air conditioners and process cooling, with the possibility of deloading peak energy consumption and making the environment pollution free. The thermal performance of the binary mixture of decanoic acid (capric acid, 65 mol%) and dodecanoic acid (lauric acid, 35 mol%) mixture, herein referred to as the C-L acid, was closely examined for thermal energy storage for cooling application. Both fatty acids are basically derived from coconut oils that are abundantly available in tropical countries.

Based on DSC analysis, C-L acid exhibits a heat of fusion value of 140.8 $kJ\cdot kg^{-1}$ and a melting point of 18.0°C. Corroborative results were obtained when the thermal energy storage performance and heat transfer behavior were experimentally elucidated from the radial and axial temperature distribution during its melting and solidification. The heat transfer characteristics of the C-L acid were determined. All experimental results were in good agreement with the numerically simulated results. Subsequent performance investigations were carried out on C-L acid with chosen organic additives to lower the melting point while retaining its substantial energy capacity without impairment of its requisite qualities.

* watanabe@chemenv.titech.ac.jp

C-L acid has sound benefits particularly on the melting congruency, no degradation, chemical stability, its behavior in relation to melting and freezing reproducibility, and reasonable energy density. Although the temperature depression of 8 to 11°C with respect to the C-L acid had not been attained in all PCM mixtures that were analyzed, C-L acid with methyl salicylate provided the most effective additive. This PCM mixture is specifically good enough for industrial applications that will focus on waste heat recovery, process cooling, impregnated construction materials and other environmental safety related concerns.

INTRODUCTION

The global anxiety on economic growth, environmental quality, human health and welfare prompted the conscientious innovations in science and technology. Foremost of these concerns is the access to basic services in response to better living conditions. Thermal energy storage presents a significant role in the energy economy without sacrificing the comfort levels. It has continued to gain considerable recognition as the most economical alternative to take advantage of lower off-peak electricity rates. Its capability to load heat energy in a short duration with a small cooling equipment and served over a long period of time has trimmed down both capital and operating costs.

Background of Latent Heat Storage

Several storage systems are today at the stage of industrial prototypes. In the process of determining the most favorable system, many researchers have opted to consider phase change materials for thermal energy storage. Investigations had been made on the different phase transitions to uncover the most practical system. Solid-gas and liquid-gas transformations provide high latent heats of fusion. But since gases normally occupy large volumes that required large containment facilities and complex structures, these kinds of phase change materials (PCM) in general, are not in use. Solid–solid transformation also offers high latent heats of fusion with corresponding high melting temperatures. However, this type of PCM has limited materials. It is noteworthy to consider the disadvantages that the previous investigators have presented such as the complexity, use of advanced materials, toxicity or corrosion of the involved components and very efficient heat exchanger requirements. Solid–liquid phase transformation involves comparatively smaller latent heats but the small change in its volume has become the subject of increased attention and use over the past decades. Early investigations by Telkes [1-3] proposed the use of solid to liquid latent heat for energy storage, rather than the sensible heat of water or rocks. The latent heat storage media densities may be considered high enough so that much smaller volumes and weights can be used for absorbing energy than are required for the same energy storage based on sensible heat. The stored energy is transferred at a fairly constant melting or solidification temperature matched to that required by the intended application.

Overview of Latent Heat Energy Storage

Thermal energy storage in the solid-to-liquid phase change (latent heat of fusion) had wide acceptance for some applications, mainly because of its higher energy density and its consequent smaller energy storage volume. Recent considerations are focused on latent heat over sensible heat storage system due to the following general advantages: (1) It represents the much needed energy obtained from phase change materials which have higher energy capacity; (2) The temperature of the phase change material remains more or less constant as it undergoes phase transformation; (3) The compactness of the storage unit allows for a greater flexibility in choosing a location for the storage unit; and (4) There is much less insulation required, thereby maintaining reasonable heat losses. Compactness allows the installation of PCM containers in locations where sensible heat storage containers are difficult and expensive to install. These advantages make the system very attractive from both technical and economical point of view. A number of studies on the theory and use of PCMs have been carried out [4-15].

The thermal energy extracted from the storage system is usually less than it previously stored, as a result of losses and irreversibilities inherent in the storage process. Thus, a very important feature that must be considered when choosing the thermal storage system is the storage temperature that depends on its application. The temperature range of its application dictates the choice of the PCM.

In a tropical and humid environment, building establishments need central air conditioning units, not to mention the refrigeration facilities needed for food and even materials preservation. Cold energy storage gained considerable recognition as a most economical alternative to take advantage of lower off-peak electricity rates, when cooling demands significantly contribute to high demand charges. For cold energy storage, the melting temperature ranges from –10 to 27°C. Studies on the cold energy storage system for air conditioning a residential or industrial space unit have indicated a pleasant environment using PCMs with considerable heat of fusion in the temperature range of 7 to 10°C specifically in a location having an ambient temperature range of 30 to 37°C. System modeling and field tests on buildings have revealed that a 7°C of cold air system obtained from PCMs provides good overall benefits as compared to a PCM having 13°C melting temperature [16]. In humid climates, the optimal supply of a more comfortable air temperature may be obtained from PCMs with melting range as low as 4°C. This system allows for smaller ductwork and air-handling equipment equating to increased space benefits, indoor air quality, and better comfort. However, the system needed to achieve 4°C is more expensive than that for a 7°C system.

Latent Heat Storage Materials

Latent heat storage media or PCMs provide high energy densities. Such energies are delivered over a very narrow range of temperature. PCMs for thermal energy storage in solar heating and cooling systems have received considerable attention due to the reduction in storage volume. **Figure 1** shows some PCMs collated for cold energy storage.

PCMs are categorized as inorganic salt hydrates, clathrates, organic materials, and eutectic mixtures. Salt hydrates are inorganic compounds of salt and water. Organic materials are classified as paraffins and fatty acids. Paraffins are mineral oil products consisting mostly of alkanes. Fatty acids are naturally occurring fats and oils. Eutectic mixtures are solutions having inorganic and organic components in such proportions that the melting point of the mixtures is the lowest with their components.

There has been a large volume of research relating to salt hydrates. Researches had been stimulated to a large degree by interest in evaluating energy transfer in buildings, solar collectors, and fluid-filled thermal storage tanks using salt hydrates the melting temperatures of which range from 0 to 150°C. Water was included for comparison. Salt hydrates' high heats of fusion are largely due to their high water content. Some have volumetric heats of fusion similar to ice, are non-toxic, non-flammable, and inexpensive. Most studied among them were sodium sulfate decahydrate (Glauber's salt, $Na_2SO_4{\cdot}10H_2O$), sodium thiosulfate pentahydrate ($Na_2S_2O_4{\cdot}5H_2O$), and calcium chloride hexahydrate ($CaCl_2{\cdot}6H_2O$) that have intermediate operating temperatures of 32.4°C, 56.0°C, and 29.7°C, respectively [1,17]. Comprehensive works had been done in both heating and cooling systems. Unfortunately, although some salt hydrates are being used for industrial applications, extensive studies revealed some problems connected with their tendency to form metastable compounds, to supercool, and to melt incongruently to a certain extent leading to erratic storage performance. The main difficulty is the incongruency in melting forming a multiphase heterogeneous system. The formation of liquid and solid phases tends to segregate with the latter settling at the bottom. Thickening agents are used to overcome this difficulty and this solution has been partially successful. The mechanical means of stirring, vibration, rolling cylinders, etc. are employed to prevent phase separation.

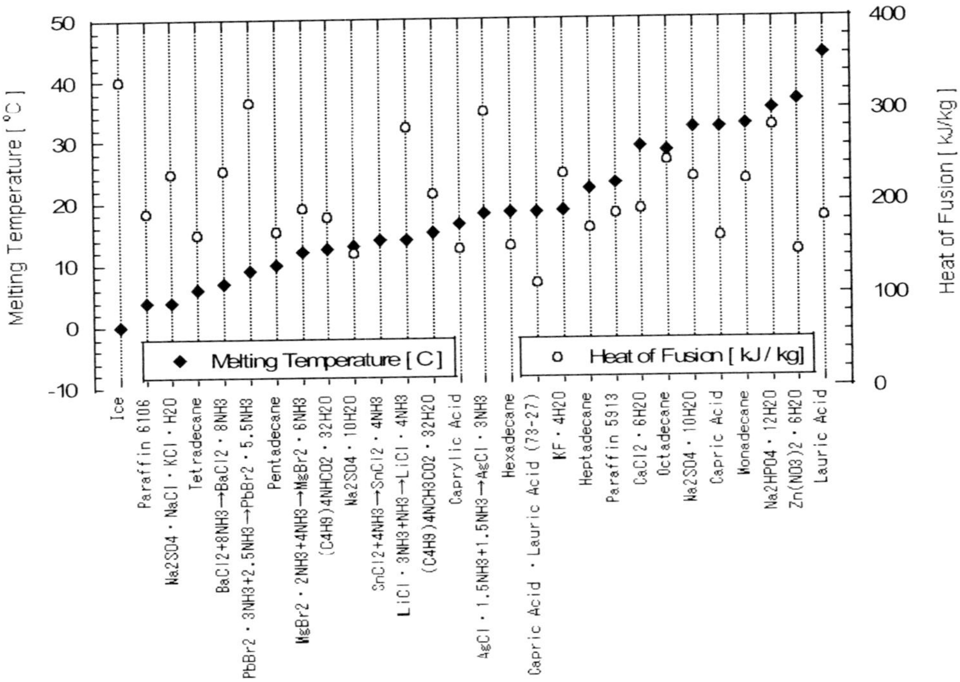

Figure 1. Latent heat storage materials for low temperature energy storage application.

Eutectics of salt hydrates were then formulated to meet the requirements of the intended applications. The unwanted phase separation due to the concentration gradients induced by the thermogravitational process nonetheless still remains as a problem connected with their tendency to form metastable compounds. Such problem showed a considerable reduction in the storage heat capacities of most salt hydrates after a certain number of cycles thus making them unfit for long term use as storage materials [18].

Organic Phase Change Materials Applied to Latent Heat Energy Storage

Realizing the strong advantages that prevail over salt hydrates, attention has been turned toward organic PCMs. A considerable amount of work has been accomplished in the last 60 years with respect to the use of organic PCMs with varied melting temperatures suitable for their projected use. In spite of their lower volumetric energy than salt hydrates, organic compounds such as *n*-alkanes, alcohols, glycols, and the esters of long chain carboxylic acids, or fatty acids are considered promising PCM candidates for energy conservation in buildings. This is because they can be impregnated or directly incorporated into conventional construction materials at some stage of their manufacture [19-36]. It has been demonstrated by Himran *et al.* [37] that this family of organic compounds has a large spectrum of latent heats, melting points, densities, and specific heats so that the heat storage designer has a good choice of storage application.

Paraffins

Paraffins are self-nucleating and exhibit no or only minor supercooling. The temperature of paraffins ranging from 6 to 80°C appears particularly well for applications to energy conservation in building constructions, having generally higher heats of fusion than salt hydrates.

Since the heat of fusion is primarily the indicator of the heat storage capacity of a storage system, the heat of fusion of paraffins increases with the increasing number of carbon chain lengths. On this respect, just from the start, great care has to be taken on exploiting the possibility of utilization of the actual paraffin one wants to use. Paraffins having an even number of carbon atoms are preferred over odd numbers, since they are cheaper, more abundant, and more stable. Another thing, paraffins have low vapor pressure thus ensuring long-term reproducibility due to its non-flammability characteristic over salt hydrates, do not supercool and solidify rapidly. Unlike salt hydrates which are corrosive in the presence of water, paraffins are stable, unreactive and non-corrosive. However, their exposure to air at elevated temperature and prolonged periods causes their slow oxidation. Although paraffins are non-corrosive, it is imperative to choose the proper containment or encapsulating material. Paraffins can permeate in and soften some plastics [15].

Paraffins have a density generally ranging from 750 to 800 kg·m^{-3}. Hence, their high mass thermal capacity is contrasted by their lower volume thermal capacity which consequently leads to poor economical usefulness. The current prices of paraffins are expensive. Due to cost considerations, only technical grade paraffins may be used as PCMs in latent heat stores. But considering the stabilizing additives and processes required of salt hydrates to deliver good thermal storage performance, the cost difference between the two types of latent heat storage materials were somewhat reduced. The difficulty on low thermal conductivity in the solid phase cannot be easily solved and many research efforts are now devoted to it.

The cheapest among the paraffins studied was Paraffin SunTech P-116 manufactured by Sun Oil Company [5]. It is basically composed of hydrocarbons of alkanes C-20 until C-32 in different weight proportions. Ghoneim [38] developed a model of phase change energy storage using this paraffin wax and found out that the storage volume required for an effective system performance is the same as required by a conventional water tank.

Many organic polymers with high heats of fusion melt at temperatures in the range of 110 to 130°C, which is suitable for hot-side storage in solar absorption or Rankine cooling systems. When high-density polyethylene is partially cross-linked, it retains almost all of its original heat of fusion, but passes through the phase transition without changing its form. Form-stable pellets of this and similar materials seem to undergo melting without flowing or agglomerating and can be used in packed beds with direct heat transfer to a gas or liquid. Inaba and Tu [27] reported on the thermophysical properties and measuring methods of shape-stabilized paraffin as a new type of latent heat storage material, which retains the same solid state shape of paraffin as it melts. This was considered essential since this type of paraffin can be used in a latent heat storage system without encapsulation.

Evaluation of results from differential scanning calorimetry (DSC) measurements made by Abhat [5] stressed that only the straight chained paraffins may be selected for low-temperature application since the *n*-paraffins experience the primary phase transition from solid to liquid in a narrow temperature range of about 2°C. Chemically pure paraffins have high latent heat, in the order of about 200 kJ·kg^{-1}, and they melt at a uniform temperature. Tetradecane and pentadecane with the corresponding melting temperatures of 5.9 and 9.9°C fall within the desired range for cooling application. However, these paraffins are more expensive and are potential fire hazards and certainly corrosive. Only technical grade paraffins, like Paraffin 5913 that is essentially a paraffin mixture and is not completely refined of oil, are suggested for thermal storage considering its least price in the group. Paraffin 5913 has a melting temperature of 22 to 24°C, not feasible for cold energy storage in tropical and humid countries.

Fatty Acids

Alternatively, fatty acids are chemically, heat and color stable, non-toxic, with reproducible melting and freezing characteristics, and have low corrosion activity just like the paraffins. They have latent heats in the same order as salt hydrates. However, some of these PCMs have an odor in their raw state. Even so, this can be neutralized by the utilization of newly developed deodorizing products, such as the family of synthetic molecular sieves that have a strong affinity for organic odor molecules [12].

Fatty acids are considered excellent heat storage substances but are somewhat too expensive for large-scale solar applications. However, they are much cheaper than research grade paraffins and are recommended for the function tests of latent heat storage materials [5, 35]. More significantly, the raw materials of fatty esters are derived from renewable vegetable and animal sources that multiply with the increasing world food production. It then guarantees a life-long environmentally friendly source that is independent of the deteriorating energy supply.

Feldman, *et al.* did extensive studies on the melting and freezing behavior of fatty acids such as stearic acid, capric acid, palmitic acid, lauric acid, oleic acid, myristic acid, pentadecanoic acid, and margaric acid for incorporation in building construction materials [12,23-25,31,35]. The performances of gypsum wallboard and concrete block that have been impregnated with the mixture of fatty acids were examined [26]. Manufacturing techniques, applications of PCM wallboard and flammability tests on wallboards and PCM concrete blocks were conducted and further supported on by Neeper [30]. Fatty acids also have received considerable attention lately as potential PCMs for diurnal heat storage in direct gain passive solar houses.

Eutectic Mixtures of Paraffins and Fatty Acids

Studies have also been made on the eutectic mixtures of organic to get the preferred melting points of suitable PCMs. Feldman, *et al.* [23-25] did wide ranged investigation on the thermophysical properties mixtures of palmitic acid, stearic acid, lauric acid, capric acid, and myristic acid in different proportions. Others like Tan and Leong [39], Py [40] and Hirasawa [41] proceeded in evaluating thermal characteristics of paraffin waxes [22]. Additional studies [42-45] dealt with storage systems of multiple paraffin waxes to enhance the heat transfer characteristics for efficient thermal storage.

Based on the studies conducted, fatty acids meet the thermodynamic and kinetic criteria for low-temperature latent heat storage. They possess the desired reproducible melting and freezing behavior, chemical stability, and physical properties. However, from among the many fatty acids examined, only caprylic acid has a melting point (16.5°C) nearest to the desired temperature range for cooling application of a space unit. For room cooling, eutectic mixtures of fatty acids had been considered as a versatile cold energy medium that can be applied to a wide variety of building conditions. Nevertheless, combinations of fatty acids with different melting points can form eutectic points that may possibly conform to the desired melting range.

A particularly interesting phase change material that is being considered for cold energy storage in a tropical country is the mixture of capric acid and lauric (C-L) acid [4,19]. These fatty acids can be derived from coconut oil [46]. An alternative source of capric acid is goat oil [47]. Both sources are abundantly available in tropical countries. In spite of the large amount of work of fatty acids, little effort has been made to evaluate the thermal performance of the C-L acid mixture. Among the fatty acid binary mixtures evaluated, only the C-L acid mixture provided the transition temperature nearest to that suitable for the cooling application of thermal energy storage.

Extensive literature reviews are available [6,10] with various simplifications and assumptions in mathematical analysis and numerical methods have been done describing the melting time and temperature distribution in the PCM. The use of materials for energy storage in cooling systems has limited acceptance because of a variety of operational problems. There are broad literatures on applications in which the cooling storage technology has been immediately successful, but the conditions under which the storage materials can produce the needed cooling characteristics or their heat transfer enhancement are not so clear. Although the possibility of using fatty acids as PCM energy storage was recognized and the theoretical works include techno-economic feasibility studies of a power plant design, no comprehensive study of these materials as energy storage media specifically for cooling application was reported in literature. This motivated the idea of pursuing in detail the C-L acid mixture.

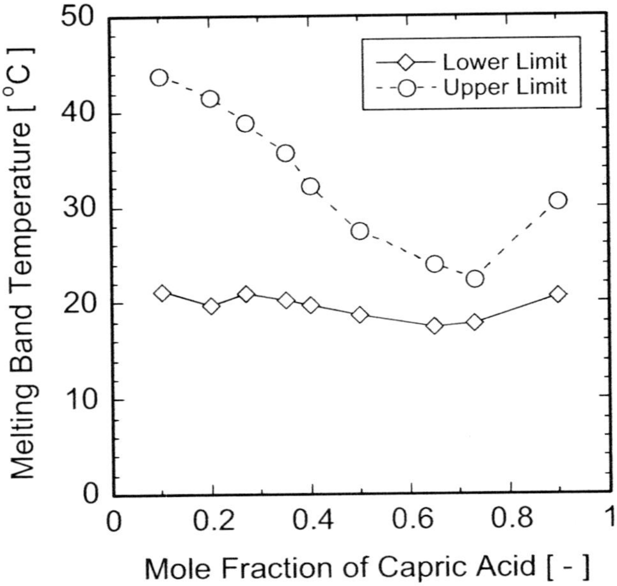

Figure 2. Melting band information of different C-L acid combinations.

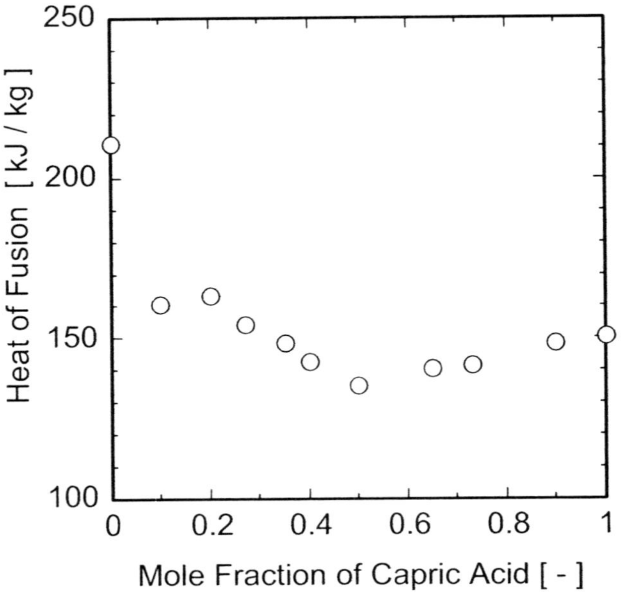

Figure 3. Heat of fusion in different C-L acid combinations.

Capric Acid and Lauric Acid Mixture as PCM

On the basis of the standard criteria set for the selection of storage materials such as the required desirable thermodynamic, kinetic and chemical properties, the behavior in relation to melting and freezing characteristics, higher energy density, reproducible melting and cooling behavior, negative property degradation and no supercooling, coupled with its low corrosion activity, non-toxicity, heat and color stability, and with definite melting temperature considered near the desired range for cooling application, the C-L acid mixture gained merits and showed as a potential PCM mixture from among several candidate PCMs.

It can easily be impregnated or directly incorporated in construction building products at some stage of their manufacture. In the liquid phase, it has a surface tension in the order of 0.02 to 0.30 $N{\cdot}m^{-1}$ [23]. This is high enough so that it can be retained in the projected construction material. It exhibits small volume change during melting and solidification with a melting dilatation of around 100 to 200 $m^3{\cdot}kg^{-1}$.

The development of the material is presently of prime importance to meet the specific requirement for an effective storage medium. Due to this end, the cost may have to be sacrificed in the interest of a better PCM performance. Changes in the materials in terms of concentrations and addition of foreign substances to meet the specific requirements are made in order to obtain the optimal performance of the C-L acid mixture.

Characterization of Capric Acid and Lauric Acid Mixture

Both capric acid and lauric acid have high melting temperatures in their individual forms. Lowering of melting point may be achieved with different mol% compositions of the binary mixtures correspondingly of capric acid and lauric acid could lower. Analytical verification of the thermal properties at different mole compositions of the mixture was confirmed employing the DSC method. The heating rate was controlled at 2.5°$C{\cdot}min^{-1}$ in the temperature range of 0 to 45°C.

The melting points and the heats of fusion of the C-L acid mixtures follow a downtrend with the decrease in concentration of capric acid in the mixture. **Figures 2 and 3** present a summary of the melting characteristics of the different capric acid and lauric acid combinations derived from DSC analysis. This behavior is seen until its composition of 65 capric acid and 35 lauric acid by mol%. With a higher concentration of capric acid, a favorable single peak endotherm is generated. Beyond the equal ratio with more of lauric acid in the mixture, two melting curves are distinctly produced obscuring the proper estimation of the latent heats of the mixture. A more apparent comparison was obtained from the DSC thermal curves of the binary mixture. The 65 and 35 by mole% composition of C-L acid was chosen as the most expedient combination. With a single peak feature indicating homogeneity in the mixture, its melting point was the lowest (18.0^{o}C) with a sufficient latent heat of fusion of 140.8 $kJ{\cdot}kg^{-1}$.

A more substantial confirmation of the C-L acid's melting congruency based on the range of stability of the phases and the phase transitions in different concentrations of the binary mixture is depicted from the phase diagram. **Figure 4** provides the T-X diagram of the temperature against the composition of the capric acid and lauric acid as a mixture. The melting point diagram provided in the figure shows that the liquid components of capric acid and lauric acid are completely miscible but the solid components are not, particularly at higher concentrations of lauric acid. An important point established is the eutectic mixture in the composition of 65 mol% capric acid. Based on the phase diagram, the 65 mol% capric acid is transformed from an all solid mixture of capric acid to an all liquid state within few degrees of temperature. Based on the thermograms of the different concentrations of the binary mixture obtained from the DSC analysis, the melting peak in this combination is the sharpest having the least melting band width and the magnitude of heat effect is largest for the isothermal arrest of this C-L acid mixture. Although microscopic examination of the eutectic mixture under high magnification discloses its heterogeneous character, solid will separate on cooling only when the eutectic temperature of 18.0°C is reached. Such a mixture is considered a simple system that forms an ideal solution in the melt. In this composition, the two solid phases of capric acid and lauric acid will deposit simultaneously and the temperature will remain constant until the liquid has disappeared. During melting, on the other hand, the solid C-L acid will melt at temperature of 18.0°C. The solution composition and its temperature will remain constant until the entire system melts completely. With this behavior, congruency in melting was assuredly maintained.

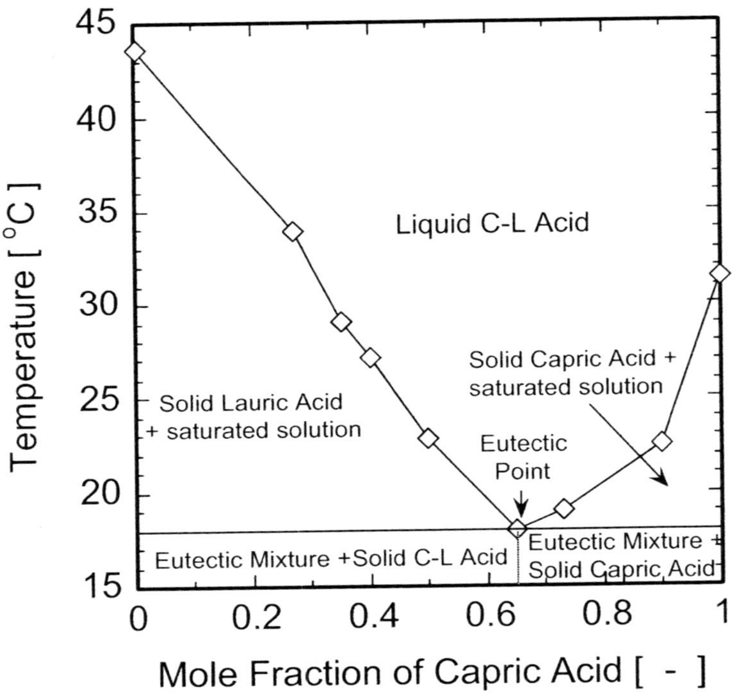

Figure 4. Phase diagram of the capric acid and lauric acid mixture. The eutectic point is established in 0.65 mole fraction of capric acid in the mixture.

Materials Significance Based on Thermal Performance

The chemical storage stability and the reproducibility of the chosen 65 mol% capric acid and 35 mol% lauric acid combination had been determined employing the classical calorimetry method using a fabricated storage cell [19]. The minimal deviation in the continuous heating and cooling operations (at a minimum of 120 cycles) of the system apparently demonstrates the chemical storage stability of the C-L acid. Physical observations have noted reproducible melting and freezing behavior with no supercooling. It did not exhibit volume change during phase transition and there is no visible separation in the components of the C-L acid mixture during charging and discharging tests. It thus indicates good chemical stability.

NUMERICAL ANALYSIS OF CAPRIC ACID AND LAURIC ACID MIXTURE DURING CHARGE AND DISCHARGE PROCESSES

A numerical analysis on the C-L acid's instantaneous temperature distribution to evaluate on its heat transfer characteristics and thermal storage capacity provides the physical understanding of the phenomena involved in phase transitions of PCMs notwithstanding the enhancement of computational capabilities appropriate to the system. A good number of works on the numerical solution of the rate of solid to liquid phase change during a unidirectional heat transfer involving finite geometries had been presented. One effective method was numerical integration of which the finite difference method is particularly useful. Previous solutions had involved heat flow with conduction as the sole transport mechanism in both solid and liquid phases disregarding the presence of convection. The inward melting or solidification of a PCM contained in a vertical tube was significantly affected by fluid motion in the liquid melt. As melting proceeds, the liquid phase had considerable lower thermal conductivities than in the solid phase. With the presence of a melted layer, heat resistance was prevented by the onset of the convective fluid motion [48-50]. Aside from this, since the fatty acid PCM mixture had an apparent melting band-width and a non-uniform temperature during phase transition, natural convection had a role to play [51,52]. Since natural convection normally existed during the melt, its significant effect must be considered. Thus, aside from the presence of solid and liquid phases, the moving interface or boundary between them was also included in the analysis. Computational studies were conducted in order to provide a better understanding of the phase change process in the C-L acid and consequently, its mixtures.

Models of PCM Behavior

The analysis took into account the conduction-based treatment of the solid and liquid phases as well as the natural convection induced by temperature difference in the liquid melt. In the analysis, the following assumptions were made:

1. The radial temperature and density distribution in the PCM tube were assumed to be axisymmetric representing the two-dimensional heat transfer by a one-dimensional heat conduction model.
2. The thermal resistance across the insulated cylindrical wall was assumed negligible.
3. The direction of heat transfer was only radial.
4. The phase change of the PCM was assumed to undergo isothermal phase change.

5. There existed at all times a sharply defined line of division indicative of the melting front between the solid and the liquid.
6. The thermophysical properties in terms of the thermal conductivities and densities were different for the solid and liquid phases but remain constant in their corresponding solid and liquid phase domains. During phase transition, their values are taken on the average of the solid and liquid phases.
7. C-L acid behaved ideally without supercooling and property degradation.

The system involved a non-linear transient heat conduction governed by a time dependent, one dimensional energy equation inside the cylindrical storage capsule expressed in terms of enthalpy, H, as

$$\rho \cdot \frac{\partial H}{\partial t} = \frac{1}{r} \cdot \left[\frac{\partial}{\partial r} \cdot \left(k \cdot \frac{r}{c_p} \cdot \frac{\partial H}{\partial r} \right) \right] \tag{1}$$

where r was the interface location. The thermophysical properties were based on the prevailing solid phase, phase transition and liquid phase of the PCMs. The energy equation in eq. (1) was modified to give the radial temperature distribution in the different phases of the PCM in a vertical cylinder.

Corresponding to the thermophysical properties in their respective domains the following simplifications were applied accordingly to a one-dimensional heat conduction approach.

For the solid phase:

$$\partial H = c_{p-S} \times \partial T \,. \tag{2}$$

For the liquid phase:

$$\partial H = c_{p-L} \times \partial T \,. \tag{3}$$

A different treatment was made during phase transition when the sharpest variation in the thermal properties occurred. The effect of natural convection was considered during phase transition using the effective thermal conductivity, k_e, in the melted layer during phase transition that also included the downward melting effect over the whole axial length of the vertical PCM containment.

During phase transition, the governing equation was

$$\partial H = L_f \times \partial \alpha \,. \tag{4}$$

The melt fraction (α) was considered as the latent heat (L_f) was absorbed during melting. Phase transition was achieved as soon as the total latent heat of fusion was completed with the melt fraction equivalent to 1.0. All the heat supplied to the control volume that experiences phase transition was used for changing the amount of latent heat content of that control volume. Liquid fractions were updated from the temperature field. The melting point of 19.0°C was considered for melting and of 20.0°C for solidification as obtained from the experimental temperature history in both processes. The phase transition region was established as a mixture of solid and liquid with the thermodynamic and transport properties assumed to be the arithmetic mean of the two phases.

The initial temperature of the PCM was assumed to be uniform throughout the radial positions from the center to the wall. For the initial condition, the radial temperature distribution was as follows:

$$T(r,0) = 35.0°C \quad 0 \le r \le R$$

for the melting process and

$$T(r,0) = 2.0°C \quad 0 \le r \le R$$

for the solidification process.

The boundary conditions of the first kind were chosen because of experimental convenience. There were two boundary conditions considered. One boundary condition to be satisfied was its outer temperature between the water and the solid interface. Implicitly, the PCM tube wall did not go through a step change in its temperature. That was

$$T = T_w + (T_w + T_i)\exp(at) \quad \text{at } r = r_{wall}. \tag{5}$$

The exponential change was considered with the value of "a" evaluated from the experimental measurements.

For the center PCM temperature,

$$\frac{\partial T}{\partial r} = 0 \quad \text{at } r = 0. \tag{6}$$

The thermophysical properties of the liquid and solid phases of the PCM mixture are different and independent of temperature. The values respective of the solid and liquid properties of the PCMs are indicated in **Table 1**. The melting points and heats of fusion of the eutectic mixture of C-L acid as well as the C-L acid in combination with the component additives were taken from the DSC analysis.

Table 1. Thermophysical properties of 65:35 C-L acid.

	C-L Acid Mixture
Melting Point [°C]	18.0 †
Heat of Fusion [$kJ \cdot kg^{-1}$]	140.8 †
Density (l) [$kg \cdot m^{-3}$]	894.9 ‡
Density (s) [$kg \cdot m^{-3}$]	900.0 ‡
Specific Heat (l) [$kJ \cdot kg^{-1} \cdot K^{-1}$]	2.24 †
Specific Heat (s) [$kJ \cdot kg^{-1} \cdot K^{-1}$]	1.97 †
Thermal Conductivity (l) [$W \cdot m^{-1} \cdot K^{-1}$]	0.139 ¶

Thermal Conductivity (s) [$W \cdot m^{-1} \cdot K^{-1}$]	0.143¶
Viscosity (l) [Pa · s]	1.65×10^{-3}¶
Prandtl Number [-]	26.7

† - measured, ‡ - (Mark *et al.*, 1978) [53], ¶ - (Humphries, 1977) [54]
‖ - (Lide, 1995) [55], l- liquid phase, s- solid phase

The problems considered here were those of the inward melting and solidification of a cylindrical PCM of which the time required for the PCM to become totally melted or solidified and the temperature profile at that moment were determined.

Charge Process

The melting temperature history in different radial positions provided the good agreement in general for all height positions, indicating the discretization validation of the numerical model. A representation to this claim is shown in **Figure 5**, which includes the experimental results that will be explained in Sec. 4.2. Further examination of the temperature profile based on the radial distribution manifested a difference in the highest layer in all wall temperatures. The temperature difference between the C-L acid and the wall influenced the temperature distribution in the liquid phase, as representatively demonstrated in **Figure 6**. Natural convection caused a recirculating flow of the melted layer flowing up along the heated side and flowing down along the layer where phase transition was occurring, therefore the application of the effective thermal conductivity is advantageous for the estimation of the temperature distribution in the melted layer.

The total heat charged and discharged from the PCM radial temperature measurements in different probe height positions were estimated from the following conservation of energy relations integrating implicitly with time and accounting for the whole process from start to finish.

Below the lower melting band limit while the PCM was still in the solid phase:

$$dQ_1 = \int_{Ts}^{T} 2\pi\, \rho_S\, Cp_S\, l\, r\, dr\, dT\,. \tag{7}$$

During the PCM phase transition within the melting band:

$$dQ_2 = \int_{Ts}^{Tm_1} 2\pi\, \rho_S\, Cp_S\, l\, r\, dr\, dT + \int_{Tm_1}^{T} 2\pi\, \rho_{ave}\, Cp_{ave}\, l\, r\, dr\, dT + \int_{0}^{\alpha} 2\pi\, \rho_{ave}\, L\, l\, r\, dr\, d\alpha\,. \tag{8}$$

Above the upper melting band limit when the PCM had melted:

$$dQ_3 = \int_{Ts}^{Tm_1} 2\pi\, \rho_S\, Cp_S\, l\, r\, dr\, dT + \int_{Tm_1}^{Tm_2} 2\pi\, \rho_{ave}\, Cp_{ave}\, l\, r\, dr\, dT + 2\pi\, \rho_{ave}\, L\, l\, r\, dr + \int_{Tm_2}^{T} 2\pi\, \rho_L\, Cp_L\, l\, r\, dr\, dT\,. \tag{9}$$

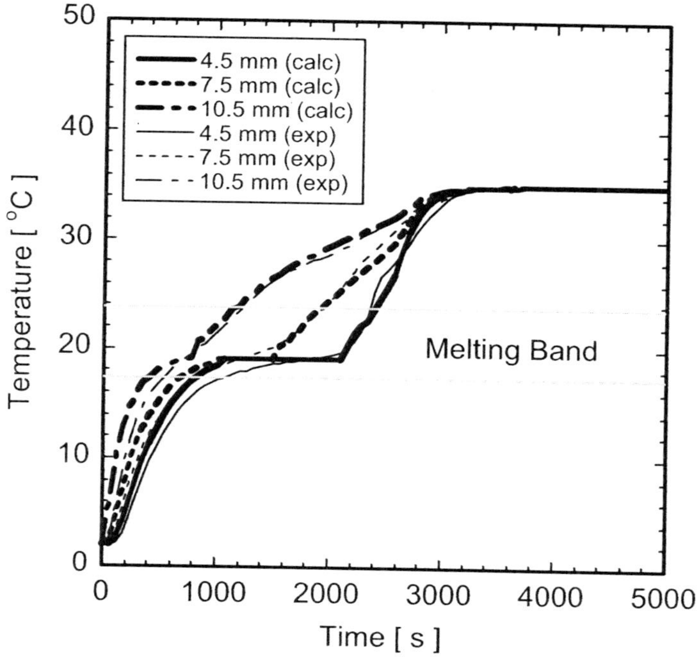

Figure 5. Numerical (calc) and averaged experimental (exp) melting temperature history of C-L acid at 35°C wall temperature.

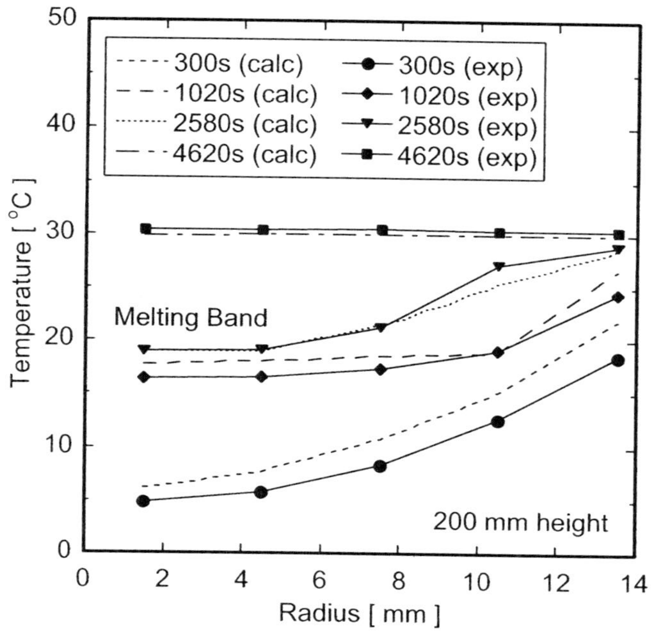

Figure 6. Radial melting temperature comparison between numerical (calc) and experimental (exp) profile of C-L Acid at 200 mm probe height position with water at 35°C.

And integrated over the radius

$$Q = \int_0^R dQ_i . \tag{10}$$

On charging, eq. (7) approximated the solid sensible heat prior to phase transition. During phase transition, the storage capsule considers not only the latent heat but also the sensible heat effects. In eq. (8), the second term furnished the small amount of sensible heat derived from the large melting band of the PCM. The third term provided a large contribution in the heat energy mainly accounted from the latent heat. The immobile phase of either the liquid or solid material, expressed in terms of the melt/solid fraction (α) that developed as the C-L acid gave up its heat of fusion, was considered in the equation. On the final stage of the charging process, the liquid sensible heat was incorporated in eq. (9). The PCM height (l), corresponds to 50 mm for (0 to 50mm), 100 mm for (50 to 150 mm), 100 mm for (150 to 250 mm), and 75 mm for (250 to 325 mm). The total amount of heat was finally integrated in every radial position using eq. (10). Taking the whole vertical column and dividing it into 4 parts corresponding to the position of the PCM probe set-up, the charge energy in each part as estimated and cumulatively considered on to the top of the capsule. A program has been developed to facilitate the computational process of determining the total amount of heat that could be generated from the charge and discharge of the PCMs.

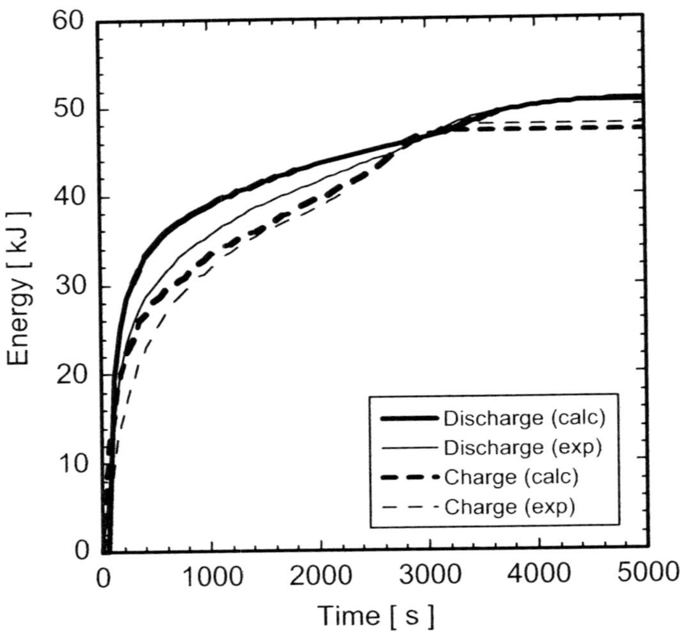

Figure 7. Charge energy at 35°C and discharge energy at 2°C taken from experimental (exp) and numerical (calc) analyses.

The charge energy of the C-L acid at different wall temperatures provided good agreement with the experiment as shown in **Figure 7**. The experiment was based on the averaged temperature history in all height positions. The charge energy was a strong function of the temperature difference between the PCM melt and wall temperatures. The greater temperature difference corresponded to the increase in charge energy resulting from an enhanced natural convection. The important information was that with 35.0°C of wall temperature, the C-L acid was capable of charging as much as 47 kJ of heat.

The validity of the model was further investigated. The numerical analysis showed the actual speed or duration of the melting process in which the highest height position in the experimental analysis also showed. The experimental melting time taken from the average of the height positions 25 to 200 mm from the base of the C-L acid tube provided a satisfactory agreement with the numerically obtained melting time. The wall temperature had a great effect on melting as shown in **Figure 8**. The greater the temperature difference, the faster the C-L acid melted.

The two-dimensional heat transfer by a one-dimensional model with an effective thermal conductivity in the melt, that took into account not only the natural convection but also the two-dimensional effect also, was considered. The effective thermal conductivity may be thought of as equivalent to the axial dispersion coefficient, usually used in a one-dimensional model to describe a two-dimensional diffusion. The unsteady state heat conduction equation in the cylindrical coordinate in both phases was used and solved numerically using the explicit finite difference method.

The model was expected to predict the average condition of heat transfer along the whole height of the cylinder, since the thermal effective conductivity provided the true average effect of natural convection. In this study, similar approach was followed to account for convection. With an increase in the melted layer width, a corresponding increase in effective thermal conductivity existed as manifested in **Figure 9**. The effective thermal conductivity was strongly dependent on the melt layer thickness and the difference between the wall and melting temperatures.

The transition from conduction to natural convection of a PCM during phase change experienced in melting and conversely in solidification involved a complicated analysis. One way of interpreting the heat transfer mechanism considering temperature distribution in the C-L acid was by means of its Grashof number and Nusselt number. Generally, the expected pattern of melting from the upper layer down to the bottom of the PCM due to the governance of natural convection was evident in the radial temperature distribution. The Grashof number

$$Gr = \frac{g \times \beta \times (T_w - T_{PCM\ melt}) \times L^3}{\upsilon^2} \tag{11}$$

accounted for the natural convection being developed in the melt layer, with an upflow adjacent to the tube wall and a downflow adjacent to the melting front. A recirculation pattern normal for natural convection was accelerated by the rates of radial inward and downward melting of the PCM. Since melting proceeded in an unsteady state, a more vivid convection behavior in the PCM was depicted by applying the correlations applicable during the phase transition.

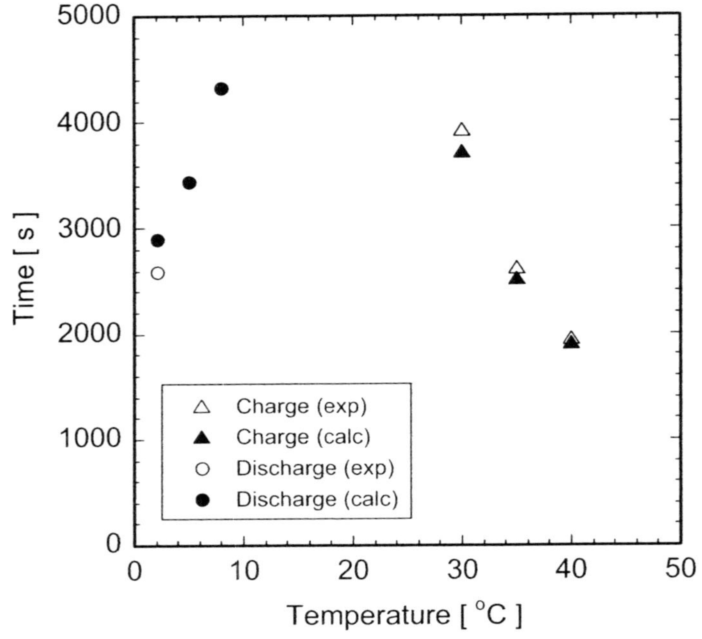

Figure 8. Melting and solidification times at different wall temperatures

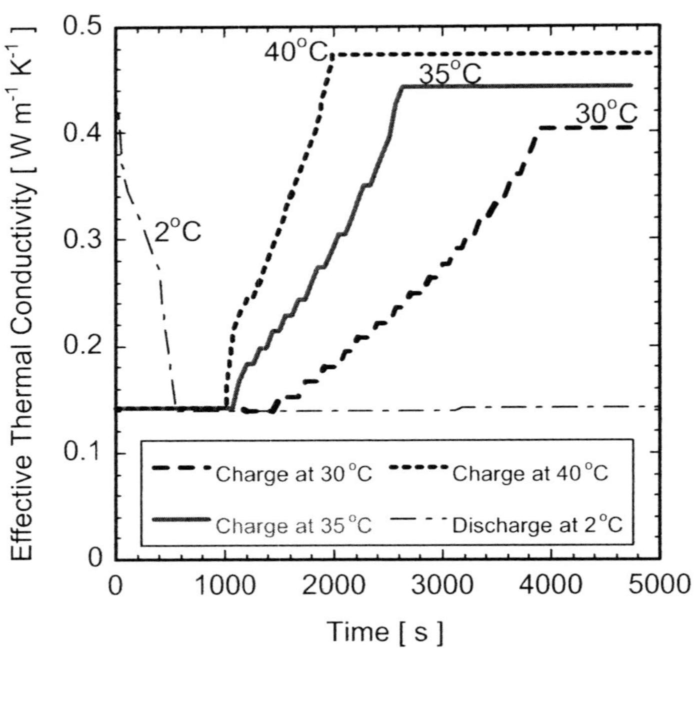

Figure 9.Effective thermal conductivity of C-L acid during phase transition at different wall temperatures.

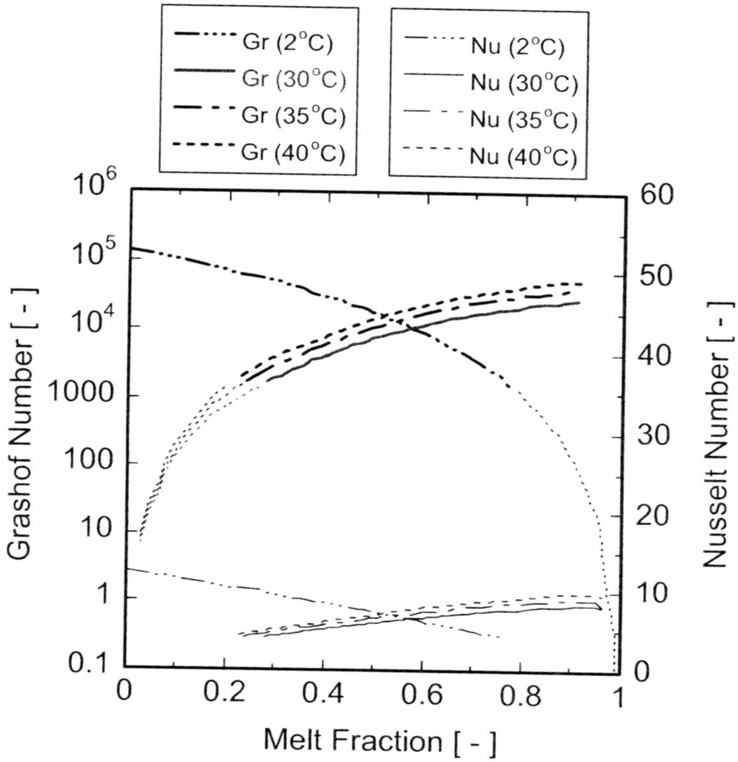

Figure 10. Grashof number and Nusselt number during phase transition of C-L acid estimated from numerical analysis in different water temperatures.

From solid to liquid phase, melting proceeds in an annular form. Considering the correlation developed by Farid *et al.* [56] for natural convection applied to a vertical annulus, the Nusselt number was estimated from the relation

$$Nu = 0.28 \times (Gr \cdot Pr)^{0.25} \tag{12}$$

with the characteristic length, D_{eq} taken from the annular radius of the melted region

$$D_{eq} = r_o - r_{i,\alpha}. \tag{13}$$

The plot of the *Gr* and *Nu* during phase transition of the C-L acid with different wall temperature is presented in **Figure 10**. *Gr* at the start indicated a very thin melting layer that implied that natural convection was insignificant. An undetermined value of Nusselt number from the start until melting set in as indicated by a gap in Figure 10 indicated the dominance of conduction heat transfer. But as time progressed and melting proceeded, *Gr* increased considerably. *Nu* on the other hand, increased linearly.

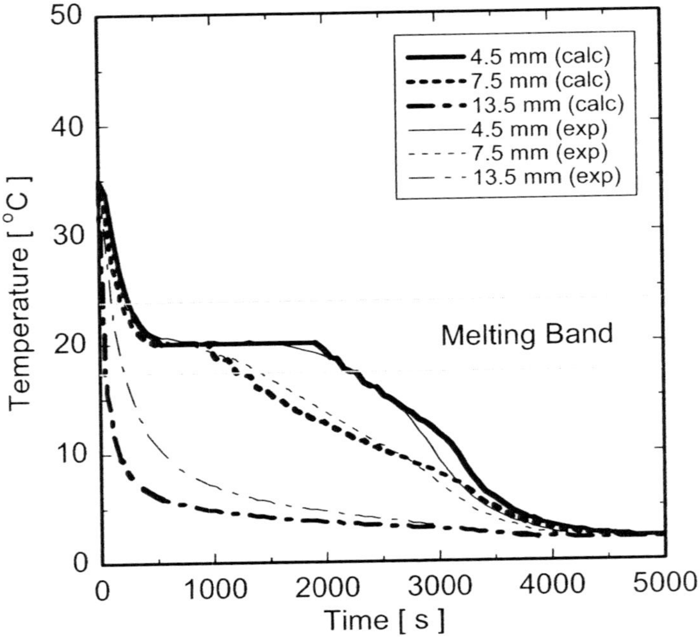

Figure 11. Numerical (calc) and averaged experimental (exp) solidification temperature history of C-L acid at 2°C wall temperature.

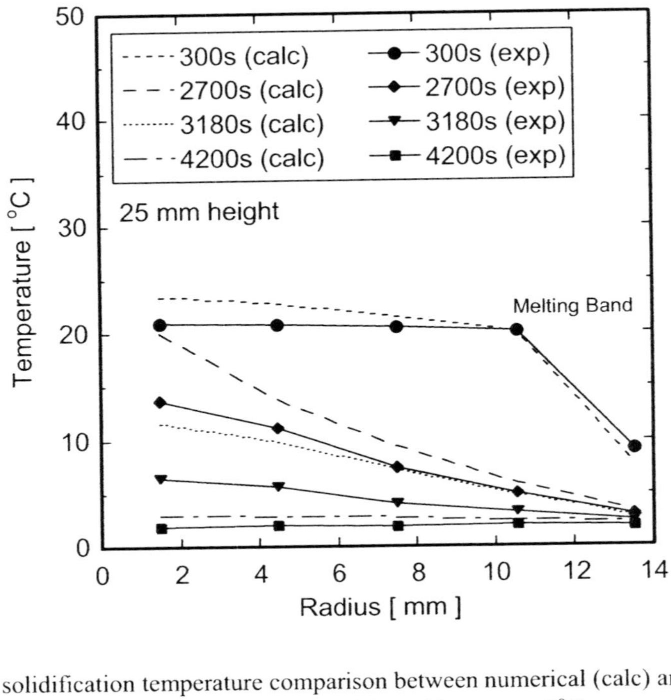

Figure 12. Radial solidification temperature comparison between numerical (calc) and experimental (exp) profile of C-L Acid at 25 mm probe height position with water at 2°C.

Discharge Process

The temperature history in different radial positions also provided the applicability of the numerical model as shown in **Figure 11**. Consistency in the radial temperature profile at the lowest layer was shown during solidification. The temperature distribution agreement in the 25 mm height position in **Figure 12** showed an ideal behavior since the lowest layer in reality solidified faster than the higher layers. The concurrence of the experimental and numerical results in different radial positions at varying times inferred the stable heat transfer controlled by conduction.

The numerically estimated discharge energy of C-L acid was likewise in good agreement with the experimental results as presented in Figure 7. The experiment was also based on the averaged temperature history in all height positions. The amount of heat discharged by the C-L acid obtained from the numerical analysis indicated a favorable energy with lower wall temperature. The greater temperature difference between the PCM melt and the wall temperature corresponded to a more effective conductive mechanism yielding higher discharge energy in the C-L acid.

Different wall temperatures were assigned to ascertain the effect of temperature to solidification time. Figure 8 shows the increase in solidification time with higher wall temperatures. With smaller temperature difference between the C-L acid solidification and wall temperatures, the heat transfer mechanism became unstable and heat conduction weakened. Nevertheless, the numerical and experimental solidification times were still in good agreement further substantiating the applicability of the model. The melting temperature difference with the wall has an insignificant effect on the liquid remaining width.

During solidification, convection mechanism was analyzed in terms of the equivalent radius (D_{eq}) of the PCM in cylinder treating it as a solid tube as

$$D_{eq} = r_{o,\alpha} \tag{14}$$

and applied in eq. (12).

The log plot of the *Gr* and *Nu* during the liquid to solid phase transition of C-L acid with different wall temperatures is presented in Figure 10. *Gr* at the start is very large indicating the dominance of natural convection. *Nu* demonstrated a high value at the start and proceeded to decrease linearly until conduction heat transfer governed.

Industrial Application

Among the various employment of PCMs for thermal energy storage application, two ways by which the C-L acid may serve as a cold energy storage material could be embedding C-L acid within floors, ceilings, and walls, and using a specialized C-L acid storage box outside the space unit to be cooled. As the C-L acid is part of the structural system of the room or space unit, it does not involve extra cost. During the daytime, the cooled or solidified mass of the C-L acid, if adequately insulated on the outside, serves as a direct heat sink within the occupied space unit. The C-L acid absorbs passively the heat penetrating into and generated inside the space unit. The main design implication of this cooling system involves the need to ensure sufficient mass, a large surface area of the mass exposed to the space unit to be cooled, and the thermal insulation at the outside of the space unit.

System Description

In this paper, the C-L acid mixture constitutes the principal storage medium. A simple scheme to demonstrate its possible employment for industrial application is shown in **Figure 13**. The material is either packaged in specialized containers such as tubes, shallow panels, contained in convectional building elements such as wallboards or ceilings, or encapsulated in plastic containers or self contained grains that are stacked or immersed in a storage tank through which cold water is circulated. Water serves as the heat transfer fluid although it has been indicated that encapsulation is better if air is the heat transfer fluid. Nevertheless, liquid systems may be used with both types of heat transfer fluid. It is incorporated into a thermal storage system. A standard chilling equipment or a refrigerating machine is used to solidify or discharge the C-L acid storage medium. Since the C-L acid has a freezing temperature greater than 10°C, its chiller solidification temperature need not be below 4°C. The cold energy is stored at night when the demand for it is high during the day. Storage tank materials include steel, concrete and plastic. Storage can be divided into multiple tanks, if necessary, in order to permit simultaneous storage for cooling.

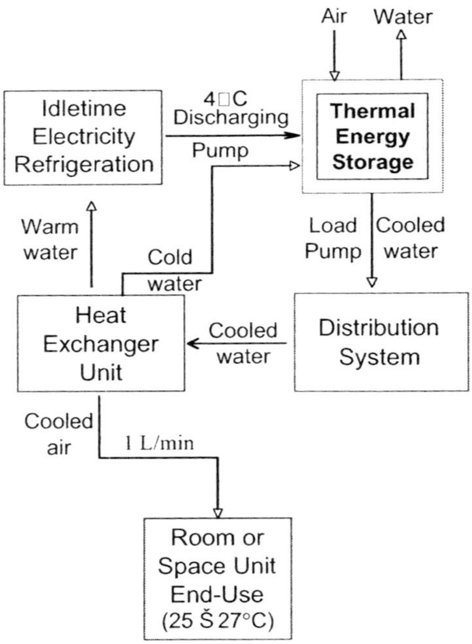

Figure 13. Concept of latent heat storage for cooling application.

Cold energy is pumped in the space unit. One option is to place a heat exchanger in between the storage tank and the space unit loop with the presence of pressure sustaining valves in order to maintain or recover the energy during the cooling load distribution. This cold energy storage system may either be fully or partially stored depending on the demand. Full storage is employed if all cooling is generated during off-peak periods and shut down during peak periods. Partial storage is implemented on a day basis or for the purpose of minimizing the electric utility needs.

Heat Transfer Characterization

Estimation is made on the cooling load for a room having a size of 20 m^2 using the C-L acid as the storage medium adapting the capability of the conventional window or split-type air conditioning system. Vertical cylindrical tubes are placed in banks in a storage box having a height of 0.30 m. The width and length depends on the number of tubes that will contain the C-L acid needed to provide the necessary cool in a space unit. Air is used as the heat transfer fluid during space cooling and water as the heat transfer fluid during the C-L acid solidification. During space cooling, the C-L acid is discharged and during solidification, the C-L acid is charged. The heat flow coming from the air conditioning storage unit is 1 kW. The capacity of the storage mass to serve as a heat sink during the following day is determined by the number of tubes that contain the required amount of C-L acid on the required temperature range.

Discharge Process

Iterations are made to come up with the right diameter of the C-L acid tube that will converge with the diameter estimated from the convective heat transfer. Analysis and validation starts from the inside heat transfer coefficient treating the C-L acid melt in annular form. Nu is estimated using the relation [56]

$$Nu = \frac{h_i \times D}{k_{eff}} = 0.15 \times Ra^{0.25} \quad \text{for } 10^3 < Ra < 10^7 , \tag{15}$$

from which the heat transfer coefficient (h_i) of the C-L acid is determined considering the effective thermal conductivity (k_{eff}) during its phase transition from

$$k_{eff} = k_L \times 0.28 \times Ra^{0.25} \times \frac{\delta}{L} . \tag{16}$$

The maximum melt layer width (δ) is taken as the tube diameter.

The conditioned air supplied from the PCM capsules must be sufficient to absorb heat from the room to maintain the comfortable condition. Cooling is accomplished by the convection of air. Forced convection in air is maintained by using fans driven by motors.

Air flow rates of 0.047 to 0.060 $m^3 \cdot s^{-1}$ per 1 kW [57] are normally recommended to achieve good performance of cooling systems. With the proposed design of the thermal energy storage system, the volumetric flow rate of 0.060 $m^3 \cdot s^{-1}$ maintains an effective cooling capacity for 1 kW. Air is treated as a cross flow in a circular cylinder.

Considering the annular distance for one tube pass, and the geometry of the tubes arrangement, the outside area provides an air velocity of 7.1 $m{\cdot}s^{-1}$. Air with this velocity is considered turbulent since Reynolds number is 2.72×10^4 obtained from

$$Re = \frac{U_{\infty,\max} \times D}{\upsilon} . \tag{17}$$

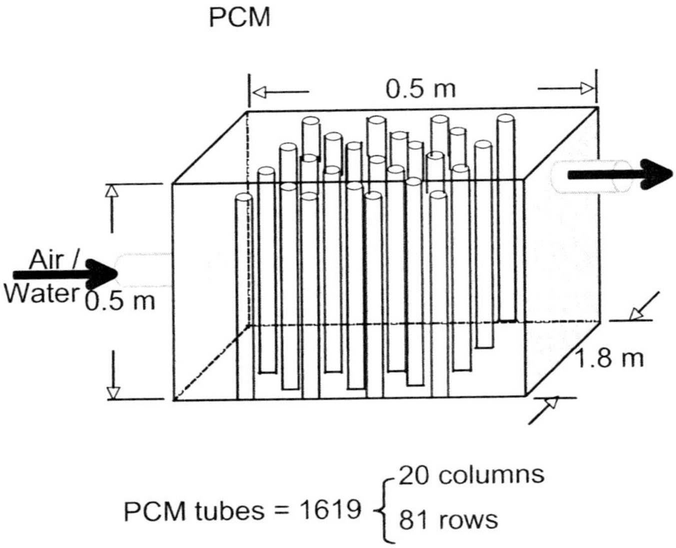

Figure 14. Thermal energy storage box.

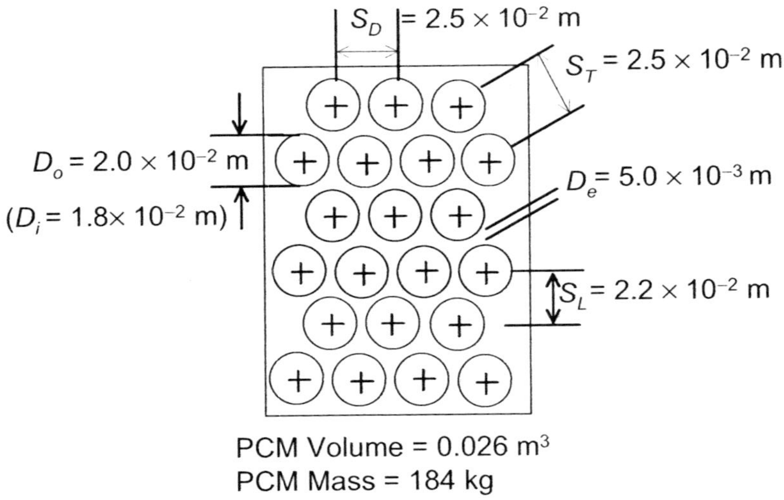

Figure 15. PCM tube geometry in the storage box.

The outside heat transfer coefficient is estimated using the relation [58]

$$\frac{h_o \times D}{k_f} = 0.193 \times Re^{0.618} \times Pr^{1/3} \text{ for } Pr > 0.6,\ 4000 > Re > 40000\,. \tag{18}$$

An h_o of 126 W·m^{-2}·K^{-1} implies that C-L acid controls the heat transfer rate.

The heat transfer rate is determined from the overall heat transfer coefficient (U) that is 11.4 W·m^{-2}·K^{-1} estimated from the heat transfer coefficients of C-L acid and air based on the relation

$$\frac{1}{U} = \frac{1}{h_i} + \frac{1}{h_o}\,. \tag{19}$$

Similarity in the heat transfer area is closely met with a PCM tube diameter of 0.018m. Considering the amount of heat absorbed by the C-L acid and estimating the amount of heat that is capable of absorbing for 8 hours, the volume associated with it is 0.21 m^3.

The heat capacity of the C-L acid that can sustain the same capacity of 1 kW using an air conditioning unit for 8 hours needs 1619 tubes. Each tube with a length of 0.50 m and diameter of 0.018 m containing 0.11 kg of the C-L acid can supplement the needed energy. A rectangular storage box measuring 1.8 m × 0.5 m × 0.5 m (**Figure 14**) containing vertical cylindrical tubes in staggered arrangement of 20 columns and 81 rows, as shown in **Figure 15**, furnishes the designed capacity.

Charge Process

Water enters the cool exchanger at 15°C and leaves at 10°C. Solidification is accomplished by means of cold water. Cold water at 12.5°C cross-flows over the C-L acid tubes inside the energy storage box at a slow rate of 9.52×10^{-5} m^3·s^{-1}.

The inside heat transfer coefficient is based on the assumption that as the C-L acid undergoes phase transition from liquid to solid, the cylinder diameter takes on its solid form. For the estimation of the heat transfer coefficient of C-L acid inside the tube, the following empirical correlation [58] applies.

$$Nu = 0.42 \times (Gr_D Pr)^{0.25} \times Pr^{0.012} \times \left(\frac{L}{D}\right)^{-0.30}. \tag{20}$$

The above equation is based on constant heat flux condition, with $10^4 < Gr_D Pr < 10^7$, $1 < Pr < 20000$, and $10 < \frac{L}{D} < 40$.

The heat transfer coefficient of water outside the tube is estimated from the relation

$$Nu_f = (0.35 + 0.56 Re_f^{\,0.52}) \times Pr_f^{\,0.3} \text{ for } 10^{-1} < Re_f < 10^5 \tag{21}$$

that better represents liquids to tubes in cross flow [59].

The minimum permissible water flow velocity is that which will produce a high rate of heat transfer which is comparable to that of air. In the discharge process, the convective heat load of 2.72 kW more than doubles the 1 kW cooling capacity that has been designated in the estimation. In the charge process, the increasing temperature difference in water corresponds with the decrease in its volumetric flow rate. A reduced flow rate of water is desired since it matches with lesser energy cost. With a much smaller heat transfer coefficient of C-L acid compared to the water, C-L acid and water provides a volumetric flow rate of 9.55×10^{-5} $m^3{\cdot}s^{-1}$, the designated cooling capacity of which equates with the charging convective heat load. Hence, it rationalizes a lesser electricity consumption with a fairly reasonable solidification at 4 hours and cooling capacity for 8 hours. **Table 2** provides the design conditions and parameters intended for the industrial application of the C-L acid for cold energy storage.

With an overall heat transfer coefficient of 8.87 $W{\cdot}m^{-2}{\cdot}K^{-1}$, convergence has also been reached with the tube diameter of 0.018 m. Estimating the time of solidification, it takes 0.5 hour for C-L acid in a tube to completely charge. In the Philippine setting, considering the current electricity rate and supplementary charges associated with it, the designed thermal energy storage that can provide the same cool using a 1 kW air conditioner for 8 hours can provide a minimum of 60% reduction in the energy cost.

Wrapping up, the cooling capacity offered by the C-L acid compensates the reduction in the energy cost proving its potential as a latent heat storage material. With a large amount of C-L acid required to make up for the cooling load delivered by an ordinary air-conditioning unit, however, an improvement in its melting characteristics by the supplement of appropriate chemical additives is necessary.

Table 2. Design parameters of the thermal energy storage system for commercial and industrial application.

Process	Temperature [°C]	Energy [W]	Heat Transfer Fluid	Inside	Outside
Discharge	Inlet = 30.0	1000	1.03×10^{-3} $m^3{\cdot}s^{-1}$	h_i=12.6 $W{\cdot}m^{-2}{\cdot}K^{-1}$	h_o=126 $W{\cdot}m^{-2}{\cdot}K^{-1}$
	Exit = 20.0			$Gr = 1.18\times10^4$	$Re = 2.72\times10^4$
	Air = 25.0			$Ra = 3.13\times10^5$	
	PCM = 18.0		1.23×10^{-5} $kg{\cdot}s^{-1}$	$Nu = 6.62$	$Nu = 94.3$
				$U = 11.4$ $W{\cdot}m^{-2}{\cdot}K^{-1}$	
Charge	Inlet = 15.0	2000	9.52×10^{-5} $m^3{\cdot}s^{-1}$	h_i=9.27 $W{\cdot}m^{-2}{\cdot}K^{-1}$	h_o=231 $W{\cdot}m^{-2}{\cdot}K^{-1}$
	Exit = 10.0			$Gr = 1.28\times10^4$	$Re = 31.0$
	Water = 12.5			$Ra = 2.98\times10^5$	
	PCM = 18.0		9.5×10^{-5} $kg{\cdot}s^{-1}$	$Nu = 3.76$	$Nu = 7.05$
				$U = 8.87$ $W{\cdot}m^{-2}{\cdot}K^{-1}$	

MELTING CHARACTERISTICS IMPROVEMENT OF CAPRIC AND LAURIC ACID SYSTEM

Although the C-L acid possesses sufficient capacity for thermal energy storage, the assurance that it is a potential phase change material was held in reservation specifically for the cooling application. Its melting point of 18°C is considered high for cooling a space unit. It has always been indicated that a temperature below 10°C must be maintained by the storage system in order to maintain a comfortable temperature of 22 to 25°C in a space unit having an ambient temperature range of 30 to 35°C. This requires a melting temperature of about 7 to 10°C for the cold side storage to achieve dehumidification.

The possibility of obtaining the desired melting temperature through the addition of foreign substances to a composite material having different melting points had been suggested several times and in fact investigated. Aside from the references that had been cited earlier in the first chapter, Sohns *et al.* [60] carried out an experimental investigation employing the dynamic difference calorimetry technique to determine the effect of impurities on PCMs such as normal paraffins and fatty acids applied for space heating. It inferred in the study that the variation of melting temperature and the latent heat of fusion due to the presence of impurities of organic components does not significantly impinge on the performance of the latent heat storage system. Although the study was meant for heating application, similar effect on materials for cooling application are likely to follow.

A variation in the melting temperature and latent heat of fusion of a PCM, through the addition of impurities, for instance, slightly affects the PCM's thermal storage performance. A temperature decrease of 10 K in the original storage system's melting point can reduce the performance by only 2%. A system having a latent heat of 200 $kJ{\cdot}kg^{-1}$ and beyond can only decrease its performance by ± 2% if the deviation in the latent heat is by ± 40%.

n-Pentadecane offers high latent heat (168 $kJ{\cdot}kg^{-1}$) with low melting point (9.9°C). Its melting point falls within the range suitable for cold energy storage. The thermal performance of the C-L acid with pentadecane as a latent heat storage was investigated [16] revealing heterogeneity in the mixture. As pentadecane is in itself an energy storage material that is currently expensive, it was proposed that other organic chemical additives that can supplement better thermal characteristics for beneficial cooling application be investigated.

Selection of Chemical Additives

PCMs with enhanced properties for cooling application are modified through the incorporation of a variety of additives. The performance of these additives is often difficult to predict from considerations of their known behavior in the separate blend constituents. However, researches conducted regarding eutectic mixtures and multiple phase change materials have set off the chances of finding the optimal chemical additive that can improve the characteristics of the C-L acid as a base PCM. In this paper, the screening of available additives that are compatible in the C-L acid prepared in different proportions was conducted to investigate on their melting characteristics and heat transfer behavior, and select the most favorable combination that would meet the desired temperature range in unison with maintaining a substantial heat of fusion.

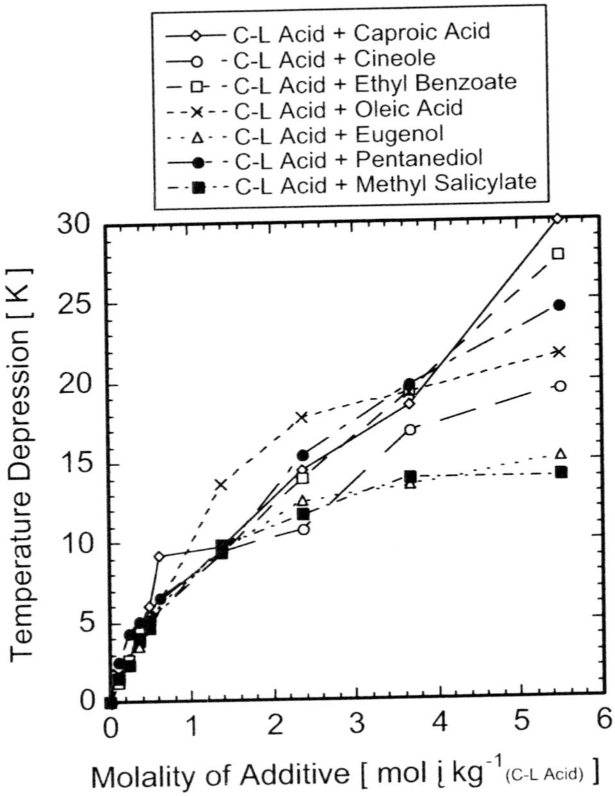

Figure 16. Melting temperature depression at different molalities of the chemical additive in the C-L acid mixture.

The chemical additives were chosen on the basis of two major factors. The first factor considered the chemical characteristics while the second covered on the safety and economic aspects of the additives. The chemical characteristics included the thermophysical properties and solubility of the additives in the C-L acid. The melting point must be 8°C to 11°C lower than that of the C-L acid with a heat of fusion sufficient enough to maintain a satisfactory thermal energy for cooling capacity. Complementary soluble additives in the C-L acid considered to promote homogeneity in the PCM mixture that would support better thermal energy storage capacity and heat transfer due to more defined phase transition. For safety considerations, carcinogens, flammable, toxic, volatile, hazardous vapor emitting and environmentally degrading additives are eliminated. Cost and availability are other features of economic concern.

Some organic chemicals that were seemingly possible materials and compatible with the C-L acid system with low melting points have low heats of fusion. Nevertheless, they were also included for analytical verification for the purpose of easily obtaining the characteristic degree of the melting point lowering of the C-L acid mixture.

Among the possible additives, caproic acid, cineole, ethyl benzoate, eugenol, oleic acid and methyl salicylate were initially tested with the C-L acid on the basis of their availability.

Different mole fractions of the additives were individually prepared and made homogeneous in the C-L acid that served as the base system. Capric acid, lauric acid (both 98.0% purity), were manufactured by Tokyo Chemicals, Inc. Caproic acid (99.0% purity), eugenol, pentanediole, and cineole were produced by Sigma Chemical Co.; ethyl benzoate (99.0% purity) was manufactured by Aldrich Chemical Co., Inc.; and the oleic acid, from D & L Industries, Inc.

The melting characteristics in terms of the heat of fusion, the melting temperature, melting peak and melting width of the C-L acid with the corresponding additives in different proportions were analyzed using the DSC821 Mettler Toledo STAR^e 2000 software. The thermodynamic quantities vary with the composition of the PCM mixture. The characteristic behavior in each system of C-L acid with the component additive was observed.

The lowering in the melting point of the C-L acid is related to the effect of the chemical additive solute on the activity of the C-L acid acting as the solvent and on its chemical potential or molality of the additive solute's Gibbs free energy. There is linearity in the melting temperature depression only at low molality concentration of the chemical additive from 0 to 0.61 mole per kg of the C-L acid (mol·$kg_{C\text{-}L}^{-1}$) as shown in **Figure 16**. The slope of the curve is independent of the identity of the chemical additive. The colligative property of a mixture is ideally demonstrated in very dilute solutions. That is why at lower concentration of additives, a more consistent decreasing trend follows a linear correlation between the temperature depression and the concentration of the mixture. From this figure, the cryoscopic constant (K_f) was determined from the simple colligative equation

$$\Delta T_{fus} = K_f \times m \tag{22}$$

in which

$$K_f = \frac{M_1 R T_{fus}^{*2}}{\Delta H_{fus}} . \tag{23}$$

The slope obtained from eq. (22) provided an average melting point depression constant of 10.7 K·$kg_{C\text{-}L}$·mol^{-1} that conveys the distinctive characteristic of the C-L acid mixture for all additives. In lower concentrations of additives in the C-L acid, the latter is particularly the solvent while the additives are the solutes. This is valid only at low concentrations since eq. (23) was derived from the Gibbs-Helmholtz equation yielding

$$\frac{d \ln x_1}{dT} = \frac{\Delta H_{fus}}{RT^2} . \tag{24}$$

For very dilute solutions, the heat of fusion (ΔH_{fus}) of the solvent C-L acid, is assumed independent of temperature over a certain range and eq. (24) may be simplified by expressing x_1 (mole fraction of solvent C-L acid) in terms of x_2 (mole fraction of additive) as

$$\ln(1-x_2) = \frac{\Delta H_{fus}}{R}\left(\frac{T-T_{fus}}{T\times T_{fus}}\right). \tag{25}$$

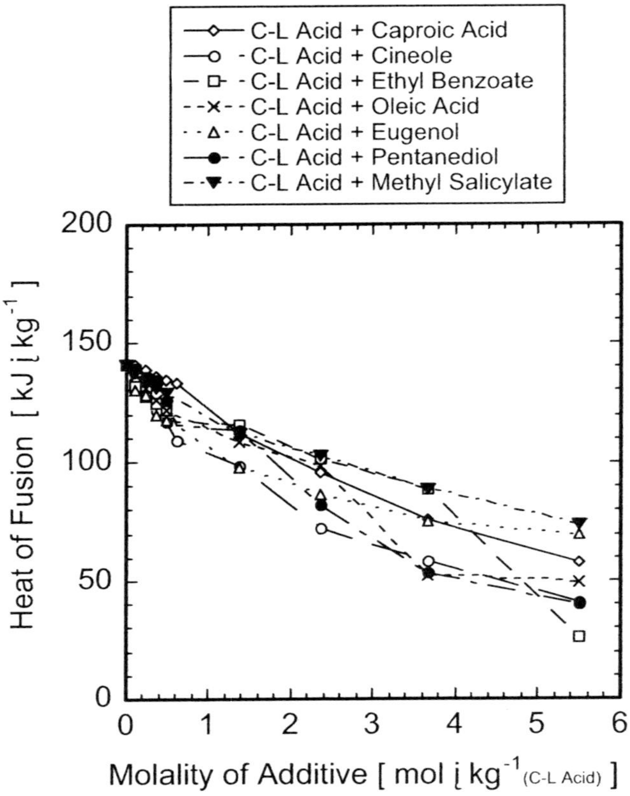

Figure 17. Heat of fusion in different molalities of chemical additive in the C-L acid mixture.

The term ln (1-x_2) is expanded in a power series and if mole fraction of the additive solute is very small, only the first term is left as

$$\ln(1-x_2) = -x_2 - \frac{1}{2}x_2^2 - \frac{1}{3}x_2^3 - \cdots \cong -x_2\,. \tag{26}$$

(T-T_{fus}) represents the freezing point depression and considered small such that $T \times T_{fus} \approx T_{fus}^2$. Hence, eq. (26) involves the assumption that ln $(1 - x_2)$ is approximately equal to the mole fraction of the solute, x_2. This assumption is valid only at low concentrations when the higher powers in the series expansion of ln $(1 - x_2)$ become low and negligible. But this is not true for higher concentrations. The slope of the melting depression against the molality sets off to deviate at higher additive concentrations. The temperatures of the mixtures drop variably, with the gradation depending on the additives' tendency to lower the melting point of the C-L acid.

A practically small change in the heat of fusion of the PCM mixture matches with the temperature depression linearity at low additive concentration (i.e., below 0.61 mol·kg_{C-L}^{-1}) as shown in **Figure 17**. The degree of the inclination of the heat of fusion with respect to higher additive concentration reveals the complex identity of the mixture. As the concentration of the additive is increased, the system becomes complex. The characteristic of the C-L acid is altered causing the deviation from its ideal condition. The real solvent cannot be distinctly identified due to the existence of more amount of additive in the PCM mixture that at higher additive concentration makes the mixture a defined ternary system.

The increase in ethyl benzoate concentration in the C-L acid furnishes the greatest decrease in ΔH_{fus} while methyl salicylate showed better ΔH_{fus} at higher concentration as shown in Figure 17. Oleic acid manifested an irregularity at higher concentrations of 3.7 and 5.5 mol·kg_{C-L}^{-1}. For an effective cooling application of a thermal storage system applying C-L acid as the base PCM medium, a melting point depression of 8 to 11°C must be furnished by the C-L acid-chemical additive combination. Caproic acid demonstrated a linear temperature depression development at higher concentrations. All the others manifested the same temperature depression at lower concentrations.

Examining their DSC derived data, all the chemical additives in the C-L acid showed a consistent trend in their upper melting band temperatures at different concentrations. Ethyl benzoate, on the other hand, demonstrated an obviously lower melting band temperature with a greater melting band width. For applicability in the present investigation, the lower melting band temperature limit of the PCM mixture must not be below 2.5°C. All PCM mixtures with the additive concentration of 0.61 mol·kg_{C-L}^{-1}, except that with oleic acid, exhibited acceptable lower melting band temperatures.

Consistency in their concentrations is needed in order to confidently decide on the optimum combination of the C-L acid with chemical additive mixture that can provide the best melting characteristics. Reliable information can be obtained from concentrations where high linearity in melting point depression exists. In such case, an additive concentration of 0.61 mol·kg_{C-L}^{-1} offers a safe basis for melting characteristics evaluation. Among the chemical additives, this concentration of methyl salicylate, cineole, and eugenol in the C-L acid provided the least melting band width [61]. On the basis of higher heats of fusion with the manifestation of the agreeable melting temperature depression in the C-L acid mixture, the selection finally boiled down to the C-L acid with methyl salicylate, cineole and eugenol as additives. Methyl salicylate offered the highest heat of fusion that provided the greatest melting point depression, followed by cineole, and eugenol. A narrow melting band provides a better thermal storage capacity in the PCM. **Table 3** presents the relevant melting characteristic information of C-L acid individually combined with 0.1 mole fraction of the chemical additives. The C-L acid with methyl salicylate presented the narrowest melting band width.

Considering their melting temperatures, the desired temperature depression in the C-L acid of 8 to 11°C was not attained in the additive concentration of 0.61 mol·kg_{C-L}^{-1} but was observed in the 1.37 mol·kg_{C-L}^{-1} (Figure 16). Nevertheless, the PCM mixtures with 0.61 mol·kg_{C-L}^{-1} can be employed for industrial process cooling applications. Other probable organic chemicals that are compatible with the C-L acid must be investigated to come up with the right additive that can suffice the required melting point and heat of fusion for the residential space unit cooling application of thermal energy storage.

Table 3. Melting band temperature information of C-L acid with 0.1 mole fraction of chemical additives taken from DSC analysis.

PCM	Melting Peak [°C]	Melting Band [°C]	
		Lower Limit	Upper Limit
C-L Acid	22.4	17.5	24.0
C-L Acid + Caproic Acid	8.8	5.7	21.0
C-L Acid + Cineole	19.7	7.5	21.5
C-L Acid + Ethyl Benzoate	13.3	8.4	22.6
C-L Acid + Eugenol	20.9	8.6	23.8
C-L Acid + Oleic Acid	12.6	1.3	21.7
C-L Acid + Methyl Salicylate	18.9	7.6	21.3
C-L Acid + Pentanediole	19.0	5.5	21.3

Experimental Set-up for Thermal Energy Storage Performance Test

The efficiency of PCMs depends on its ability to store the needed amount of energy and consequently change the temperature of an adjacent space unit or any subject intended for cooling application. Optimum effects obtained from careful evaluation of storage medium performance rely heavily on a good working system. Thus, the storage capsule design is vital in response to a better performance test evaluation. Theoretical analysis corresponds to experimental verification. Hence, they form the basis of the experimental design.

Heat transfer in latent heat storage materials is usually complicated by the phase change that occurs during the absorption or release of energy. Several latent heat thermal energy storage systems used in practical applications made use of PCMs in varied designs such as spherical capsules, vertical, horizontal or even inclined concentric cylinders, rectangular flat plates or slabs, and storage capsules with fins or continuous stirrer. Various efforts included an array of geometric and fluid flow designs to provide effective heat transfer [33,62-70]. Works that focused on heat transfer enhancement in thermal energy storage systems were reviewed. Among the geometries considered, the vertical tube and cylinder was employed in this paper because of its simplicity in structure and easy handling. The PCM was placed in the inner tube in order to discern its effective and steady phase change characteristic.

Temperature Measurement System

In pursuit of an effective storage performance evaluation, the efficiency of the fabricated temperature measurement system has to be confirmed and its adequacy has to be characterized [71,72]. A temperature measurement system was assembled to monitor the temperature during the various stages of phase transition. A sensitive method for temperature measurement during phase change has to be integrated with an efficient storage capsule design providing a basis for a competent evaluation of the performance of the heat storage system.

Base metal K-type chromel and alumel wires each having 0.28 mm diameter and coated with insulator was joined in one end by gas welding. The sensing junction beads were formed in diameters ranging from 0.6 to 0.9 mm. Accuracy of the thermocouples were tested using ice point 0°C as the reference temperature and water at different controlled temperatures. Each thermocouple (TC) was individually calibrated for higher accuracy assurance, showing a ± 0.68% deviation against established K-type standard. The thermocouples were further subjected to heating (0→45.0°C) and then to cooling (45.0→0°C) processes, with the temperatures simultaneously read every 10 s.

A heat transfer process was associated with all temperature measurements. The temperature probe set-up was arranged such that the radial and axial temperature distribution can provide a quantitative energy storage performance evaluation of the C-L acid. The thermocouples were held in 2 rectangular acrylic slabs and the acrylic pipe supported the thermocouple wires. The probe beads were distanced 1.5, 4.5, 7.5, 10.5, and 13.5 mm on both sides from the center of the slab. Individual thermocouple sensors were placed along the inlet and outlet water ports in order to monitor the temperature of water.

The temperature measurement data indicated the capability of the designed system to provide adequate quality. The repeatability of the data at the same temperature range and probe positions ruled out the possible errors on time delay which are difficult to predict. Problems on the analysis of the system behavior were dealt with on the heat transfer between the PCM and the heat transfer fluid, considering factors that should not affect the temperature and eventually, the quantities of heat to be measured. These problems were dealt with and looked into through the estimation and evaluation of the system's heat flux, and the heat transfer coefficient. The heat flux in the C-L acid at varying height positions during melting and solidification indicated the confident level of the experimental set-up.

Quantified estimation showed the negligible effect on the possible enhancement to heat conduction of the thermocouple wires and the acrylic pipe support. The cross-sectional areas accounted for the thermocouple wires and the acrylic pipe support were immaterially small. The probe set-up insignificantly affected the rate of heat transfer in the PCM. Natural convection was not suppressed by the probe set-up. Estimation of the heat transfer coefficients with and without the probe set-up presented negligible effect on heat convection. In unsteady conduction, the temperature of the surrounding water was not only changed but the magnitude of the overall heat transfer coefficient was also significantly changed until a steady condition was achieved. The irregularity in the axial height values as time proceeded suggested complications arising from the mechanism of natural convection and dropping of solid melt during phase transition until melting completion along the PCM column. This depicted the added complexity of temperatures that was a function of time as well as a function of radius and height.

During solidification, the thermal resistance was smaller with lesser difference in the heat flux in different probe height positions. This was due to the fact that the conduction heat transfer during solidification was more stable.

As inward solidification took place, the inner PCMs were dominated by the required latent heat supply which in turn controlled the flow of heat through the rest of the material along the axial height region. A distinctly clearer variation in the overall heat transfer coefficient at varying height positions implied a smoother heat transfer proceeding during heat conduction.

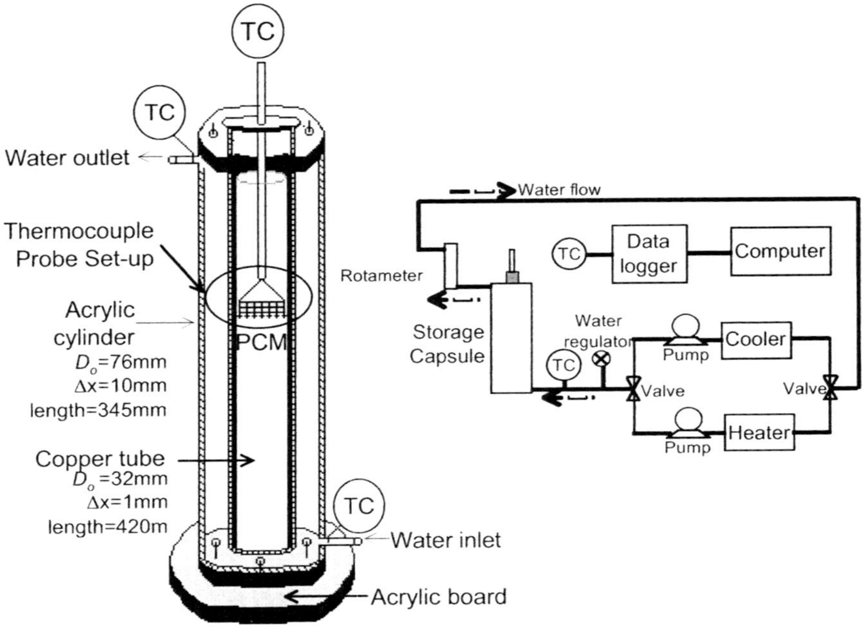

Figure 18. The vertical tube and cylinder storage capsule used in the thermal storage performance tests of C-L acid and with additives. The inset schematic diagram indicates the flow of the charging and discharging processes.

Thermal Energy Storage Facility

A simple experimental set-up was designed to evaluate the melting and solidification characteristics of the C-L acid and its mixture with the prospective and compatible component additives. The heat storage module shown in **Figure 18** consisted of an annular acrylic cylinder and a copper tube. Both ends of the storage module were securely fixed and sealed with acrylic boards. The vertical copper tube was fixed vertically in the center of the cylinder. The PCM was placed inside the copper tube together with the temperature probe sensor. The PCM volume of 2.25×10^{-4} m^3 was set at a length of 325 mm in the storage capsule corresponding to 0.21 kg of C-L acid. Water continuously flowed through the acrylic cylinder which had an inside diameter of 56 mm, thickness of 10 mm, and length of 345 mm. Inlet and outlet provisions provided for the continuous flow of water. The position of the thermocouple probe assembly can be varied at different height positions of 25, 100, 200, and 300 mm from the base of the copper tube. One cycle of melting and solidification operation provided the radial temperature distribution in one height position.

A schematic flow of the experiment is presented in Figure 18. Water served as the heat transfer fluid with its temperature regulated at a constant temperature. Heater and cooler water baths with pumps and temperature controllers were used correspondingly for the charging (melting) and discharging (solidification) processes. During charging, inlet water was maintained each time at 30.0°C, 35.0°C, and 39.8°C for the C-L acid and with its component additives, with the flow rate controlled at 1.67×10^{-5} $m^3 \cdot s^{-1}$. During the discharge process, inlet water was kept at an average temperature of 2.1°C for both storage media at a flow rate of 2.51×10^{-5} $m^3 \cdot s^{-1}$. Water entered at the bottom of the acrylic cylinder and exited at the top of the storage capsule on both melting and solidification processes.

Performance Evaluation of C-L Acid with Chemical Additives as Latent Heat Store for Cooling Application

An additive concentration of 0.61 $mol \cdot kg_{C\text{-}L}^{-1}$ offered a safe basis for melting characteristics evaluation. Among the chemical additives, this concentration of methyl salicylate, cineole, and eugenol in the C-L acid provided the least melting band width. The melting behavior at 0.10 mol fractions of methyl salicylate, cineole, and eugenol was evaluated subjecting each PCM combination to a series of charging and discharging processes. The radial temperature distribution at varying probe height positions provided an image of the solid to liquid interfacial transition during the charging and discharging processes. The melting band elucidated the phase transition region evaluated from the DSC analysis. Below the melting band is the solid phase and above the melting band is the liquid phase.

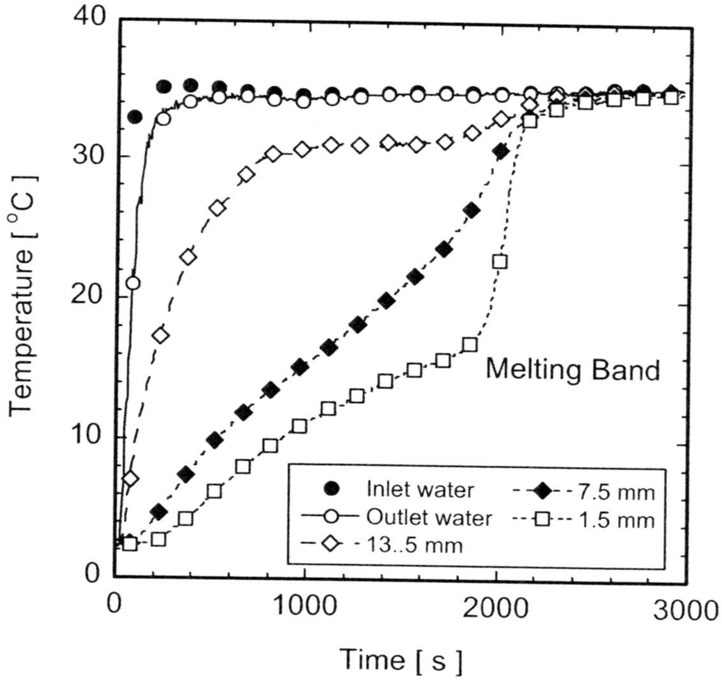

Figure 19. Temperature history of methyl salicylate (0.10 mol fraction) in C-L acid mixture during melting monitored at 100 mm probe height position. Inlet water was maintained at an average temperature of 35°C.

The characteristic feature of the PCM blend was manifested in the melting temperature histories represented in **Figure 19** by the C-L acid mixture with methyl salicylate at one height position during its charging. The phase transition in the PCM mixture did not occur at the same temperature but over a certain range of temperature within its melting band.

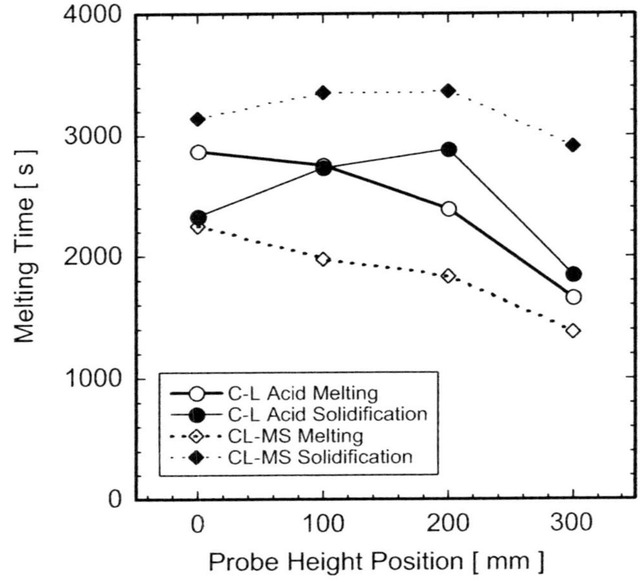

Figure 20. Melting and solidification times of C-L acid and 0.10 mol fraction of methyl salicylate in C-L acid at different probe height positions. Inlet water was maintained at 35°C during melting and at 2°C during solidification

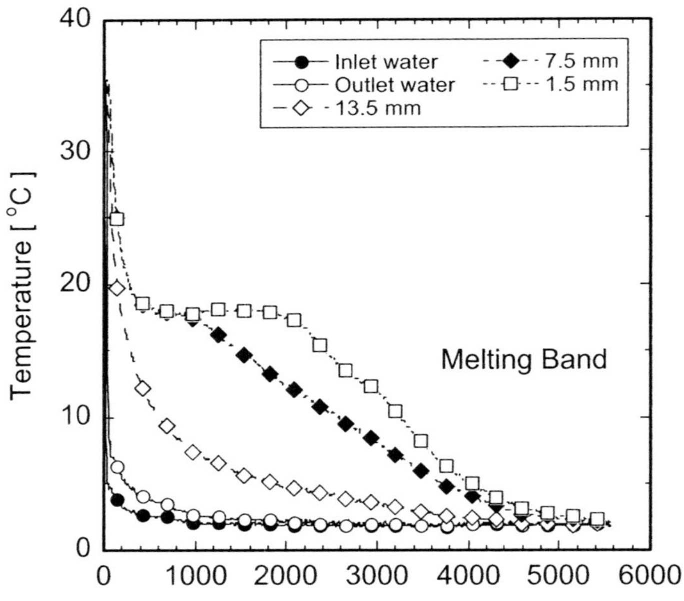

Figure 21. Temperature history of the methyl salicylate (0.10 mol fraction) in C-L acid mixture during solidification monitored at 100 mm probe height position. Inlet water was maintained at an average temperature of 2.1°C.

Consistent with the latent heats of PCM mixtures at 0.1 mol fraction of additives indicated in Table 3, the C-L acid and methyl salicylate combination demonstrated a progressively increasing charged energy followed by the C-L acid and eugenol combination. The upper melting band temperature limit affected the liquid sensible heat yield of the PCM mixtures during melting. The lower melting band temperature affects the sensible heats generated from phase transition up to complete solidification. The broader melting band of the C-L Acid and eugenol combination and its slightly greater heat of fusion caused its greater transferred energy compared to the C-L acid and cineole combination. The choice finally zeroed in on the 0.10 mole fraction of methyl salicylate as the best combination of the C-L acid and additive blend based on the negligible variation with its higher heat of fusion and narrower melting band width.

An efficient storage system for industrial and residential applications requires a thermal energy load completed within a relatively short time period and delivered using small cooling equipment over a longer period of time. The results depicted in **Figure 20** showed the improvement of the C-L acid in terms of higher latent heat with lower melting point in relation to the lesser melting time expended during the charging of the PCM mixture. The melting time of the C-L acid was decreased by addition of methyl salicylate.

The flattening of the radial temperatures within the melting band in **Figure 21** indicated the dominance of freezing. The slow rate of heat transfer provided enough time for the PCM to solidify and the temperature remained almost constant until solidification was complete in all radial positions. From the temperature histories presented, it was implied that solidification took longer time than melting.

The more diluted C-L acid manifested greater melting time among the different concentrations of the C-L acid and methyl salicylate combinations. However, the more concentration of methyl salicylate in the C-L acid signified a longer solidification time manifesting the characteristic ternary blend of the mixture. The melting times shown in Figure 20 shows that melting proceeded from the highest to the lowest layer of the melting band. Solidification took longer time as it proceeded from the bottom to the upper layers along the vertical column of the PCM mixture. Based on the progression of the different PCM mixture solidification times, the lower temperature limit of the melting band showed a strong effect on solidification time.

The C-L acid with methyl salicylate manifested the favorable heat transfer characteristics based on the shorter charge and discharge rates corresponding to a narrower melting band width and a higher heat of fusion value. This ensured an improvement in the melting characteristic of the C-L acid system for the cooling application of thermal energy storage.

CONCLUSION

The latent heat thermal energy storage system employing the C-L acid system as the storage medium offers considerable potential as a viable energy conversion alternative for cooling in industrial and commercial applications. It has sound benefits particularly on the chemical properties, its behavior in relation to melting and freezing characteristics, reasonable energy density, reproducible melting and cooling behavior, congruency in melting, negative property degradation, no supercooling, non-toxic, non-corrosive, and having definite melting temperature.

The identification of the C-L acid as a promising phase change material with its melting characteristics offers strong possibilities for industrial application. It is applicable in process cooling, waste heat recovery or in waste water treatment plants, domestic water systems, or cold convective storage that store the cold at night for daily savings management although the stores are almost always connected to heat pumps. It may be used to support passive, hybrid and low energy cooling techniques needing to rely on the possibility to use some type of short term cold energy storage or diurnal thermal energy storage. They may be implemented on underground thermal energy system in the building, industrial and agricultural sector. Conventional roof mounted flat plate solar panels may serve as solar collectors and the thermal energy is transferred to the C-L acid mixture through a series of pipes under the floor. C-L acid may be impregnated or directly incorporated in construction or building materials as wallboards or ceilings or encapsulated in self-contained grains. Air that would pass over wallboards or concrete incorporated with the C-L acid mixture may be cooled down due to heat absorption when the mixture melts. The wallboards or concrete may consequently be recharged when the C-L acid mixture in it solidifies as its stored latent heat is released to the cool air during the night. In this direction, the thermal characteristics of the C-L acid may further be improved for a more efficient storage system that can provide the necessary comfort in a building or even residential facility. In this context, operating temperature ranges, the thermal storage performance, and the improvement of the performance covering on several courses of evaluation are sure to evolve. Such are the selection of the wallboard or concrete for the C-L acid mixture compatibility, their modifications through addition of appropriate components, their curing effects, duration effects, examination on the various aspects of corrosion, the effects of possible contamination on the thermal properties and many more.

In air conditioners for residential application, a melting point lower than 10°C is a rather ambitious aim. However, the founded decrease in melting point in the C-L acid with the addition of compatible organic additives have offered conviction of its attainment provided further investigation on other component additives will be carried out in the future. In due time, a design may also be developed for air conditioning employing the C-L acid mixture having a melting point of 12.5°C.

With the information provided on the C-L acid's melting characteristics and thermal performance, future works dealing with its application that seem to have more suitability as far as its melting point is concerned, will be of great help to interested research bodies.

REFERENCES

[1] Telkes, M. (1974). Solar energy storage. *ASHRAE J.*, *16* (9), 38-44.

[2] Telkes, M. (1947). Solar house heating- a problem of heat storage. *Heating and Ventilating, 44* (1), 68-75.

[3] Telkes, M., & Raymond, E. (1949). Storing solar heat in chemicals – a report on the Dover house. *Heating and Ventilating*, *46* (1), 80-86.

[4] Dimaano, M. N. R. & Watanabe, T. (2002). Performance investigation of the capric and lauric acid mixture as latent heat energy storage for a cooling system. *Solar Energy, 72* (3), 205-215.

[5] Abhat, A. (1983). Low temperature latent heat thermal energy storage: heat storage materials. *Solar Energy, 30* (4), 313-332.

[6] Zalba, B., Marin, J. M., Cabeza, L. F., & Mehling., H. (2003). Review on thermal energy storage with phase change: materials, heat transfer analysis and applications. *Applied Thermal Engineering, 23*, 251-283.

[7] Hawlader, M. N. A., Uddin, M. S., & Khin, M. M. (2003).Microencapsulated PCM thermal-energy storage system. *Applied Energy, 74* (3), 195-202.

[8] Kauranen, P., Peipopo, K., & Lund, P. D. (1991). An organic PCM storage system with adjustable melting temperature. *Solar Energy, 46* (5), 275-278.

[9] Cui, H., Yuan, X., & Hou, X. (2003). Thermal performance analysis for a heat receiver using multiple phase change materials. *Applied Thermal Engineering, 23* (18), 2353-2361.

[10] Farid, M. M., Khudhair, A. M., Razack, S. A. K., & Al-Hallaj, S. (2004). A review on phase change energy storage: materials and applications. *Energy Conversion and Management, 45* (9-10), 1597-1615.

[11] Gu, Z., H. Liu, & Li, Y. (2004). Thermal energy recovery of air conditioning system – heat recovery system calculation and phase change materials development. *Applied Thermal Engineering, 24* (17-18), 2511-2526.

[12] Feldman, D., Banu, D., & Hawes, D. W. (1995). Low chain esters of stearic acid as phase change materials for thermal energy storage in buildings. *Solar Energy Materials and Solar Cells, 36* (1) 311-322.

[13] Valkilaltojjar, S. M.& Saman, W. (2001). Analysis and modeling of a phase change storage system for air conditioning applications. *Applied Thermal Engineering, 21* (3), 249-263.

[14] Sari, A. (2004). Form-stable paraffin/high density polyethylene composites as solid-liquid phase change material for thermal energy storage: preparation and thermal properties. *Energy Conversion and Management, 45* (13-14), 2033-2042.

[15] Hamdan, M. A. & Elwerr, F. A. (1996). TES using a PCM. *Solar Energy, 56* (2), 183-189.

[16] Dimaano, M. N. R. & Watanabe, T. (2002). The capric-lauric acid and pentadecane combination as PCM for cooling applications. *Applied Thermal Engineering, 22* (4), 367-377.

[17] Jin, Z., Asako, Y., Yamaguchi, Y., & Yoshida, H. (2000). Thermal and water storage characteristics of super-absorbent polymer gel which absorbed aqueous solution of calcium chloride. *Int. J. Heat Mass Transfer, 43* (17), 3407-3415.

[18] Hidaka, H., Yamazaki, M., Yabe, M., Kakiuchi, H., Ona, E., Kojima, Y., & Matsuda, H. (2004). New PCMs prepared from erythritol-polyalcohol mixtures for latent heat storage between 80 and 100°C. *Journal of Chemical Engineering of Japan, 37* (9), 1155-1162.

[19] Dimaano, M. N. R. & Escoto, A. (1998). Preliminary assessment of a mixture of capric acid and lauric acids for low-temperature thermal energy storage. *Energy, 23* (5) 421-427.

[20] Sari, A. & Kaygusuz, K. (2001). Thermal energy storage system using stearic acid as a phase change material. *Solar Energy, 71* (6), 365-376.

[21] Sari, A. (2003). Thermal characteristics of a eutectic mixture of myristic and palmitic acids as phase change material for heating applications. *Applied Thermal Engineering, 23* (8), 1005-1017.

[22] Farid, M., Kim, Y., Honda, T., & Kanzawa, A. (1989). The role of natural convection during melting and solidification of PCM in a vertical cylinder. *Chem. Eng. Comm., 84* (1), 43-60.

[23] Feldman, D, Banu, D., & Hawes, D. W. (1995). Development and application of organic phase change mixtures in thermal storage gypsum wallboard. *Solar Energy Materials and Solar Cells, 36* (1), 147-157.

[24] Feldman, D., Banu, D., Hawes, D.W., & Ghanbari, W. (1991). Obtaining an energy storing building material by direct incorporation of an inorganic phase change material in gypsum wallboard. *Solar Energy Materials. 22*, 231-242.

[25] Feldman, D., Khan, M. A., & Banu, D. (1989). Energy storage composite with an organic PCM. *Solar Energy Materials, 48* (3), 333-341.

[26] Shapiro, M. M., (1988). PCM wallboard tempers passive-solar temperature swings *Sun World, 12* (1), 46-47.

[27] Inaba, H. & Tu, P. (1997). Evaluation of thermophysical characteristics on shape-stabilized paraffin as a solid-liquid phase change material. *Trans. Journal of Heat and Mass Transfer, 32* (2), 307-312.

[28] Hawlader, M. N. A., Uddin, M. S., & Khin, M. M. (2003). Microencapsulated thermal energy storage system. *Applied Energy, 74* (1-2), 195-202.

[29] Hasan, A., (1994). Phase change material energy storage system employing palmitic acid. *Solar Energy, 52* (1), 143-154.

[30] Neeper, D. A., (2000). Thermal dynamics of wallboard with latent heat storage. *Solar Energy, 68* (5), 393-403.

[31] Banu, D., Feldman, D., Haghighat, F., Paris, J., & Hawes, D. (1998). Energy-storing wallboard: flammability tests. *J. Materials in Civil Engineering, 10* (2), 98-105.

[32] Paris, J., Falardeau, M., & Villeneuve, C. (1993). Thermal storage by latent heat: a viable option for energy conservation in buildings. *Energy Sources, 15*, 85-93.

[33] Cho, K. & Choi, S. H. (2000). Thermal characteristics of paraffin in a spherical capsule during freezing and melting processes. *Int. J. Heat Mass Transfer, 43* (17), 3183-3196.

[34] Aboul-Enein, S. & Olofa, S.A. (1991). Thermophysical properties of heat-of-fusion storage materials for cooling applications. *Renewable Energy, 1* (4), 791-797.

[35] Hawes, D. W., Feldman, D., & Banu, D. (1993). Latent heat storage in building materials. *Energy and Buildings, 20* (1), 77-86.

[36] Ames, D. E., (1990). Eutectic cool storage: current developments. *ASHRAE Journal, 32* (4), 46-53.

[37] Himran, S., Suwono, A., & Mansoori, G. A. (1994). Characterization of alkanes and paraffin waxes for application as phase change energy storage medium. *Energy Sources, 16* (1), 117-128.

[38] Ghoneim, A. A., (1989). Comparison of theoretical models of phase change and sensible heat storage for air and water-based solar heating systems. *Solar Energy, 42* (3), 209-220.

[39] Tan, F. L. & Leong, K.C. (1996). Freezing inside an enclosure using *n*-paraffin materials : an experimental study. *J. Materials Processing & Manufacturing Science, 5* (1), 45-55.

[40] Py, X., Olives, R., & Mauran, S. (2001). Paraffin/porous-graphite-matrix composite as a high and constant power thermal storage material. *Int. J. Heat Mass Transfer, 44* (14), 2727-2737.

[41] Hirasawa, Y., Saito, A., Sato, T., Takegoshi, E., & Imura, S. (1991). A study on phase changes of heterogeneous composite materials. *JSME Inter. Journal II, 34* (2), 189-194.

[42] Choi, E., Cho, Y. I., & Lorsch, H.G. (1992). Thermal analysis of the mixture of laboratory commercial grades hexadecane and tetradecane. *Int. Comm. Heat Mass Transfer, 19* (1), 1-15.

[43] Py, X., Olives, R., & Mauran, S. (2001). Paraffin/porous-graphite-matrix composite as a high and constant power thermal energy storage. *Int. J. Heat Mass Transfer, 44* (14), 2727-2737.

[44] Watanabe T., Kikuchi, H., & Kanzawa, A. (1993). Enhancement of Charging and Discharging Rates in a Latent Heat Storage System by Use of PCMs with Different Melting Temperatures. *Heat Recovery Syst. & CHP, 13* (1) 57-66.

[45] Sari, A., H. Sari, & Onal, A. (2004). Thermal properties and thermal reliability of eutectic mixtures of some fatty acids as latent heat storage materials. *Energy Conversion and Management, 45* (3), 365-376.

[46] Banzon, J. A. & Resurreccion, A. P. (1979). Fatty acid distribution in coconut oil obtained by four processing methods and secured from four Philippine Types of Coconuts. *The Philippine Journal of Coconut Studies, 4* (2) 1-8.

[47] Mark H. F., Othmer, D. F., Overrberger, C.G., & Seaborg, G.T. (1985). Kirk-Othmer Encyclopedia of Chemical Technology. 3rd ed., 4, New York, John Wiley and Sons, 86-187.

[48] Bareiss, M. & Beer, H. (1980). Influence of natural convection on the melting process in a vertical cylindrical enclosure. *Letters in Heat and Mass Transfer 7*, 329-338.

[49] Sparrow, E. M. & Broadbent, J. A. (1982). Inward melting in a vertical tube which allows free expansion of the phase-change medium. *Trans. ASME, Journal of Heat Transfer, 104*, 309-315.

[50] Sparrow, E. M., Patankar, S.V., & Ramadhyani, S. (1977). Analysis of melting in the presence of natural convection in the melt region. *Trans. ASME, Journal of Heat Transfer, 99*, 520-526.

[51] Farid, M. M. & Mohammed, A. K. (1987). Effect of natural convection on the process of melting and solidification of paraffin wax. *Chem. Eng. Comm., 47*, 297-316.

[52] Dimaano, M. N. R. & Watanabe, T. (2003). The Transient Behavior Analysis of The Capric Acid and Lauric Acid Mixture in a Vertical Tube. *Journal of Chemical Engineering of Japan, 36* (12), 1421-1431.

[53] Mark, H. F., Othmer, D .F., Overrberger, C. G., & Seaborg, G. T. (1985). Kirk-Othmer Encyclopedia of Chemical Technology 3rd ed., 4, New York, John Wiley and Sons, 86-187.

[54] Humpries W. R. & Griggs, E. I. (1977). A Design Handbook for Phase Change Thermal control and Energy Storage Devices. NASA Technical Paper 1074, 35.

[55] Lide D. R. (1995). CRC Handbook of Chemistry and Physics. 76th ed., Boca Raton, Florida, CRC Press.

[56] Farid, M., Kim, Y., Honda, T., & Kanzawa, A. (1989). The role of natural convection during melting and solidification of PCM in a vertical cylinder. *Chem. Eng. Comm., 84*, 43-60.

[57] Parsons, R. A., (1987). ASHRAE Handbook: HVAC Systems and Applications. GA, American Society of Heating, Refrigerating and Air Conditioning Engineers, Inc.

[58] MacGregor, R. K. & Emery, A. P. (1961). Free convection through vertical plane layers: moderate and high Prandtl number fluids. *Trans. ASME, Journal of Heat Transfer, 91*, 391-403.

[59] Fand, R.M. (1965). Heat transfer by forced convection from a cylinder to water in crossflow. *Int. J. Heat Mass Transfer, 8* (7) 995-1010.

[60] Sohns, J., Seifert, B., & Hahne, E. (1981). The effect of impurities on the melting temperature and the heat of fusion of latent heat storage materials. *International Journal of Thermophysics, 2 (1)*, 71-87.

[61] Dimaano, M. N. R. & Watanabe, T. (2002). The capric and lauric acid mixture with chemical additives as latent heat storage materials for cooling application. *Energy, 27 (9)*, 869-888.

[62] Gong, Z. X. &. Mujumdar, A. S (1996). Enhancement of energy charge-discharge rates in composite slabs of different phase change material. *Int. J. Heat Mass Transfer, 39* (4), 725-733.

[63] Farid, M. M., Kim, Y., & Kanzawa, A. (1990). Thermal performance of a heat storage module using PCMs with different melting temperature: Experimental. *Trans. ASME, Journal of Solar Energy Engineering, 112*, 125-131.

[64] Kim, W. S. & Song, H. O. (1988). A study of the melting heat transfer characteristics of energy storage system utilizing the PCM. *Journal of Chemical Engineering of Japan, 21* (1), 69-74.

[65] Kemink, R. G. & Sparrow, E. M. (1981). Heat transfer coefficients for melting about a vertical cylinder with or without subcooling and for open or closed containment. *Int. J. Heat Mass Transfer, 24* (10) 1699-1710.

[66] Brosseau, P. & Lacroix, M. (1994). Numerical simulation of a compact multi-layer phase change material energy storage system. *ASME Fundamentals of Phase Change: Sublimation and Solidification, 286*, 13-20.

[67] Aceves, S. M., Nakamura, H., Reistad, G. M., & Martinez-Frias, J. (1998). Optimization of a class of latent thermal energy storage systems with multiple phase change materials. *Trans. ASME, Journal of Solar Energy Engineering, 120* (21), 14-19.

[68] Sari, A. & Kaygusuz, K. (2001). Thermal energy storage system using stearic acid as a phase change material. *Solar Energy, 71* (6), 365-376.

[69] Saitoh, T. & Hirose, K. (1986). High performance phase-change thermal energy storage using spherical capsules. *Chem. Eng. Comm., 41*, 39-58.

[70] Takeda, S., Nagano, K., Mochida, T., & Shimakura, K. (2004). Development of a ventilation system utilizing thermal energy storage for granules containing phase change material. *Solar Energy, 77* (3), 329-338.

[71] Dimaano, M. N. R. & Watanabe, T. Temperature measurement system for the mixture of capric Acid and lauric acid as a low-temperature thermal energy storage medium. *Acta Manilana, 48* (1), 1-10 (2000).

[72] Segall, A. E. (2001). Solutions for the correction of temperature measurements based on beaded thermocouples. *Int. J. Heat Mass Transfer, 44* (15), 2801-2808.

In: Solar Energy: New Research
Editor: Tom P. Hough, pp.43-70
ISBN 1-59454-630-4

Chapter 2

SMART MATERIALS FOR TEMPERATURE REGULATION IN PASSIVE SOLAR BUILDINGS

R. Nikolić*, J. Savović,
Institute of Nuclear Sciences "Vinča", Belgrade, Serbia and Montenegro
S. Gadžurić, I. Zsigrai
Faculty of Sciences, University of Novi Sad, Novi Sad, Serbia and Montenegro

ABSTRACT

In this chapter we present a review of our results on thermal energy storage in passive solar buildings. Thermophysical properties of a large number of phase change materials were studied, particularly those undergoing solid/liquid transition near ambient temperature. The materials studied can be applied in two typical, so called "intelligent" solar objects: a) in transparent walls of agricultural greenhouses and b) incorporated with building materials of wallboards in residential buildings.

The materials studied for thermal storage in transparent walls are considered to be an original approach, with the combination of isothermal latent heat storage and the thermochromic behavior of cobalt (II) chloride complex compounds dissolved in selected low melting mixtures. In this way self-adaptive control of temperature and light intensity can be attained through physico-chemical processes.

For the purpose of possible incorporation with various building materials, to be used in passive solar buildings with storage in walls, we studied a large number of organic compounds and their mixtures. The following materials showed almost ideal behavior for reversible heat storage and easy incorporation in building materials: hexadecane, poly (ethylene glycol-400) distearate, methyl stearate, methyl palmitate, cetyl stearate, cetyl palmitate and their mixtures of defined composition.

The value of the available stored energy in walls impregnated with studied phase change materials (PCMs) exceeds all known systems studied to date.

* Corresponding author: Tel.: +381 11 2453967; fax:+381 11 4447207.

Introduction

The main priorities of sustainable developments are closely connected to energy demands and environmental concerns. The prospect of energy future based on renewable energy sources and additional efforts in energy savings enable us to contribute significantly in solving both problems. It is estimated that heating of buildings makes up more than one third of primary energy consumption and is a major source of urban air pollution. By promoting the use of solar thermal energy for heating purposes it is possible to cut energy demands to a very high degree, not only in the civil sector but also in public services, as well as for industrial processes and agriculture. Solar thermal energy can be utilized in:

a) Active solar systems with the use of solar collectors, heat exchangers, heat storage units, pumps, ventilation systems and equipment for automatic control of temperature, lights, humidity and other relevant parameters;
b) Passive solar systems in which heat absorption and regulation is attained by the special construction of buildings, enabling maximum heat gain and by the use of special materials, serving as heat storage and/or self-regulation systems of the relevant parameters.

Efficient application of solar energy both in active and passive systems requires efficient thermal storage. The most convenient thermal storage of low grade heat can be accomplished as latent heat of phase change, usually solid-liquid or solid-solid phase transition. In this way high storage density can be attained. Furthermore, heat is stored almost isothermally, in a narrow temperature range in which phase change occurs. The phase change temperature should correspond with the specific application, thereby enabling minimization of heat losses.

The study of phase change materials for possible use as latent heat storage units in active solar systems was pioneered by M. Telkes' work on Glauber salt as a latent heat storage material [1]. These studies provoked further screening of inorganic and organic materials for potential use in low grade thermal storage. Comprehensive lists of PCMs for possible application as latent heat storage media, their thermal properties, and the most important criteria for their possible application in energy storage units were published [2-6]. However, a large number of the materials listed showed very serious disadvantages (i.e. incongruent melting, toxicity, thermal and chemical instability, low thermal conductivity, high price e.t.c.) making them unavailable for practical use. Therefore, studies of new materials and their thermophysical properties were continued in an attempt to obtain a better choice of materials with a good balance of the necessary criteria to be fulfilled for particular applications.

Characterization of Phase Change Materials

A considerable number of new PCMs have been thoroughly explored according to following methodology:

- Selection of materials for desirable temperature range;
- Characterization of thermophysical properties;

- Reversibility, short term and long term studies;
- Thermal cycling;
- Compatibility with construction materials (for passive application).

Selection of materials according to phase transition temperatures is only a first step, which is closely associated with climatic conditions and the purpose of application. For heating purposes the chosen PCM should have a transition temperature slightly above comfortable ambient temperature, enough to allow an efficient heat withdrawal rate. In order to minimize heat losses it is preferable to minimize the difference between the phase change temperature and desired ambient temperature in the solar object. An efficient heat exchange rate is attained with PCMs, which have high thermal conductivity and heat storage density.

Therefore it is important to define all the relevant thermophysical properties of chosen PCMs, as enthalpy of phase change, specific heat, density and volume change and thermal conductivity. The most suitable and largely used experimental method for measurement of enthalpy and specific heat is differential scanning calorimetry (DSC). This method was useful for laboratory scale studies of reversibility, supercooling, short and long term stability of a large number of inorganic and organic PCMs [7-11]. DSC measurements proved to be useful for the studies of PCMs by our research group, particularly in tailoring acetamide based eutectics with various salts [12-16].

Other experimental methods and simplified calculations were also used for the characterization of thermophysical properties, long and short-term stability and reversibility phase change transition [17-22]. Supercooling and incongruent melting was shown to be the main problem occurring in salt hydrates and a large number of seeding substances were used to improve the crystallization properties of salt hydrates [3, 22-23]. However, low thermal conductivity is a limiting factor in the use of organic phase change compounds for practical applications. Compensation of poor thermal conductivity in active solar heating systems can be achieved with appropriate design of the container and choice of container material. In passive solar systems mixing of PCMs with construction materials helps in improving heat exchange.

Selection of PCMs for Passive Solar Applications

Passive solar systems are highly economic and well suited for large scale use, particularly in the areas with fluctuating climatic conditions. Appropriate design of buildings, based on the long term study of climatic conditions and well chosen building materials with good storage capacity and efficient insulation can contribute to significant energy savings. It is of particular importance to attain on-peak energy savings and 24 hour storage of solar thermal energy. For these purposes, incorporation of suitable phase change materials in wallboards, roofs, ceilings and floors offers the greatest benefit. Passive solar systems can also be integrated into agricultural greenhouses to reduce their energy consumption for heating.

It has been claimed that energy savings of up to 15 % of the annual energy cost can be attained in buildings with a PCM wall [24]. Energy savings in agricultural greenhouses with solar thermal storage can reach 20-75 % of the annual energy costs [25]. However, parametric simulation studies of PCM wallboards as energy load managing device showed an almost limitless possibility to increase the energy efficiency of a building with the appropriate choice of PCM [26] and reasonable glazing area. Using a Transient System Simulation Program (TRNSYS) to test a PCM wallboard containing a PCM with a variable transition temperature, amount of PCM present and under variable climatic conditions, it has been shown that the payback of investment costs resulting from annual energy savings in a passive solar building is three to five years [26-29]. Simulations and model calculations performed in the study showed that the energy savings were directly related to the concentration of PCM in the wallboard, but model calculations did not reach general conclusions regarding optimal PCM properties. Parametric analysis can only identify the sensitivities of storage capacity to a number of sub optimum properties and details of the passive solar system. For instance, increasing insulation levels or highly insulated glazing areas may alter the effectiveness of a certain storage material and favor one storage material over another.

The most obvious factors influencing the effectiveness of a passive solar building are the thermophysical properties of the heat storage material. Therefore it is essential to choose the appropriate phase change materials for possible use in corresponding climatic conditions, architectural designs and for that particular application.

The following parameters are decisive in the choice of PCMs for passive applications:

- Temperature range of phase change (depending on climatic region);
- High enthalpy of phase change;
- Compatibility with building materials.

The most convenient transition temperature range is 22-28 ^{o}C, depending on the climatic conditions and in particular on the architectural design. Although a large number of PCMs are available for possible practical applications, there are a limited number of PCMs with phase transitions in the desired temperature range. Therefore there is a need for further studies of new materials to be used in walls with thermal storage.

Materials for Simultaneous Control of Solar Radiation and Temperature

Passive solar objects are usually designed with large windows or with wide glazing areas (sun space), oriented to the south or south-west. Thus, in some periods of the year overheating is often observed, unless a high mass of PCM is contained within the wall, or an automated regulation system is provided [30]. To overcome this problem, and to avoid additional energy costs for air conditioning, studies of sophisticated glazing materials have recently been carried out [31]. Switchable or "smart windows" change their optical properties in response to a signal produced by an electric field or by temperature and/or light intensity. Some of the existing technologies are commercially available but found only limited applications in building and vehicle windows due to high price [32]. Therefore further studies of smart glazing materials have been carried out by several research groups [33-35].

Most extensively applied passive solar objects in past few decades were agricultural greenhouses. The main purpose of their use is to produce agricultural products outside their usual cultivation season. In other words, the application of agricultural greenhouses increases cultivation time, quality and yield of most agricultural products. In addition, growing of delicate plants can be achieved in areas with normally unsuitable climatic conditions for that particular plant.

In order to fulfill the demand for satisfactory products quality and yield it is necessary to achieve suitable indoor conditions, especially in periods and areas with extreme climatic fluctuations. Thus, the following parameters must be controlled:

- temperature
- humidity
- light intensity
- indoor air quality.

Many of the parameters listed above are very much dependent on the type of energy source used for maintenance of the required temperature range. The application of solar energy as much as possible in agricultural greenhouses and use of other energy sources as a supplement, only when unavoidable in order to maintain a desired temperature range, is inevitable. This can be achieved with latent heat storage in a PCM, with phase transition temperature range varying from 20– 35 °C, depending on the cultivating plants.

Overheating is a significant problem in agricultural greenhouses in hot or fluctuating climate areas. Therefore extreme outdoor conditions must be balanced with both heating and cooling systems, special design and orientation, moveable isolation, roof or shading [25, 36-38].

Our research group developed an original method of simultaneous control of temperature and light intensity. In order to balance the thermal capacity of the solar heating and storage system, the possibility of using a phase change heat storage material in combination with thermochromic substances was proposed, to be placed in the south side wall or on the south part of the roof. The thermochromic substance is a transition metal complex with mixed ligands, dissolved in the phase change material. The complex undergoes reversible changes of optical properties in response to a change of temperature, acting thus as a self-regulating shading [39]. The phase change material may be placed in the south wall or on the south part of the roof, between two transparent plates, made from simplest and inexpensive plastics or glass.

The materials applied for the above procedure are required to satisfy the following criteria:

a) Thermophysical properties:
   - The melting point close to ambient temperature
   - High value of the enthalpy of fusion
   - Reversible phase change process, minimized supercooling
b) Spectroscopic properties
   - The change of molar absorption coefficient of the thermochromic complexes in the visible spectral range from low values at room temperature to high values at temperatures elevated up to 60 °C.

The concentrations of the additives for thermochromic complex formation are very low and therefore they should not change the thermophysical properties of the phase change material. They have no significant influence on the total costs of the heat storage system.

Diagram presented in Figure 1 schematically describes the process of phase change and thermochromic auto shading.

The phase change material should be transparent in the molten state. The metal complex compound should exhibit low absorbance in the visible spectral range at the melting point. When the melting process is completed (at T_{l1}) the phase change material acts as sensible heat storage system, and its temperature increases (till T_{l2}). As a result of thermochromic auto shading the temperature inside the object does not increase above T_{l3}. Studies by Hamdan showed the temperature effect of mechanical shading on the thermal gains inside a solar object with large glazing areas [40]. Hence, as a result of shading $T_{l3} < T_{l2}$.

From earlier studies it is known that such complexes are formed in solvents containing mixed ligands like chloride, nitrate, water or an organic polar compound [41, 42].

We utilized the fact that cobalt (II) halide complexes in various media with mixed ligands undergo changes of the coordination and rearrangements of the ligands with temperature [43], as the basis for the study of the changes of the absorption spectra in the visible spectral range.

The first step in our studies was to select the phase change materials undergoing solid-liquid transition above the average ambient temperature in the summer period and to define their thermophysical properties. The second criterion was that the material in molten state is a good solvent for complex formation of thermochromic species.

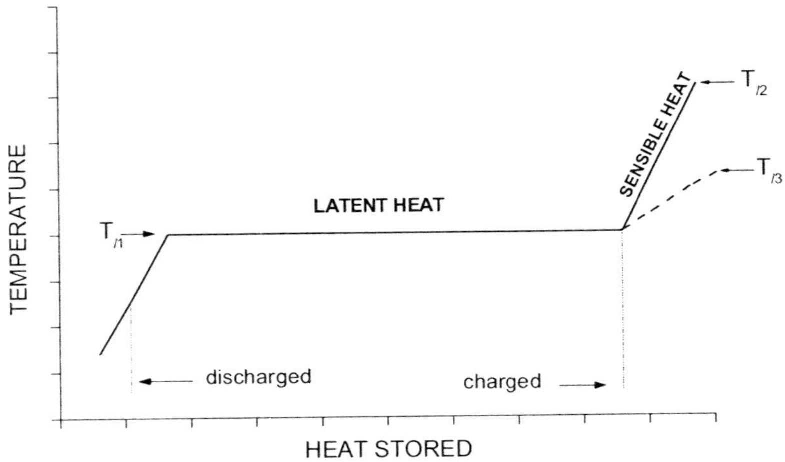

Figure 1. Schematic presentation of combined latent heat storage and auto shading.

Two binary mixtures with variable compositions were thoroughly studied by DSC measurements [44-45]:

1. Binary mixture of calcium nitrate tetrahydrate and calcium chloride hexahydrate
2. Binary mixture of acetamide with calcium nitrate tetrahydrate.

The compositions of the mixtures were adjusted so that their melting temperatures were suitable for passive solar applications and they were appropriate solvent media for thermochromic cobalt (II) halide complexes. The following compositions of PCMs satisfying both thermophysical and spectroscopic criteria were selected as solvents for Co (II) thermochromic complex (TC) species:

PCM TC (1): $0.925\ Ca(NO_3)_2 \cdot 4.06\ H_2O + 0.075\ CaCl_2 \cdot 6.11\ H_2O$
PCM TC (2): $0.9\ CH_3CONH_2 + 0.1\ Ca(NO_3)_2 \cdot 4\ H_2O$

THERMOPHYSICAL PROPERTIES OF PCMS WITH THERMOCHROMIC COMPLEXES (TC)

The most important thermophysical properties: melting and crystallization temperature (T_m and T_c), enthalpy of melting and crystallization (ΔH_m and ΔH_c) and heat capacity (c_p) at the melting/crystallization temperature, obtained by DSC measurements, are presented in Table 1.

T_m and T_c are temperatures of melting and crystallization, respectively, ΔH_m and ΔH_c are heat of fusion and heat of crystallization, respectively and c_p is the heat capacity obtained in the solid state measured at the beginning of the deviation of the melting curve from the baseline.

Table 1. Thermophysical properties of PCM TC(1) and PCM TC(2) [44]

Mixture	T_m	T_c	ΔH_m	ΔH_c	c_p (m.p.)	Storage density
	(°C)	(°C)	(kJ/kg)	(kJ/kg)	($J/K^{-1}g^{-1}$)	(MJ/m^3)
PCM TC(1)	35.6	32.4	155	130	2.96	265
PCM TC(2)	27.7 (22.3)[a]	25.1	141	127	2.51	168

[a)] Value in brackets corresponds to the melting temperature of the metastable crystal form.

The most frequently used phase change material for these purposes is $CaCl_2{\cdot}6H_2O$, because it is a low cost material with a melting point of 29.8 °C and has a fairly high heat storage capacity of 150 kJ/kg or 230 MJ/m^3 [25]. However, supercooling and sensitivity towards moisture are serious disadvantages for long-term use of the storage unit with $CaCl_2{\cdot}6H_2O$. In most cases special thickening and seeding agents are added to the hydrated salt [46-47]. Calcium nitrate tetrahydrate is also considered as potentially good latent heat storage material but with somewhat higher melting point (43 °C) [48]. None of these hydrates are suitable as solvents for the chromogenic behavior of cobalt (II). From the work of Pacak and Slama [49] it is known that in the mixture of $CaCl_2$ - $Ca(NO_3)_2$ - H_2O there is a large concentration range of the components with crystallization temperatures ranging between 29 °C and 43 °C. DSC measurements of the compositions with variable concentrations of the two hydrates were followed by spectroscopic measurements in those mixtures, in order to obtain a composition with the best thermochromic behavior. Additions of calcium chloride hexahydrate to calcium nitrate tetrahydrate caused the lowering of the melting temperature, but also broadening of the sharp melting peak of calcium nitrate tetrahydrate. In Figure 2 the melting thermogram of a sample of PCM TC(1) compared with the melting thermogram of a sample of pure calcium nitrate tetrahydrate are presented. The melting point of PCM TC(1) at 35.6 °C obtained from DSC measurements agreed with the temperature calculated from the literature data [49].

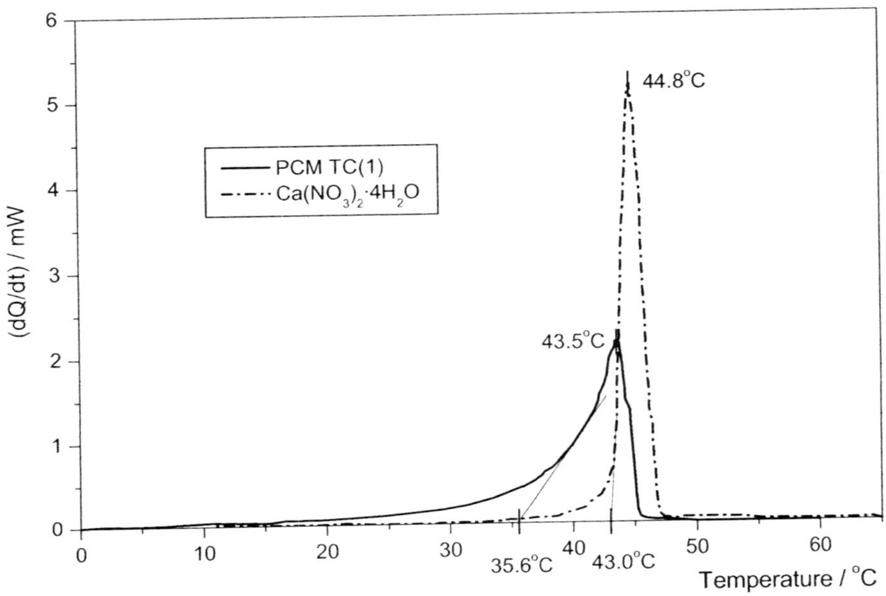

Figure 2. The melting thermograms of PCM TC(1) and of pure $Ca(NO_3)_2 \cdot 4\ H_2O$, scanned at heat rate 5 °C/min, [45].

Studies of the thermophysical properties of PCM TC(2) were based on earlier DSC measurements in acetamide based binary mixtures [12, 13] and further detailed spectroscopic measurements and evidence of the chromogenic behavior of Co (II) mixed complexes in pure acetamide and some of the studied mixtures [42, 49, 50].

DSC measurements of the melting temperatures in the binary mixture of acetamide and calcium nitrate tetrahydrate showed that there is a mutual solubility of the components in the entire concentration range, some mixtures having melting points below 0 °C. A pronounced thermochromic behavior of cobalt (II) chloride complexes at a defined excess of chloride ions was also observed in the mixture containing 0.8 CH_3CONH_2 + 0.2 $Ca(NO_3)_2 \cdot 4\ H_2O$, but this mixture was liquid below 0 °C [42]. In order to obtain a mixture with a melting temperature in the desirable range gradual addition of calcium nitrate tetrahydrate to acetamide was made. The changes of the melting thermograms were followed by DSC. Following additions of calcium nitrate tetrahydrate the sharp melting peak of acetamide, with melting point at 80.9 °C [12] was smeared and the peak was gradually moved towards lower temperature until the composition of the mixture was 0.9 CH_3CONH_2 - 0.1 $Ca(NO_3)_2 \cdot 4\ H_2O$. At this composition a sharp melting peak was obtained at 27.7 °C, with an inflection at ≈ 22 °C, indicating formation of an eutectic [51]. Similarly to other acetamide - nitrate salt mixtures [12, 13] in some subsequent DSC runs the melting peak appeared at 22 °C, behavior typical for the metastable crystal formation of acetamide and its mixtures. In this mixture supercooling was ≤ 3 °C. In Figure 3 typical melting and crystallization thermograms of a PCM(2) sample are shown. The two distinct peaks at the crystallization scan show co-existence of the two crystalline forms, which also have different enthalpies of fusion. It has been shown that the stable crystalline form usually appeared in samples crystallized from solution with mass larger than 10 grams. Moreover, the stable form was always obtained from the mixture 0.9 CH_3CONH_2 + 0.1 $Ca(NO_3)_2 \cdot 4\,H_2O$ when it was cooled to at least 10 °C below the crystallization temperature.

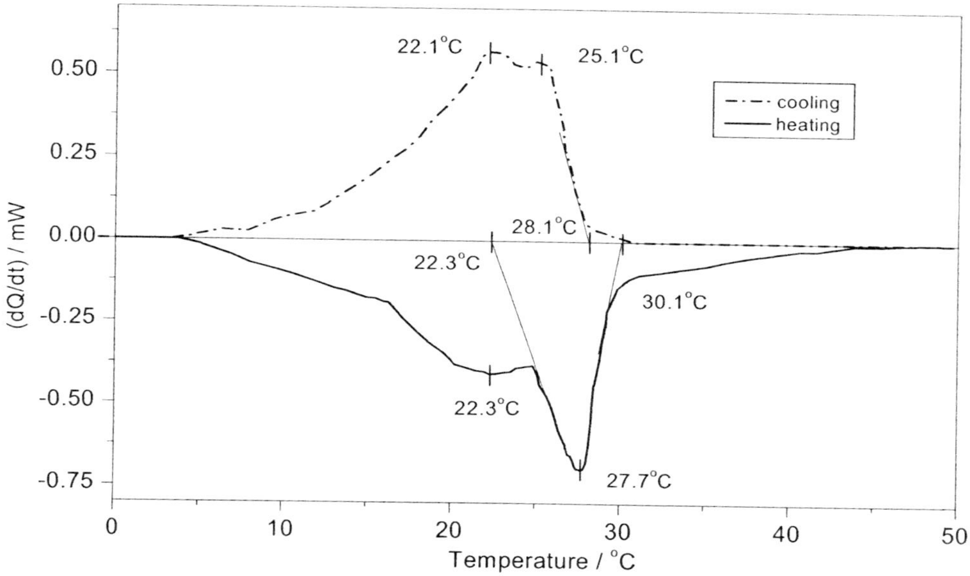

Figure 3. DSC thermograms of a sample of PCM TC(2), with two co-existing crystalline forms, scanned at heat rate 5 °C/min [45].

Phase transition in both PCM TC(1) and PCM TC(2) does not occur at a sharp temperature, which is not a great disadvantage for application in solar heated greenhouses. It is always preferable to have single compound or pure eutectic, with sharp melting point. However, none of the PCMs with sharp phase transition temperatures, studied so far, showed satisfactory optical properties in the desired temperature range.

Spectroscopic Properties of PCM TC(1) and PCM TC(2)

In order to satisfy the required spectroscopic properties and to obtain the desired thermochromic behavior in the desired temperature range (relevant for passive solar applications), it is necessary to add the appropriate amount of a transition metal salt together with the appropriate amount of ligand. Thus, chemical equilibria between various complex species are established in the phase change material. The change of color is a result of the shift of chemical equilibria with a change of temperature. As a first step, absorption spectra of cobalt (II) chloride complexes in liquid or supercooled $CaCl_2 \cdot 6H_2O$ and in $Ca(NO_3)_2 \cdot 4H_2O$ were recorded in the temperature range 28–60 °C and visible wavelength range (400-750 nm). Cobalt species dissolved in these pure hydrates do not show color changes with a change of temperature. Cobalt chloride species in $CaCl_2 \cdot 6H_2O$ are blue and absorption spectra show three distinct maxima between 580-750 nm, with high absorption coefficients, which do not change significantly with temperature. In the pure $Ca(NO_3)_2 \cdot 4H_2O$ spectra of cobalt (II) chloride species show different co-ordinations and consequently different colors (pink) and lower absorption coefficients [44, 45].

In Table 2 the values of the molar absorption coefficient (ε) in pure calcium nitrate tetrahydrate and pure calcium chloride tetrahydrate at the characteristic absorption wavelengths (λ) are given at various temperature.

Table 2. Temperature dependence of the maximum molar absorption coefficient of Co (II) chloride in pure components of PCM TC(1) [45]

Compound	T (°C)	ε_λ ($cm^{-1}mol^{-1}l$)			
		510 nm	625 nm	665 nm	685 nm
$Ca(NO_3)_2 \cdot 4H_2O$	28	6.59			
	40	6.95			
	60	7.79			
$CaCl_2 \cdot 6H_2O$	28		164.9	226.0	252.6
	40		167.3	237.9	268.6
	60		182.9	266.4	290.8

When Co (II) species are dissolved in molten mixture $CaCl_2 \cdot 6H_2O + Ca(NO_3)_2 \cdot 4H_2O$ melts formation of mixed ligand species of Co (II) occurs. Stability of mixed ligand Co (II) complexes is strongly dependent on temperature. Consequently, the absorption spectra of the species show a pronounced change with temperature [42, 52].

In order to obtain the most efficient thermochromic behavior in the desired temperature range gradual addition of $CaCl_2 \cdot 6H_2O$ into the $Ca(NO_3)_2 \cdot 4H_2O$ containing $CoCl_2 \cdot 6H_2O$ was made. In this way the composition of PCM TC(1) was obtained. The chloride ligand concentration corresponding to this composition is 0.6476 mol/kg. The concentration of dissolved Co (II) was 0.0107 mol/kg. In Figure 4 the molar absorption coefficient of cobalt (II) chloride in this mixture, in the wavelength range 400-800 nm is presented at various temperatures in the range from ambient to 58.3 °C. The pronounced increase of the molar absorption coefficient with the increase of temperature is evident and the changes are proven to be reversible. At room temperature the molar absorption coefficient shows 4 maxima, one of them at a similar wavelength as in molten calcium nitrate tetrahydrate and the other three at wavelengths corresponding to those observed in calcium chloride hexahydrate. The color changes from light pink with low molar absorbance through purple to dark blue.

The most indicative thermochromic behavior in PCM TC(1) was obtained with cobalt (II) chloride solution containing 0.6476 mol/kg excess chloride added as $CaCl_2$. The attempts to change the composition of the mixture PCM TC(1) towards a higher calcium chloride concentration changed the shape of the spectra to be similar to that in pure calcium chloride hexahydrate and less sensitive to the temperature changes. With a lower concentration of calcium chloride in the mixture the absorption spectra were similar to those in pure calcium nitrate tetrahydrate. However, for the auto shading of a layer of PCMs it would have been better to obtain lower absorbance at the melting point in the entire spectral range.

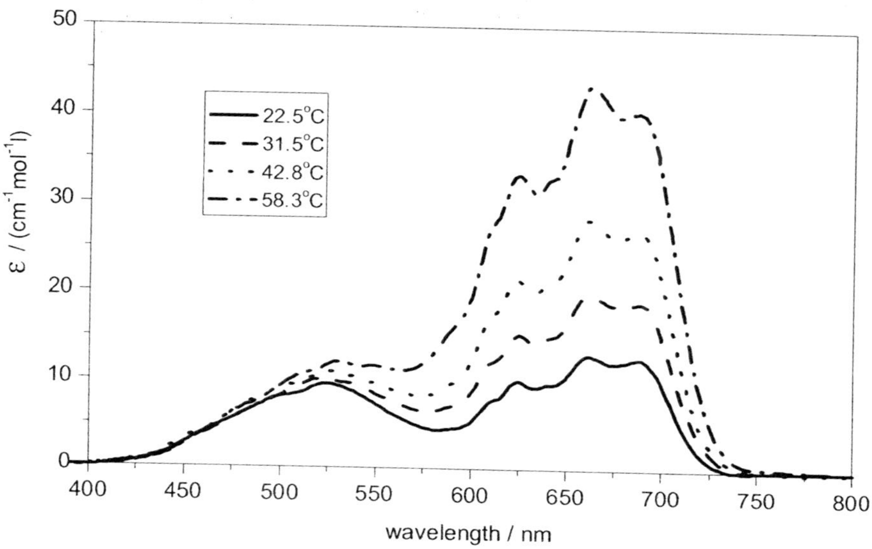

Figure 4. Variation of spectra of Co(II) chloride in PCM TC(1) with temperature [45].

Therefore another phase change material has been tested as a solvent for cobalt (II) chloride thermochromic behavior and a new thermochromic phase change material, PCM TC(2), was obtained [44, 45]. The absorption spectra of Co (II) chloride complexes dissolved in this mixture were studied with variable excess chloride concentrations added as anhydrous calcium chloride and at various temperatures, ranging from room temperature to about 60 °C. The most temperature sensitive spectra were obtained in 0.9 CH_3CONH_2 + 0.1 $Ca(NO_3)_2 \cdot 4\ H_2O$ melts with chloride concentrations between 0.22 - 0.3 mol/kg. The temperature dependence of the absorption spectra of the two solutions of cobalt (II) chloride (0.0047 mol/kg) containing two different concentrations of chloride ions, 0.2233 mol/kg and 0.2939 mol/kg, in 0.9 CH_3CONH_2 + 0.1 $Ca(NO_3)_2 \cdot 4\ H_2O$ melts are presented in Figures 5 and 6, respectively.

The two sets of spectra represent clearly the influence of the chloride ion concentration on the thermochromic effect in the mixture studied. From Figures 5 and 6 it can be seen that at room temperature these species have lower molar absorption coefficients at higher wavelengths than the corresponding spectrum in the mixture PCM TC(1). The melts are pink at room temperature and their color changes through violet to blue. The solution with 0.2939 mol/kg total chloride concentration shows a more pronounced increase in the molar absorption coefficient and more pronounced change of the spectra with temperature than the solution containing 0.2233 mole/kg chloride concentration. Both solutions contained a much lower concentration of chloride than the solution of Co (II) chloride in PCM TC(1). Further increase of chloride to the melt solution causes an increase of the absorption coefficient at room temperature, and therefore is less suitable for the purpose of autoshading.

In both thermochromic mixtures in PCM TC(1) and PCM TC(2) the molar absorption coefficient in the visible wavelength range 380-740 nm increased significantly when the temperature was raised from room temperature to about 60 °C.

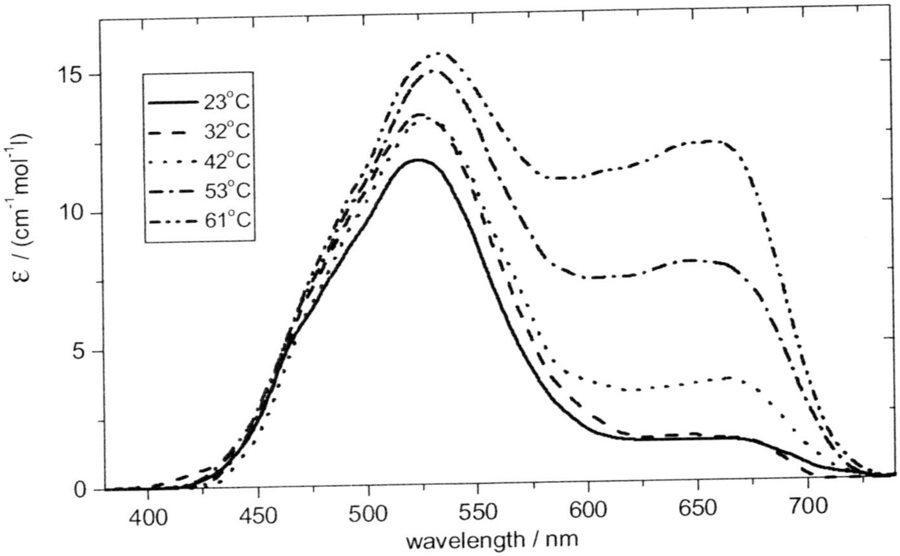

Figure 5. Variation of absorption spectra of Co (II) chloride in PCM TC(2) with O.2233 mol/kg total chloride concentration [45].

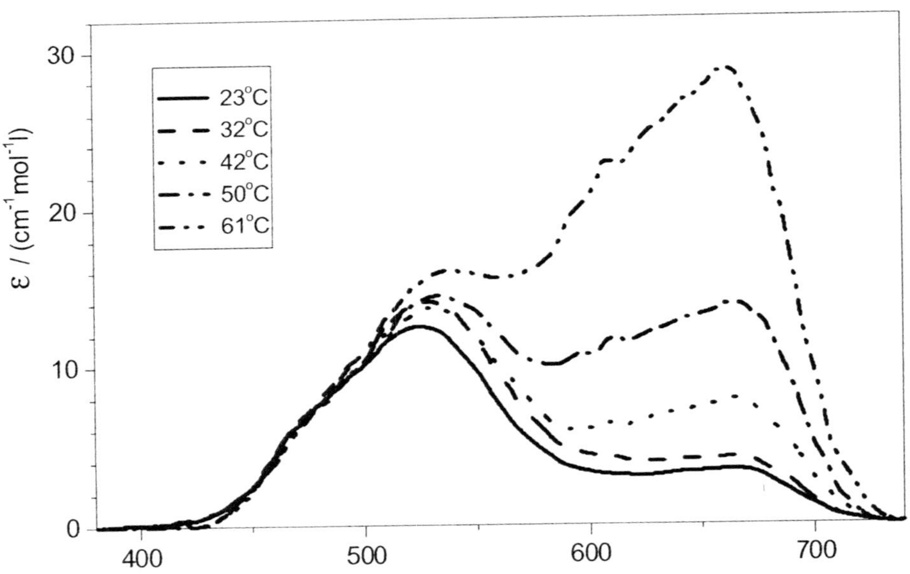

Figure 6. Variation of absorption spectra of Co (II) chloride in PCM TC(2) with 0.2939 mol/kg total chloride concentration [45].

The effectiveness of thermochromic behavior in passive solar applications can be presented in more realistic way. The overall solar energy transmitted through a wall made from a transparent material with a gap of 1 cm filled with a phase change and thermochromic material can be calculated as follows. The spectral transmittances of the wall filled with absorbing medium are multiplied by solar spectral irradiances (in steps of 5 nm) for the spectral range studied. The area bordered by these values is compared with corresponding area without absorbing media. The results expressed as integrated opacity/shading ($O = 1 - T$, where T is transmittance) in the wavelength range 380-740 nm, are presented in Table 3.

Table 3. Shading effect of thermochromic complex compounds [45]

Material	$[Co^{2+}]$ (mol/kg)	$[Cl^{-}]$ (mol/kg)	T °C	O (%)
PCM TC(1)	0.0107	0.648	23	22
	...	...	58	38.8
PCM TC(2)	0.0047	0.294	23	5.7
	...	...	61	3.8

From Table 3 it is evident that a remarkable shading effect is attained in both materials.

Our results are only a phenomenological presentation of an auto shading and temperature controlling system, based on physico-chemical processes. However, measurements of relevant thermotechnical parameters on a model of a greenhouse with an appropriate design and orientation are necessary in order to make conclusions on the energy efficiency in such a device.

Very recently, another research group made measurements on a model system using a PCM-fasade–panel for daylighting and room heating [54]. They made an experimental set-up with a south oriented layer of PCM placed in a gap between double glazed surfaces. They made tests with five PCM, having a solid/liquid transition temperature between 25-30 °C and with variable layer thickness. With sun rising and heating it takes a few hours to complete the melting of PCM and reach full transparency. During the melting period the temperature remained constant. To prevent overheating in the test room due to further sensible heat storage they used inner blinds. In summer periods it can be problematic when stored heat is released after crystallization in the evening. The results of their experiments and calculations showed that a fasade panel with PCM shows about 30 % less heat loss than without PCM but about 50 % reduced solar heat gain. Therefore, it is necessary to optimize the choice of PCM and its layer thickness with climatic conditions and with architectural design solutions.

The problem of reduced heat gain is less pronounced in agricultural greenhouses with large transparent wall surfaces. In these objects it is more important to prevent overheating and strong midday lighting, which can easily be attained with an automatic regulation, using thermochromic PCM in the south wall.

We propose a model design of a greenhouse, with a layer of thermochromic phase change material in the south wall, insulated north wall (with or without phase change material) and heat exchange through the ground (with or without phase change material).

In Figure 7 (a, b, c) schematic presentations of an intelligent greenhouse are given, with thermochromic light intensity regulations and latent heat temperature regulations. Thermochromic complex mixture in the phase change material (1) is placed in the south wall (2). Heat exchanger tubes (3) are placed through the soil deep in the ground. Another phase change material without thermochromic additives (4) is placed in the north wall (5). In order to obtained stabilized temperature range inside the object it is proposed to place some capsulated phase change material (6) in the tubes for heat exchange. It is preferable that the phase change materials placed in the ground and in the north wall have lower transition temperature than thermochromic PCM used in the south wall. Simulations of a passive solar object and calculation using TRNSYS program showed that using two different melting temperatures within one object can improve indoor temperature regulation, buffering variations in temperature requirements and improving overall energy savings [26, 27]. Moreover it is very important for root growing of the most of plants that the soil temperatures are kept bellow 20 °C.

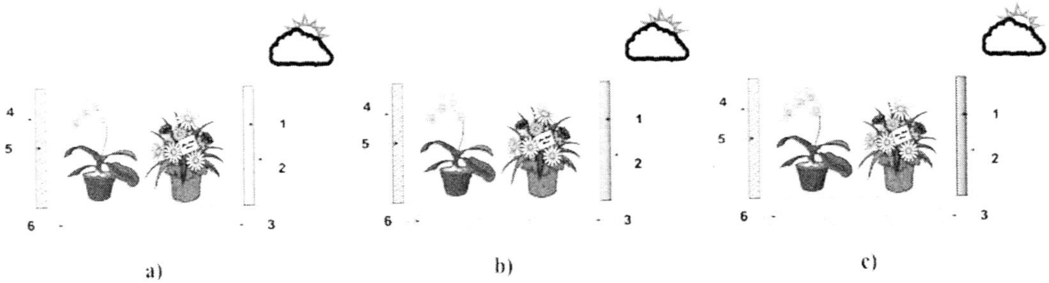

Figure 7. Schematic presentation of an "intelligent" greenhouse with thermochromic PCM. a) At the melting point of PCM TC; b) medium outside temperature increase; c) midday highest outside temperature.

INCORPORATED PCMs IN BUILDING ELEMENTS

Solar heated buildings receive solar thermal energy as direct gain through windows or through south oriented walls containing thermal mass as storage material. A combination is possible to have a sun-space or a seasonal glazed room which acts as heating system or as a buffer space, reducing heat losses [55]. With incorporated PCM in the wall it is possible to attain more favorable temperature variations and time lags than with direct gain of solar energy. In that way comfortable indoor air conditions can be ensured.

For practical use in building construction it is preferable to have light weighted elements that can be easily installed in the walls, roof, floors or ceilings. In addition to a very limited number of PCMs with phase transition temperature in the desired temperature range a serious problem is how to obtain an effective and economical containment of PCM in liquid phase. To avoid this problem, one of the ideas was to search for a material undergoing solid/solid phase transition in the required temperature range [56-58]. Some polyhydric alcohols were subject of these studies. Among a large number of compounds and mixtures studied, only a solid solution containing 60 wt % neopentylglycol and 40 wt % pentaglycerol showed solid-solid phase transition in the desired range for solar thermal storage, at 26 °C. Though it is much more convenient to prepare construction elements by mixing solid PCMs with building materials, two serious disadvantages were limitations for their practical use:

1. The enthalpy of solid/solid transition was much lower than observed in PCMs with solid-liquid transitions
2. Significant supercooling was also observed in this solid solution [58].

Therefore, for thermal storage in walls consideration of incorporation of PCMs with solid/liquid phase transition remained to be solved. When salt hydrates are to be used as PCMs in wall components they must be contained in leak proof containments, usually in polyethylene capsules, and with thickening agents [23, 59] to prevent segregation, precipitation and supercooling, though it is possible that concrete or other building material can act as stabilizing or seeding agent for a salt hydrate [60]. It has been shown that encapsulation of PCMs decreases efficiency of heat exchange and increases the price of the storage modules [61]. Microencapsulation is one of the proposed means of PCM packing in the building modules. The process of microencapsulation is based on interfacial polymerization in an emulsion system, followed by membrane formation around particles [62]. The heat exchange is much better than with macrocapsules, but the chosen PCM must be compatible with the encapsulated film, which limits already poor choice of available PCMs for passive applications and increases the price of the building modules [63-65].

Direct incorporation of PCMs appeared to be the most suitable and most economical procedure for preparation of thermal storage building elements. It can be achieved either by previous mixing of appropriate amount of a PCM with building material to make a construction element or by immersing blocks of already made building elements into a molten PCM. The price of the installation of impregnated wallboard elements is the same as for the ordinary building blocks. The impregnation procedure is more applicable to organic PCMs that can be retained in the pores of the building materials by surface tension forces.

In past two decades research has concentrated on the following materials: paraffines, paraffine waxes and their mixtures, fatty acids and esters. Experimental and parametric analysis of wallboard containing paraffine mixtures (n-octadecane based) with phase transition temperature adjusted to be between 19 and 25 °C were made [26, 29] and results confirmed possibility of benefit in energy savings, which is dependent on combinations of geometry, arrangement of components and choice of PCM properties. Detailed laboratory study and pilot tests of the wallboard impregnated with commercial butyl palmitate (49 %)-butyl stearate (48 %) (Emerset 2326) were performed by Feldman and coworkers [66-69]. Compatibility and stability of gypsum and concrete blocs impregnated with Emerset 2326, 1-dodecanol, 1-tetradecanol, 45/55 % capric-lauric acid mixture, paraffines and propyl palmitate were also studied, as well as the influence of various fire retardants and additives for stabilization [63, 70-72]. The best stability was obtained with Emerset 2326 and its uptake by concrete was up to 25 wt %. From these studies it is concluded that additional work is necessary to improve stabilization and prevent risk of flammability with the organic chemicals, particularly paraffines. In addition, it becomes obvious that it is necessary to tailor new PCMs whose transition temperature, other thermophysical properties and compatibility with building materials can suit a particular application and in particular climatic region.

Two types of organic materials have been studied by our research group: long chain fatty acid esters and their mixtures and mixtures of a paraffine (hexadecan) with poly-hydroxyl ester (polyethilen glycol-400 distearate) [74-78]. The systems showing suitable thermophysical properties were studied for compatibility with building materials and long-term stability.

Fatty Acid Esters and their Mixtures

Thermal properties of only few esters were studied so far [68, 73]. Fatty acid esters show a solid-liquid transition in a narrow temperature range, contrary to paraffins and paraffin waxes. Their mixtures can form eutectics, like many inorganic salt mixtures. They are easily mixed with building materials and in molten state held in the pores by surface tension. Solid/liquid transitions within the pores do not cause any leaking of materials in wallboard. In addition, most of the fatty acid esters are commercially available, since they are already produced in large amounts for plastics, cosmetics, textile and other industries.

In this chapter we present the results of the studies obtained by our research team, with the aim to tailor new eutectics of fatty acid esters, suitable for possible use in passive solar building elements. For that purpose solid/liquid phase equilibria were studied in methyl stearate, methyl palmitate, cetyl stearate, cetyl palmitate and their mixtures [74, 75] and their incorporation in gypsum and brick. The precise tailoring of the eutectic mixtures was made on the basis of DSC measurements. As a first step, DSC scans of each component were made and then various combinations between esters were tried in an attempt to obtain binary mixtures with sharp phase transitions in the desired temperature range [51]. Preliminary studies showed that there was a mutual solubility of the components in the entire concentration range. From DSC thermograms, scanned in temperature range (-10)-60 ^{o}C, the melting temperature T_m, the crystallization temperature T_c, the latent heat of melting ΔH_m and the latent heat of crystallization ΔH_c were determined for pure components and their mixtures in the entire concentration range.

DSC thermograms of high purity methyl stearate and methyl palmitate showed sharp peaks, corresponding to the solid/liquid transition temperatures at 37.8 ^{o}C and 29.0 ^{o}C, respectively. Additions of methyl stearate into methyl palmitate resulted in lowering of the melting temperature and widening of the endothermic peak until the composition closed to the eutectic was reached [51]. The formation of the eutectic was recognized by the appearance of the sharp endothermic peak. The melting point was 23.9 ^{o}C, lower than the values obtained for pure components. The composition of the eutectic mixture was 84.4 wt % (85.7 mol %) methyl palmitate + 15.6 wt % (14.3 mol %) methyl stearate. The freezing points coincided with melting temperatures, no supercooling was observed at low heat/cool rate (< 5 ^{o}C). The solid-liquid phase transitions were reversible, i.e. the heat absorbed as latent heat of fusion was released as heat of solidification. The endothermic DSC thermograms of both pure components and of the eutectic mixture are shown in Figure 8.

From Figure 8, it can be seen that the eutectic mixture shows the most sharp phase transition at the lowest temperature, 23.9 ^{o}C, which has been almost an ideal value for passive solar applications. The melting temperature of pure methyl palmitate is also in the range of possible interest for thermal storage applications. Therefore, samples of commercial gypsum and brick blocks, 1 cm thick layers, were impregnated with the eutectic mixture of methyl palmitate and methyl stearate and with pure methyl palmitate. The absorbed mass of methyl palmitate in gypsum was 28 mass % and in brick 12 mass %, while 25.2 mass % of the eutectic mixture was absorbed in gypsum and 10.9 mass % in brick. The solid/liquid transition temperature and the enthalpy of fusion of the absorbed substances did not change in more than 700 thermal cycles. The thermal properties of the samples, kept in an air atmosphere and at room temperature over 24 months period, were checked occasionally. No changes in thermal properties were observed by DSC.

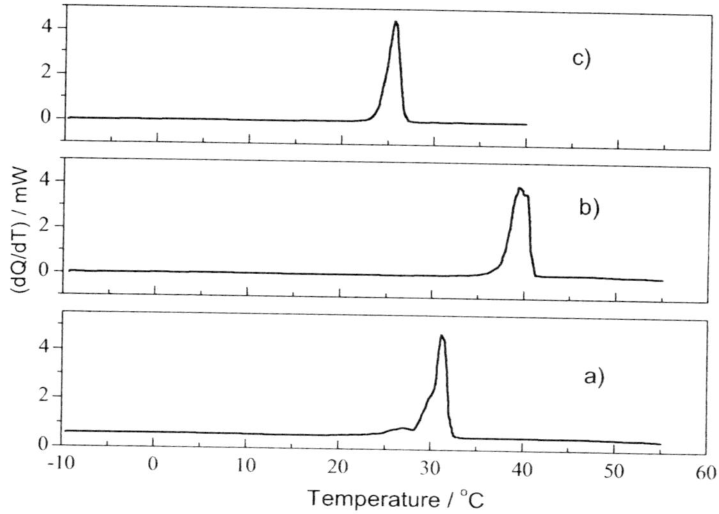

Figure 8. The endothermic DSC thermograms of a) methyl palmitate, b) methyl stearate and c) their eutectic mixture [74].

Figure 9 shows a thermal cycle, registered by DSC at heat/cool rate 5 °C/min, of a sample of gypsum impregnated with the eutectic mixture of methyl palmitate and methyl stearate.

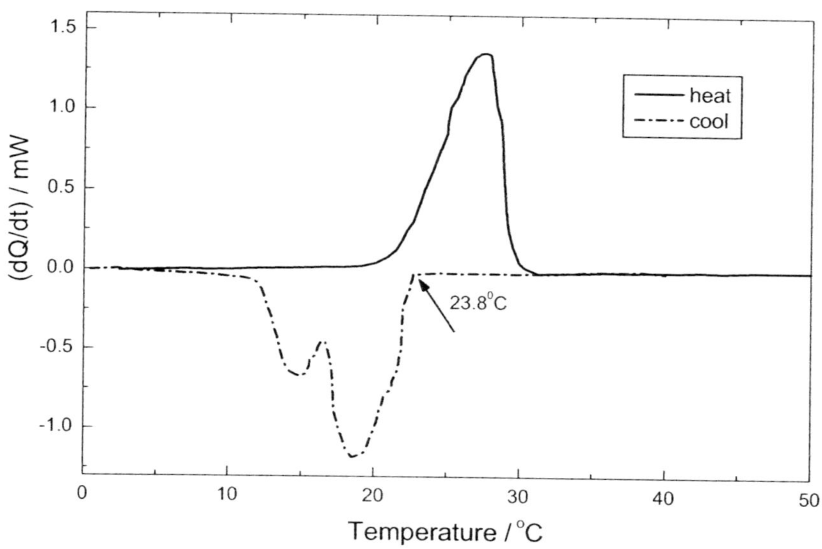

Figure 9. DSC thermogram of cycle no. 20 of gypsum sample impregnated with methyl stearate - methyl palmitate eutectic (MSMP) [75].

For the large scale preparation of building blocks impregnated with PCMs it is preferable to use materials of commercial (technical) grade purity. When the esters of technical purity (95 %) were used (methyl stearate, cetyl stearate, cetyl palmitate) their thermograms showed a number of small peaks in addition to the main sharp peak corresponding to the solid/liquid transition of the related substance. The area of small peaks did not exceed 15 % of the area of the main peak. Using the procedure that was described above, eutectic mixtures were also tailored from these chemicals [75]. Two of the mixtures showed a sharp main peak in the temperature range desired for passive solar applications.

The composition and the melting points of the mixtures were as follows: 86 mass % (90.8 mol %) methyl stearate + 14 mass % (9.2 mol %) cetyl palmitate, (MSCP), with melting point at 28.2 °C and 85 mass % (90.6 mol %) methyl stearate + 15 mass % (9.4 mol %) cetyl stearate, (MSCS), with melting point at 22.2 °C. Both of these mixtures were used for the impregnation of commercial gypsum wallboard. The mass of the mixture adsorbed was 26.2 %. The DSC thermogram of a sample of MSCP mixture is plotted in Figure 10, together with the thermogram of the gypsum sample impregnated with the same mixture. Both samples were scanned with the same heating rate (5 °C/min).

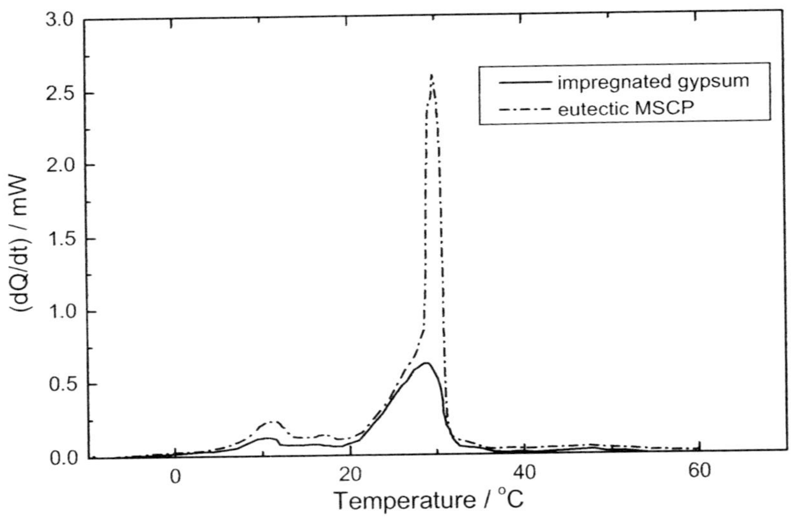

Figure 10. DSC thermograms of eutectic mixture of methyl stearate and cetyl palmitate (MSCP) and the impregnated gypsum [75].

From Figure 10, it can be seen that the solid/liquid phase change of the mixture absorbed in gypsum occurs at the same temperature as in MSCP, indicating that heat conductivity of the gypsum sample remains the same after impregnation with MSCP. The peak registered in gypsum is slightly broader. Earlier studies with gypsum impregnated with butyl stearate showed that thermal conductivity is slightly lower in impregnated samples than in the starting PCM [70]. A small peak at ≈ 10 °C is attributed to the impurities contained in methyl stearate. Three eutectic mixtures of the esters were selected for long term studies of their thermophysical properties, together with pure methyl palmitate. The samples of gypsum and brick, impregnated with these materials were also scanned after 700 thermal cycles. The thermophysical properties of the selected fatty acid ester are summarized in Table 4.

Table 4. Thermophysical properties of the studied esters [75]

Material	T_m	T_c	ΔH_m		ΔH_c	
	(°C)	(°C)	(kJ/mol)	(k J/kg)	(kJ/mol)	(kJ/kg)
MP[a]	29.0	28.7	58.1	215	57.9	214
0.86 MS[b] – 0.14 MP (MSMP)	23.9	23.8	64.8	220	64.8	220
0.91 MS – 0.09 CP[c] (MSCP)	28.2	27.9	59.6	189	58.3	185
0.91 MS – 0.09 CS[d] (MSCS)	22.2	21.8	57.1	180	55.6	175

a) MP - methyl palmitate; b) MS - methyl stearate; c) CP - cetyl palmitate; d) CS - cetyl stearate

From Table 4 it can be seen that metyl palmitate and all three mixtures have melting points in desirable range for passive solar applications and show no supercooling. They have high value of the enthalpy of melting and crystallization.

Table 5. Thermal storage capacity of building materials impregnated with PCMs [74, 75]

Building material	Impregnated with	Δm	ΔQ
		(mass %)	(kJ/kg)
brick	MP	12.5	26.9
	0.86 MS + 0.14 MP	11.7	25.7
gypsum	MP	27.9	59.8
	0.86 MS + 0.14 MP	25.2	55.0
	0.91 MS + 0.09 CP	25.7	48.6
	0.91 MS + 0.09 CS	25.6	46.0

The impregnation was performed by simple immersion of a rectangular sample of the composite (gypsum and brick) in the molten substance at ≈ 40 °C for at least 30 minutes until homogeneous absorption was attained. About 25 mass % of each mixture was absorbed by blocks of gypsum and about 12 weight % by blocks of brick. The thermal storage capacity of the impregnated composite material was calculated and presented in Table 5.

Comparison of the results presented in Table 5 with heat capacity of gypsum (≈ 30 kJ/kg) impregnated with Emerset 2326 [70] for the uptake of ≈ 20 mass % of Emerset 2326, it is seen that almost twice more energy can be stored in gypsum wallboard impregnated with these esters.

Latent heat capacity of impregnated brick, which is much less porous than gypsum or concrete, is again higher than for concrete impregnated with Emerset 2326 [71].

Mixtures of Hexadecane (HD) and Poly(ethylene glycol-400) Distearate (PGDS)

Paraffines of various compositions were extensively studied for the purpose of latent heat storage [2, 6, 1, 18, 26]. Varying composition of paraffin mixtures it is possible to adjust melting temperature [79]. Most of the studies on paraffines incorporated in wallboard materials were octadecane based. The most widely studied single paraffin is hexadecane or eicosan, with the formula $C_{16}H_{34}$, melting point at 18 °C and a high value of the enthalpy of fusion (237 kJ/kg). However, a bit too low melting point for latent heat storage and supercooling observed as in other paraffines are disadvantages for its use in combination with construction materials. On the other hand, there are no data on thermal properties of fatty acid esters with polyalcohols though they are commercially available at lower price than long chain fatty acid esters. Possibility of their blending with alkyl hydrocarbons has not been studied.

Therefore we selected to study a new organic phase change system by mixing hexadecane (HD) with poly(ethylene glycol-400) distearate (PGDS). In our studies [76-78] a commercial product, poly(ethylene glycol-400) distearate (Aldrich), was chosen to be analyzed by DSC as single compound and in combination with hexadecane (Aldrich), in an attempt to obtain a PCM which is compatible with building materials.

It was shown that there is a mutual solubility of the two compounds in the whole concentration range. Additions of HD into the melts of PGDS lowered the melting temperature of PGDS. The mixture containing 76 mass % (42 mol %) of poly(ethylene glycol-400) distearate showed narrow melting/freezing temperature range (22.5-26 °C), no supercooling and good thermal stability. However, homogeneous mixture was obtained after very slow heating or after at least three heat/cool cycles, indicating that a new polymeric compound may be formed. Once homogenization is attained no change in DSC thermogram was observed. From DSC thermograms, scanned in the temperature range (-10)-50 °C, the melting temperature T_m, the crystallisation temperature T_c, the latent heat of fusion ΔH_m and the latent heat of crystallisation ΔH_c were determined for pure compounds and their mixtures. In Table 6 thermophysical properties in the system HD + PGDS, obtained from DSC measurements, are shown.

Table 6. Thermophysical properties in the system HD + PGDS [77, 78]

Material	T_m	T_c	ΔH_m		ΔH_c	
	(°C)	(°C)	(kJ/mol)	(kJ/kg)	(kJ/mol)	(kJ/kg)
HD	18.2	14.8 - 15.2	53.8	237.5	50.8	225
PGDS	32 - 36	30 - 31	133.6	135	132.7	134
HDPGS	22.5 - 26	25.5	87.3	130	87.3	130

Melting thermograms of HD, PGDS and mixture of 0.58 HD + 0.42 PGDS (HDPGS) are shown in Figure 11. The samples were scanned at heat rate 5 °C/min. From data presented in Table 6 and in Figure 11 it can be seen that the mixture HDPGS has melting point between the values for pure HD and pure PGDS. The solid/liquid phase transition occurs in the desired temperature range, almost ideal for passive solar applications. The mixture does not supercool, i.e. melting and solidification occurs at the same temperature, as shown in Figure 12.

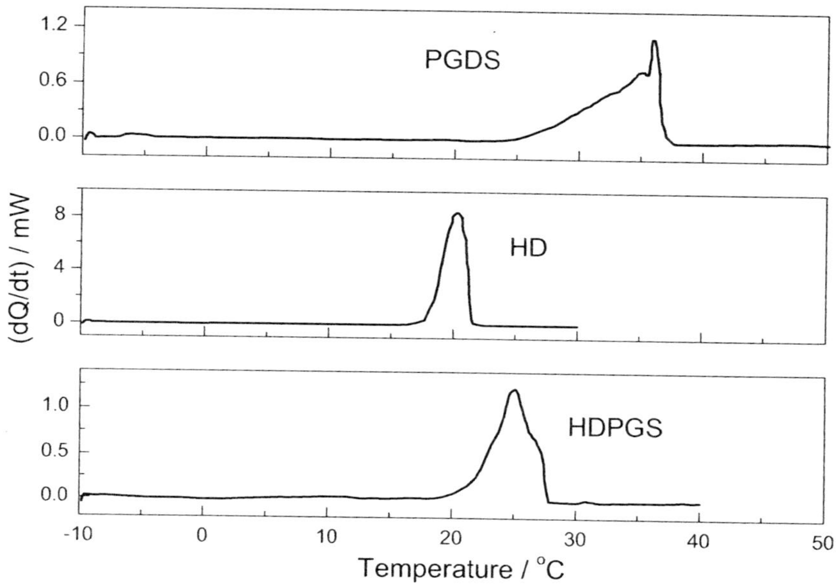

Figure 11. Melting thermograms of HD, PGDS and mixture HDPGS [77].

The value of the enthalpy of melting/crystallization is much lower than the values obtained with the eutectic mixtures of long chain fatty acid esters, but similar to the value obtained for Emerset 2326 [70].

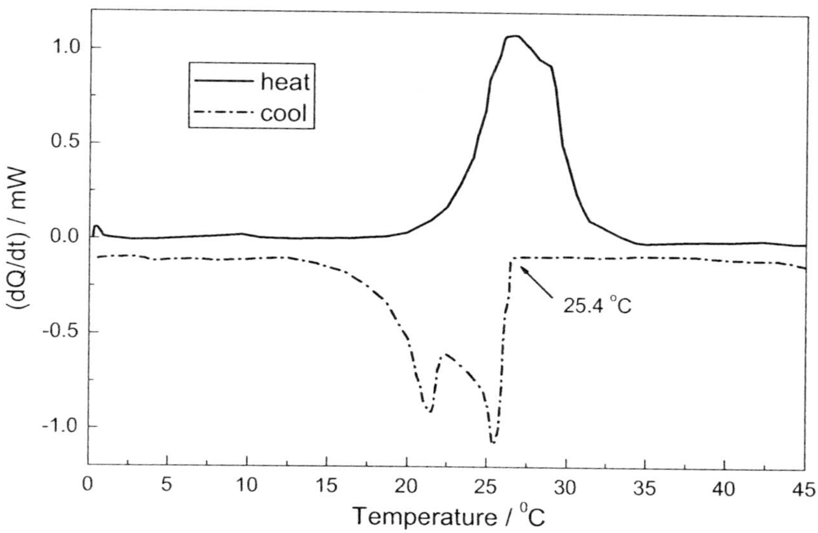

Figure 12. Melting and solidification thermograms of HDPGS [77].

The mixture HDPGS was used for impregnation of three composite materials used in buildings: brick (Vinča-Belgrade), plaster (Knauph-Austria) and tarolit (mineralized wood filings, Ostruznica, Belgrade). The thermal and the chemical stability of the impregnated samples were tested by DSC, over a period of 24 months, in 60 accelerated thermal cycles between 0 – 60 °C and additional 500 cycles from 15 –35 °C. No changes in DSC curves were observed. Figure 13 shows DSC curves scanned after 20 heat/cool cycles.

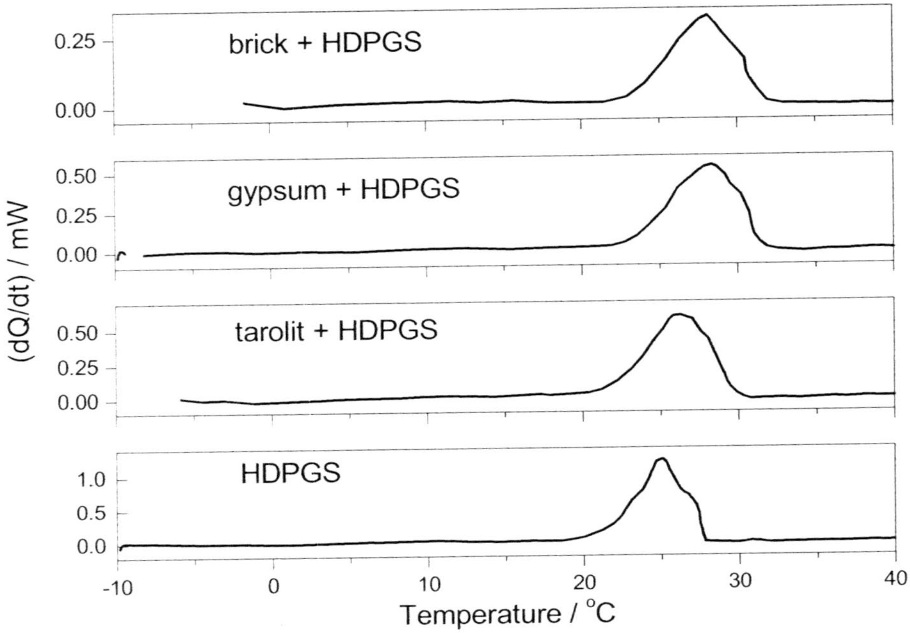

Figure 13. Melting thermograms of: a) HDPGS and composite materials, b) tarolit, c) gypsum, d) brick, impregnated with HDPGS [76].

Moreover, building materials studied showed high uptake of HDPGS. In that way relatively low value of the enthalpy of phase transition was compensated by high absortivity. Thus, high latent heat storage capacity in the impregnated building materials was attained, again almost twice higher than with Emerset 2326 and very similar to the capacity obtained by fatty acid ester mixtures [75].

In Table 7. the mass % of the absorbed HDPGS by the composite materials, are presented together with T_m, T_c and the thermal storage capacity, ΔQ, in the building materials studied.

Table 7. Thermal storage capacity of the materials impregnated with HDPGS [78]

Composite	Δm	ΔQ	T_m	T_c
	(mass %)	(kJ/kg)	(°C)	(°C)
brick	13.2	17	22.5 - 28	25
gypsum	39.0	50	22.5 - 28.2	25.2
tarolit	40.8	53	22.5 - 26.1	25.1

Figure 14. Schematic presentation of "tarolit' combinations.

Very good results were obtained in gypsum and "tarolit". Commercial blocks of "tarolit", schematically presented in Figure 14, available as two layered or three layered blocks ("tarolit kombi"), with polystyrene foam between the layers of mineralized wood, are particularly suitable for passive solar walls impregnated with PCMs, because it can be a component of a wallboard with heat storage and heat isolation properties.

CONCLUSION

In this chapter the studies, by our research group, of phase change materials that can be applied in two typical, so called "intelligent" solar objects are presented: a) in transparent walls of agricultural greenhouses and b) incorporated with building materials of wallboards in residential buildings.

The results on the studies of the first types of PCMs show that combination of spectral properties of metal complexes dissolved in a suitable phase change material and its latent heat of fusion can be of potential use for simultaneous control of temperature and the light intensity. This is of importance for application in agricultural greenhouses used for cultivation of some delicate plants in fluctuating climate conditions. The choice of metal complex, its nature and characteristic absorption spectral range may be adjusted according to the requirements for the special plant cultivation needs.

Calculations of the overall solar energy transmitted through the wall made of a transparent material with a gap of 1 cm filled with phase change and thermochromic material have been done for the three sets of experimental spectra obtained at room temperature and the highest temperature (≈ 60 °C). The results, expressed as % of the decrease of transparency (shading) in the integrated wavelength range 400-750 nm showed remarkable shading effect of thermochromic materials used.

The results need to be tested in an experimental model system with a thermochromic complex compound. However, an experimental model with a PCM placed in a gap between double glazed surfaces has recently been tested and heat losses were reduced by 30 %, though inner blinds were used for light intensity regulations [54].

The results of tailoring new phase change materials for passive solar applications in wallboard, floor or ceiling, obtained by our team, are also presented in this chapter. Three mixtures of fatty acid esters and a mixture of hexadecane and poly(ethylene glycol-400) distearate, of defined compositions, were tested in laboratory conditions, by DSC. The transition temperatures were in the range 22-28 °C, and the latent heat of phase change were in the range 130 - 220 kJ/kg. The selected four phase change materials with suitable thermophysical properties were used for impregnation of passive solar wallboard components: brick, gypsum and "tarolit."

Highest thermal storage capacity (45-55 kJ/kg) was obtained with "tarolit" and gypsum blocks impregnated with the studied PCMs. The values obtained are nearly twice higher from those known so far [66-68].

Thermal properties of the materials were not changed when they were absorbed in the composite materials studied. They remain stable during the studies in a period of 24 months. Accelerated thermal cycle tests were also made.

Although it seems that it is difficult to make longer list of candidate PCMs suitable for passive solar applications we conclude that there is still an open field of search for various combinations of organic compounds. It is essential to have a good choice of phase change material to obtain the best heat gains and other performances in different climate areas, consequently reducing additional costs and making thus reasonable short investment cost pay back. It is sometimes more efficient to use more than one PCM in building components (walls, floors, ceilings) [26]. In this way comfortable indoor air conditions can be attained. It would have been desirable to create a chameleon type of building mechanism as an additional mechanism of temperature control in the passive solar building. The intention of an initial study in this field was to mix paint with a thermochromic pigment that could absorb solar energy below a switching temperature and reflect above this temperature [80]. However, the pigment substance with a suitable switching temperature has not yet been found.

References

[1] Telkes, M. "Heat Storage Unit"; US Patent 2677367, 1954.

[2] Lorsch, H. G.; Kauffman K. W. "Residential and Commercial Thermal Storage" *Energy from the Sun Symposium*; Chicago, Ill., 1978; pp 309-324.

[3] Telkes, M. "Latent Heat Storage Techniques" *Energy from the Sun Symposium*, Chicago, Ill., 1978; pp 325-336.

[4] Abhat, A. "Low Temperature Latent Heat Thermal Energy Storage: Heat Storage Materials" *Sol. Energy* 1983, 30(4), 313-332.

[5] Lane, G. A. "Solar Heat Storage: Latent Heat Materials"; Vol. I - Background and Scientific Principles, 1983; CRC Press: Boca Raton, Fl.

[6] Marcus, Y. "Thermal Energy Storage" In Molten Salt Technology; Lovering D. G.; Ed.; Chapter 14; Plenum: New York, 1982; pp 457-498.

[7] Elder, J. P. "Thermal Energy Storage Materials – A DSC Study" *Thermochim. Acta* 1980, 36, 67-77.

[8] Srivastava, S. P.; Tandon, R. S.; Pandey, D. C.; Madwall, D. C.; Goyal, S. K. "Phase Transitions in Petroleum Waxes: Correlation With Properties" *Fuel* 1993, 72(9), 1345-1349.

[9] Sharma, S. D.; Buddhi, D.; Sawhney, R. L. "Accelerated Thermal Cycle Test of Latent Heat-Storage Materials" *Sol. Engergy* 1999, 66(6), 483-490.

[10] Sharma, A.; Sharma, S. D.; Buddhi, D. "Accelerated Thermal Cycle Test of Acetamide, Stearic Acid and Paraffin Wax for Solar Thermal Latent Heat Storage Applications", *Energy Convers. Manage.* 2002, 43, 1923-1930.

[11] He, B.; Martin, V.; Setterwall, F. "Phase Transition Temperature Ranges and Average Density of Paraffin Wax Phase Change Materials" *Energy* 2004, 29, 1785-1804.

[12] Nikolic, R.; Tripkovic, J.; Kerridge, D. H. "Phase Changes in Acetamide-Salt Systems: Melting Points and Latent Heat of Fusion of Pure Acetamide and Acetamide-Ammonium Chloride" *Thermochim. Acta* 1989, 146, 353-360.

[13] Nikolic, R.; Ristic, G.; Todorovic, M. "Binary Eutectics of Acetamide with Inorganic Nitrates: Thermophysical Properties Relevant for Heat Storage" *Sol. Energy Mater. Sol. Cells* 1992, 28, 56-69.

[14] Nikolic, R.; Marinkovic, M.D. "Cryoscopic Behavior in Acetamide-Zinc Acetate Dihydrate Melts" *Thermochim. Acta* 1995, 265, 55-62.

[15] Nikolic, R.; Marinkovic, M.; Pavlovic, M. "Cryoscopic Behavior in The Binary Mixtures of Acetamide-Sodium Acetate and Acetamide-Sodium Acetate Trihydrate" *Thermochim. Acta* 1996, 276, 17-25.

[16] Nikolic, R.; Marinkovic, M.; Savovic, J.; Zsigrai, I. "Binary Eutectics of Acetamide with Inorganic Acetates: Possible Application for Latent Heat Storage" In Molten Salt Chemistry and Technology; Wendt, H.; Ed.; Molten Salt Forum; Trans Tech Publications: Zuerich, Switzerland, 1998; Vol. 5-6, pp 521-524.

[17] Yinping, Z.; Yi, J. "A Simple Method, The T-History Method, of Determining The Heat of Fusion, Specific Heat and Thermal Conductivity of Phase Change Materials" *Meas. Sci. Technol.* 1999, 10, 201-205.

[18] Himran, S.; Suwono, A.; Ali Mansoori, G. "Characterization of Alkanes and Paraffin Waxes for Application as Phase Change Energy Storage Medium" *Energy Sources* 1994, 16, 117-128.

[19] Marcus Y, Minevich A, Ben- Dor L, " Differnetial Drop Calorimetry for the Determination of Enthalpy of Fusion", *J. Chem. Thermodyn.* 2003, 35, 1009-1018.

[20] Nikolic, R.; Kelic, K.; Neskovic, O. "The Thermal Conductivities of Some Low Melting Materials Relevant to Energy Storage" *Appl. Phys.* A 1984, 34, 199-203.

[21] Nikolic, R.; Tripkovic, J. "Measurements of Thermal Conductivities of Some Low Melting Materials in a Concentric Cylinder Apparatus" *Appl. Phys.* A 1987, 44, 293-297.

[22] Schröder, J.; Gawron, K. "Latent Heat Storage" *Energy Research* 1981, 5, 103-109.

[23] Carlsson, B.; Stymne, H.; Wetermark, G. "An Incongruent Heat-of-Fusion System $CaCl_2 \cdot 6H_2O$ – Made Congruent Through Modification of the Chemical Composition of the System" *Sol. Energy* 1979, 23, 343-350.

[24] Peippo, K.; Kauranen, P.; Lund, P. D. "A multi-component PCM wall optimized for passive solar heating" *Energy and Buildings* 1991, 17, 259-270.

[25] Santamouris, M.; Balareas, C. A.; Dascalaki, E.; Villindras, M. "Passive Solar Agricultural Grenhouses: A Worldwide Classification And Evaluation of Technologies and Systems Used for Heating Purposes" *Sol. Energy* 1994, 53(5), 411-426.

[26] Stowall, T. K.; Tomlinson, J. J. "What are the Potential Benefits of Including Latent Storage in Common Wallboard?" *Trans. ASME* 1995, 117, 318-325.

[27] "A Transient System Simulation Program" TRNSYS 1983, Solar Energy Laboratory, University of Wisconsin, Madison, WI.

[28] "A Transient System Simulation Program", Engineering Experiment Station Report" TRNSYS 1990, 38-13, Solar Energy Laboratory, University of Wisconsin, Madison, WI.

[29] Tomlinson, J. J.; Heberle, D. "Analysis of Wallboard Containing a Phase Change Material" In Proc. of the 25th Intersociety Energy Conversion Engineering Conference 1990, Reno, NV, Vol. 4, 230-235.
[30] Guillemin, A.; Morel, N. "Experimental Results of a Self-Adaptive Integrated Control System in Buildins: A Pilot Study" *Sol. Energy* 2002, 72(5), 397-403.
[31] Granqvist, C. G. "Electrochromic Tungsten Oxide Films: Review of Progress 1993 – 1998" *Sol. Energy Mater. Sol. Cells* 2000, 60, 201-262.
[32] Lampert, C. M. "Large Area Smart Glass and Integrated Photovoltaics" *Sol. Energy Mater. Sol. Cells* 2003, 76, 489-499.
[33] Cui, H. N.; Costa, M. F.; Teixeira, V.; Porqueras, I.; Bertran, E. "Electrochromic Coatings for Smart Windows" *Surf. Sci.* 2003, 532-535, 1127-1131.
[34] Jaksic, N.; Salahifar, C. "A Feasibility Study of Electrochromic Windows in Vehicles" *Sol. Energy Mater. Sol. Cells* 2003, 79, 409-423.
[35] Pehlivan, E.; Tepehan, F. Z.; Tepehan, G. G. "Comparison of Optical, Structural and Electrochromic Properties of Undoped and WO_3-doped Nb_2O_5 Thin Films" *Solid State Ionics* 2003, 165, 105-110.
[36] Turkay, O. S. "An Expert System to Control Humidity and Temperature of a Laboratory Scale Greenhouse" *Mechatronics* 1994, 4, 607-615.
[37] Santamouris, M.; Argiriou, A.; Vallindras, M. "Design and Operation of A Low Energy Consumption Passive Solar Agricultural Greenhouse" *Sol. Energy* 1994, 52(5), 371-378.
[38] Kürklü, A. "Energy Storage Applications in Greenhouses by Means of Phase Change Materials (PCMs): A Review" *Renewable Energy* 1998, 13(1), 89-103.
[39] Nikolic, R.; Yu Patent Application P-734/95, Patent No. 48969, Serbia & Montenegro.
[40] Hamdan, M. A. "Thermal gains through windows" *Energy Convers. Manage.* 1994, 35(6), 501-501.
[41] Kerridge, D. H.; Nikolic, R.; Stojic, D. *J. Chem. Soc. Dalton Trans.* 1986, 1663-1668.
[42] Tripkovic, J.; Nikolic, R.; Kerridge, D. H. *J. Serb. Chem. Soc.* 1989, 54 (9-10), 527-534.
[43] Ballhausen, C.J.; Liehr, A.D. "Intensities in inorganic complexes, Part II. Tetrahedral complexes" *J. Mol. Spectrosc.* 1958, 2(1-6), 342-360.
[44] Nikolic, R.; Savovic, J.; Marinkovic, M.; Zsigrai, I.J. "Thermochromic behavior of cobalt(II) chloride complexes in low melting binary mixtures" In Molten Salt Chemistry and Technology; Wendt, H.; Ed.; Molten Salt Forum; Trans Tech Publications: Zuerich, Switzerland, 1998; Vol. 5-6, pp 621-624.
[45] Marinkovic, M.; Nikolic, R.; Savovic, J.; Gadzuric, S.; Zsigrai, I. "Thermochromic complex compounds in phase change materials: possible application in an agricultural greenhouse" *Sol. Energy Mater. Sol. Cells* 1998, 51, 401-411.
[46] Deki, S.; Kajinami, A.; Kanaji, Y.; Mizuhata, M.; Nagata, K. "Properties of $CaCl_2$ Hydrate with an inorganic powder. Part 2. - Melting behavior and thermodynamic properties of $CaCl_2 \cdot nH_2O$ (n=6.00-7.35) with α-Al_2O_3 or SiC powder" *J. Chem. Soc. Faraday Trans.* 1993, 89(20), 3811-3815.
[47] Abate, C.; Scheetz, B. E. "Aqueous Phase Equilibria in the system CaO-Al_2O_3-$CaCl_2$-H2O: The Significance and Stability of Friedel's salt" *J. Am. Chem. Soc.* 1995, 78(4), 939-944.

[48] Y. Xu, Y.; Hepler, L.G. "Calorimetric investigations of crystalline, molten, and supercooled $Ca(NO_3)_2 \cdot 4H_2O$ and of concentrated $Ca(NO_3)_2(aq)$" *J. Chem. Thermodyn.* 1993, 25(1), 91-97.

[49] Savovic, J.; Nikolic, R.; Kerridge, D.H. "Cobalt (II) Chloride Complexes in Molten Acetamide" *Fluid Phase Equilib.* 1996, 118, 143-151.

[50] Savovic, J.; Nikolic, R.; Veselinovic, D. "Cobalt(II) Chloride Complex Formation in Acetamide-Calcium Nitrate Tetrahydrate Melts" *J. Solution Chem.* 2004, 33(3), 287-300.

[51] Franzosini, P.; Sanesi, M. "Thermal Methods" In Molten Salt Techniques; Vol. 1, Lovering, D. G.; Gale, R. J.; Eds.; Plenum: New York, 1983.

[52] Zsigrai, I.J.; Gadzuric, S.B.; Nikolic, R. "Cobalt(II) Chloride Complexes in Calcium Nitrate Tetrahydrate Melt" *High Temp. Mat. Proc.* 2001, 5, 573-581.

[53] Alen, C.W. "Astrophysical Quantities" The Athlone Press: London, 1955.

[54] Weinläder, H.; Beck, A.; Fricke, J. "PCM-Facade –Panel for Daylighting and Room Heating" *Sol. Energy* 2005, 78, 177-186.

[55] Nikolic, R.; Pucar, M.; Zsigrai, I. "Greenhouses Biult on Flat Roofs with Solar Heating and Temperature Regulation With Two Phase Change Materials" *KGH* 1996, 1, 95-99 (in Serbian).

[56] Christensen, C.; Benson, D. "Advanced Phase-Change Storage: Analysis of Materials and Configurations" SERI/TP-255-2018, Solar Energy Research Institute, Golden, Co., 1-6, 1983.

[57] Benson, D.K.; Burrows, R. W.; Webb, J. D. "Solid State Phase Transitions in Pentaerithritol and Related Polyhidric Alcohols" *Sol. Energy Mater.* 1986, 13, 133-152.

[58] Barrio, M.; Font, J.; Muntasell, J.; Navarro, J.; Tamarit, J. L. I. "Applicability for Heat Storage of Binary Systems of Neopentylglycol, Pentaglycerine and Pentaerithritol: A Comparative Analysis" *Sol. Energy Mater.* 1988, 18, 109-115.

[59] Telkes, M. "Solar Thermal Energy Storage" In Applications of Solar Energy for Heating and Cooling of Buildings, Chapter 7; Jordan, R.G.; Liu, Y.H.; Ed.; 1977; ASHRAE GRP-170.

[60] Hadjieva, M.; Stoykov, R.; Filipova, Tz. "Composite Salt-Hydrate Concrete System for Building Energy Storage" *Renewable Energy* 2000, 19, 111-115.

[61] Neeper, D.A. "The Thermal Dynamics of Wallboard with Latent Heat Storage" *Sol. Energy* 2000, 68, 393-403.

[62] Cho, J. S.; Kwon, A.; Cho, C. G. "Microencapsulation of Octadecane as a Phase-Change Material by Interfacial Polymerization in an Emulsion System" *Colloid Polym. Sci.* 2000, 280, 260-266.

[63] Hawes, D.W.; Banu, D.; Feldman, D. "Latent Heat Storage in Concrete II" *Sol. Energy Mater.* 1990, 21, 61-80.

[64] Hawes, D. W.; Feldman, D.; Banu, D. "Latent Heat Storage in Building Materials" *Energy and Buildings* 1993, 20, 77-86.

[65] Khudhair, A.M.; Farid, M. M. "A Review on Energy Conservation in Building Applications with Thermal Storage by Latent Heat Using Phase Change Materials" *Energy Convers. Manage.* 2004, 45, 263-275.

[66] Feldman, D.; Banu, D. "DSC Analysis for Evaluation of an Energy Storing Wallboard" *Thermochim. Acta* 1996, 272, 243-251.

[67] Banu, D.; Feldman, D.; Hawes, D. "Evaluation of Thermal Storage as Latent Heat in Phase Change Material wallboard by Differential Scanning Calorimetry and Large Scale Thermal testing" *Thermochim. Acta* 1998, 317, 39-45.
[68] Scalat, S.; Banu, D.; Hawes, D.; Paris, J.; Haghighata, F.; Feldman, D "Full Scale Thermal Testing of Latent Heat Storage in Wallboard" *Sol. Energy Mater. Sol. Cells* 1996, 44, 49-61.
[69] Athienitis, A. K.; Liu, C.; Banu, D.; Feldman, D. "Investigation of the Thermal Performance of a Passive Solar Test-Room with Wall Latent Heat Storage" *Building and Environment* 1997, 32(5), 405-410.
[70] Feldman, D.; Banu, D.; Hawes, D.; Ghanbari, E. "Obtaining an Energy Storing Building Material by Direct Incorporation of an Organic Phase Change Material in Gypsum wallboard" *Sol. Energy Mater.* 1991, 22, 231-242.
[71] Hawes, D.; Banu, D.; Feldman, D. "The Stability of Phase Change Materials in Concrete" *Sol. Energy Mater. Sol. Cells* 1992, 27, 103-118.
[72] Banu, D.; Feldman, D.; Haghighat, F.; Paris, J.; Hawes, D. "Energy Storing Wallboard: Flammability Tests" *J. Mater. Civ. Eng.* 1998, 10(2), 98-105.
[73] Feldman, D.; Shapiro, M. M.; Banu, D. "Organic Phase Change Materials for Thermal Energy Storage" *Sol. Energy Mater.* 1986, 13, 1-10.
[74] Nikolic, R.; Gadzuric, S.; Zsigrai, I. J. "Phase Equilibria in Methyl Stearate - Methyl Palmitate Mixtures" In Proceedings of the 5th International Conference on Fundamental and Applied Aspects of Physical Chemistry, Belgrade 2000, 51-53.
[75] Nikolic, R.; Marinovic-Cincovic, M.; Gadzuric, S.; Zsigrai, I.J. "New Materials for Solar Thermal Storage-Solid/Liquid Transitions in Fatty Acid Esters" *Sol. Energy Mater. Sol. Cells* 2003, 79, 285-292.
[76] Nikolic, R.M.; Gadzuric, S.B.; Zsigrai, I. J. "Studies of Binary Mixtures of Hexadecane and Polyethylene Glycol Distearate for Passive Solar Thermal Storage" In Proceedings of the 6th International Conference on Fundamental and Applied Aspects of Physical Chemistry, Belgrade 2002, 1, 43-45.
[77] Nikolic, R.; Zsigrai, I.J.; Gadzuric, S. "Studies of Binary Mixtures of Hexadecane and Polyethylene Glycol Distearate for Passive Solar Thermal Storage" *Mag. Kem. Fol. (Hung. J. Chem.)* 2002, 108(11), 508-511 (in Hungarian).
[78] Nikolic, R.; Gadzuric, S.; Zsigrai, I. J. "Smart Materials in Building Construction for Efficient Energy Consumption and Clean Environment" In Proceedings 2nd International Conference on Ecological Protection of the Planet Earth Bio-Environment and Bio-Culture, 5-8 June Sofia 2003, 499-504.
[79] Espeau, P.; Robles, L.; Cuevas-Diarte, M. A.; Mondieig, D.; Haget, Y. "Thermal Cycling of Molecular Alloys and Eutectics Containing Alkanes for Energy Storage" *Mater. Res. Bull.* 1996, 31(10), 1219-1232.
[80] Ma, Y.; Zhang, X.; Zhu, B.; Wu, K. "Research on Reversible Effects and Mechanism between the Energy-Absorbing and Energy-Reflecting States of Chameleon-Type Building Coatings" *Sol. Energy* 2002, 72(6), 511–520.

In: Solar Energy: New Research
Editor: Tom P. Hough, pp.71-95

ISBN 1-59454-630-4

Chapter 3

A CONCRETIZATION OF A PROPOSAL FOR AN INNOVATIVE DESIGN OF A SOLAR FLAT-PLATE THERMAL COLLECTOR

C. Cristofari*and G. Notton
Centre de Recherches "Energie et Systèmes", Université de Corse,
U.M.R. CNRS 6134, Route des Sanguinaires, F-20000 AJACCIO, France

ABSTRACT

On an international market where industries of traditional production are successfully developing, our purpose is to present a solar flat-plate thermal collector integrating a production technology which is different from the industrials processes.
The technological benefit from this new sensor resides in dividing by four the number of components, by means of the reduction of assembling steps, the thermoformed fabrication of two pieces of the totally symmetric absorber (key component, patent protected). There is no more box. For equivalent performances, the weight and the cost are strongly reduced. Thus this is not an incremental improvement of a product, but a new technology which allows industrial production.
Industrials from the solar sector are not able to lower their production costs since they have reached a limit of productivity gain for their technologies.
Using a thermal model of finite differences, the performances of this solar flat-plate thermal collector wholly manufactured in a copolymer material is studied in low flow conditions. The influences of the flow rate and the stratification of the tank are analysed. Thermal performances, productivity and efficiency of such a solar system (collector with surface of 2 m^2 and a storage tank of 150 litres) are presented for a Mediterranean site. Such a system presents a yearly mean efficiency of about 54 % and 52.0 % for respectively high thermal stratified tank and fully mixed tank.

---

* Authors for correspondence. Tel: 33.4.95.45.01.54 ; 33.4.95.52 41.52 / 31 / 51 .Fax 33.4.95.52.41.42 ; e- mail: cristofa @univ-corse.fr ; notton / poggi / louche @vignola.univ-corse.fr

These results show that a stratified tank has much higher performances than a fully mixed tank. With high degree of stratification saving energy is higher (5.25% over one year of use) than fully mixed tank. The use of polymer materials reduces the collector weight by 50% in comparison with a traditional metal collector, what implies an easier production and installation.

Key words: Solar flat-plate collector / copolymer / stratified tank / flow rate / modeling / performance.

NOMENCLATURE

A	surface area	m^2
b_0	numerical constant	
B	collector control function	
CC	thermal capacity	$W.K^{-1}$
C	specific heat capacity	$J.kg^{-1}.K^{-1}$
D	hydraulic diameter	m
DT	temperature difference: T_f-T_{ae}	K
$DT1$	temperature difference: T_{2f}-T_{1f}	K
e	thickness	m
E	energy gain by the fluid	kWh
G	total solar radiation (β=45°)	$W.m^{-2}$
h	heat transfer coefficient	$W.m^{-2}.K^{-1}$
I	instantaneous incident solar radiation on the horizontal plane	$W.m^{-2}$
K	incidence angle modifier	
L	length	m
$\dot{m}$	collector mass flow rate	$kg.s^{-1}.m^{-2}$
M	material or fluid weight	kg
Pr	Prandtl number	
q	heat carried by the heater fluid	W
r	resistance	$K.W^{-1}$
R	geometric factor	
Re	Reynolds number	
S	solar energy absorption rate	$W.m^{-2}$
t	time	s
T	temperature	K
U	heat loss coefficient	$W.m^{-2}K^{-1}$
V	wind speed	$m.s^{-1}$

Subscripts:

a	inside air
ae	outside air
b	beam

Subscripts continued:

c	collector	
cd	conductive	
cv	convective	
d	diffuse	
dp	dew point	
eb	effective beam	
ed	effective diffuse	
eg	effective ground	
f	fluid	
g	ground	
i	node (variables denoting certain parts of the system)	
iso	insulation	
j	node (variables denoting certain parts of the system)	
l	inlet fluid in the tank from load	
m	mean	
n	normal	
p	absorber	
pb	bottom half of the absorber	
pt	top half of the absorber	
r	radiative	
s	tank	
sky	sky	
w1	fluid in pipe *collector-tank*	
w2	fluid in pipe *tank-collector*	
0	total incident at the horizontal plane	
1	inlet fluid in the collector from the tank	
2	outlet fluid from collector	
22	inlet fluid in the tank from collector	

Greek letters

α	absorptivity	
β	inclination of the collector	°
ε	emissivity	
Δ	inter node spacing	m
ϕ	Potential sources or wells	W
η	Instantaneous or Daily mean of efficiency	%
λ	Thermal conductivity	$W.m^{-2}.K^{-1}$
θ	incidence angle	°
ρ	Density	$kg.m^{-3}$
σ	Boltzmann constant	($5.6697.10^{-8}$ $W.m^{-2}.K^{-4}$)
τ	transmittance	
υ	volume	m^3

INTRODUCTION

Solar water heating systems have reached technical maturity and are used in many countries. After the first oil crisis in 1973, the strategies used by industrialised and developing countries to reduce their oil dependence have been numerous. A diversification of energy import, a structural change of the large domestic product (industrial development of activities using a low energy expenditure) or an increase of the national supply have been the essential measures taken by the countries with various degree of importance.

The island character of Corsica in Mediterranean sea (particularly the fact that this island is not connected to the French continental electrical grid) caused this region to opt for an energy sources diversification [1]. Natural resources of hydroelectric energy provide about 30% of the island electrical's demand, two fuel thermal power plants meet 50% of demand and the remaining 15% are brought to Corsica from Italy via the SACO electrical cable.

This particularly character also induces a great gap between the production and distribution cost of electricity (identical in all over the French territory, 0.192 €/kWh) against 0.12 €/kWh duty-free. So, each kWh sold in Corsica increases the deficit of the French utility EDF (66 M€ deficit for a turnover of 105 M€). To reduce this deficit, EDF favours in Corsica and in French overseas territories (with the same EDF deficit) actions to reduce electricity consumption.

In this context, since the early 80's, the French Agency for Environment and Energy Management (ADEME) and the Corsican Territorial Collectivity have implemented a common policy consisting of financial aids for the development of renewable energies and actions to control and reduce energy cost.

This policy of aids (FCME) allows to develop the thermal solar energy in individual and collective applications. Today, respectively 1200 m^2 and 3000 m^2 of solar collectors have been installed in individual and collective houses.

The low integration of thermal solar energy in Corsica and more generally in France shows that a real environmental policy and more competitive prices of installations are necessary for a more extensive use of these systems.

In such way, we developed a copolymer solar collector whose thermal behaviour has been modeled and optimised using a finite difference model [2].

In this paper, we analyse more accurately the influence of the flow rate and consequently the stratification degree of the tank on the water heating system performances. The interest of this study resides in the approach used to model the tank and in the analyses of the number of the nodes used on the gained energy.

LITERATURE REVIEW

Choice of the Collector Configuration

Various researches, developments and demonstration works on large-scale low cost plastic bag solar collectors have been realised since the seventies [3, 4, 5]. The system consisted in a plastic film water bag which rested on layer of thermal insulation. In 1997,

Tsilingiris [6] proposed a solar collector based on the same principle but including a collector enclosure made with hard structural thermal insulation material.

Solar domestic hot water (SDHW) heaters are widely used today for collection of low temperature solar thermal energy. This type of collectors have been the subject of numerous studies and developments since 1950. Bliss [7] and Willier [8], Hottel and Willier [9] studied for the first time these solar collectors through the first thermal analyses for a parallel tubes collectors which served as a basis for many further developments.

An empirical equation to calculated the top heat loss coefficient from the collector plate was developed by Klein [10] following the basic procedure of Hottel and Woertz [11].

Hottel [9], Willier [8] and Bliss [7] elaborated relationships available for most collector designs (tube and sheet case) using an appropriate form of the collector efficiency factor.

Matrawy and Farkas [12] compared a two parallel plate collectors (TPPC), with respectively a parallel tubes collectors (PTC) and a serpentine tube collectors (STC). Under the same ambient and performance conditions, efficiency of the TPPC is respectively 6% and 10% greater than STC and PTC efficiency.

The authors explain performance of the TPPC by the uniform temperature distribution over the absorbed plate surface and the uniform distribution of the working fluid between the two plates. However, this comparison study has been performed for a plate thermal conductivity of different collectors equal to 211 $W.m^{-1}.K^{-1}$ corresponding to the use of expensive materials, for a configuration STC and a distance between tubes of 0.167 m.

The use of a polymer absorber has been studied by Van Nierkerk et al. [13] with the aim of evaluating the performances of parallel tubes collectors in south of Africa. The variation of geometrical variables such as tube diameter, tube spacing and pitch have been studied for optimising the collector performances. Van Nierkerk et al. [13] concluded that the best configuration is obtained for an inter tube spacing equal to zero. It appears that the configuration of the flat plate collector is the most important parameter which affects collector performance.

We propose in this paper to study the collector configuration which presents the highest performance similar to what proposed Hottel [9], Willier [8] and Bliss [7] and Matrawy and Farkas [12] but which differs mainly:

- in the choice of the low conductivity material used with the aim of having a solar collector which is corrosion resistant and not prone to scaling, of decreasing its weight and cost price;
- in its rectangular flow passageways conception in order to increase its toughness.
- in the calculation, heat capacities of the various components are taken into account and allows to make dynamical simulation of system behaviour.

Choice of the Low Flow SDHW Design

Low flow SDHW systems differ from traditional SDHW system by having low volume flow rates in the solar collector loop of the systems. This configuration has several advantages comparing with a traditional operation (high flow) at various levels:

- *Thermal stratification:* using low flow operation results in an increased outlet temperature from the solar collector and consequently induces a higher degree of thermal stratification inside the heat storage; Moreover the temperature at the top of the storage will be closer to the desired load temperature. Therefore, the auxiliary energy consumption will be decreased which increases the solar fraction. Further, with highly stratified heat storage the return temperature to the solar collector will be lowered and the working periods for the solar collector will be longer, which implies in an increased output energy from the solar collector [14, 15].
- *Piping in the solar collector loop:*with low flow systems it is possible to use smaller pipes, in this way at first less material is used for pipes and insulation and secondly the heat losses are reduced.
- *Pump:*the energy consumption of the circulation pump is decreased.

The design of low flow SDHW systems varies from one country to another, which is due to differences in [16]:

- Regulatory issues concerning hot water systems
- Design of conventional hot water systems
- Available conventional energy sources and energy prices
- Local conditions and traditions
- Hot water consumption and patterns
- The climate variations from one country to another

The Canadian and North American low flow SDHW systems are normally based on pre-heating tank and an auxiliary tank.

In northern of Europe, low flow systems are often based on vertical mantle tank. The heat exchange between the solar collector fluid and the consumption water takes place in the mantle.

In central Europe, to obtain a very good stratification, a manifold diffuser is often used but this type of system is very expensive also compared to the increased performance initiated by the fine thermal stratification. However, in order to study the influence of thermal stratification in the tank, we will opt for this last configuration.

DESCRIPTION OF THE SOLAR SYSTEM

We consider a classical solar water heating installation with a flat-plate solar collector and a water tank for storage as shown in *figure 1*. A bypass is required which is controlled by a differential controller turning on the inlet fluid in the tank when the temperature T_{22f} is greater than the temperature of the fluid in the tank $T_{s,i}$ corresponding to the lowest temperature (at the bottom of the tank). If T_{22f} is lower than $T_{s,i}$, the fluid return to the collector [17].

The studied flat plate water collector is composed of a transparent cover and an absorber-exchanger which transform the solar radiation to heat. This "absorber-exchanger" has back and side insulations (in expanded polyurethane, ρ_{iso}=80 kg.m^{-2}, λ_{iso}=0.022 W.m^{-1}.K^{-1}), as shown in *figure 2*.

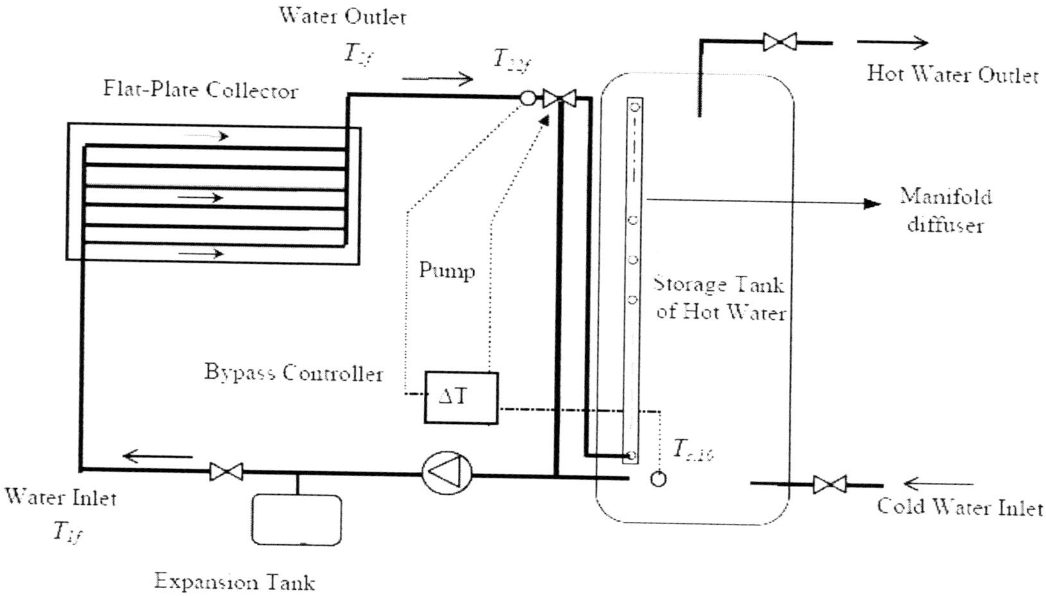

Figure 1. Layout of the solar system installation.

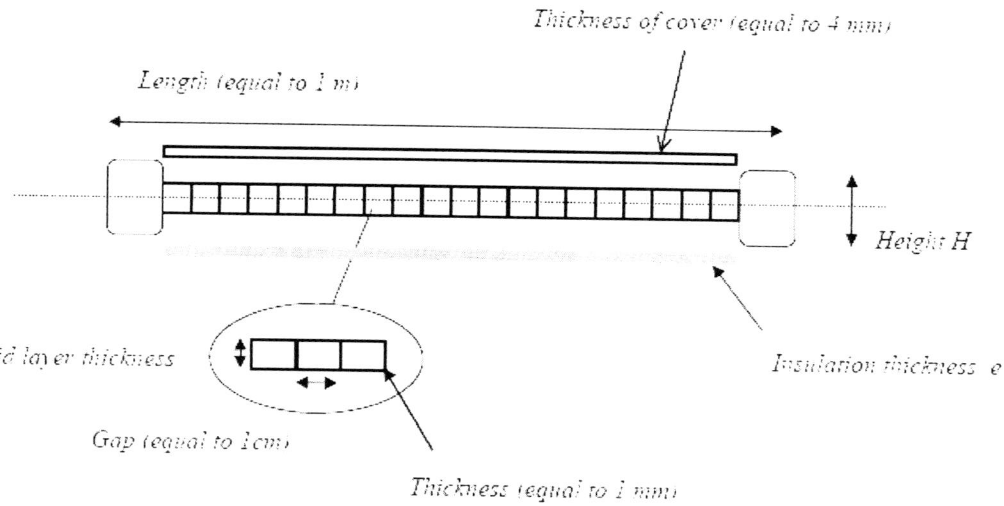

Figure 2. Transversal section of the polymer collector. Schematic section of the pipes (in the flow direction).

The fluid is distributed uniformly under the surface of the absorber. The water flows are parallel. A header pipe supplies each pipe and another one collects the warm fluid. The pipes are connected to the system users as shown in *figure 1*.

The absorber-exchanger, made of copolymer material, must satisfy the following constraints: UV protected, high thermal conductivity, water-resistant and glycol-resistant (anti-freezing solution), good thermal range of utilisation (-10°C ↔ 150°C), a good mechanical strength and to be chemically stable.

Various anti-UV treated copolymers such as polycarbonate, polyphenylen ether or methyl polymethacrylate satisfy all these criteria. However, for our study, we selected the anti-UV treated polycarbonate which is the cheapest polymer and which satisfies chemical, mechanical and thermal constraints. All the properties of this copolymer are summarised in table 1. We chose a single glass cover plate with a thickness equal to 4 mm.

Table 1. Properties of polycarbonate

Properties	**Units**	**Polycarbonate copolymer**
Physical		
Density	$kg.m^{-3}$	1.18
Water Absorption	%	0.15
Rate of cristallinity	%	0
Mechanical		
Breaking Constraint	MPa	65
Breaking elongation	%	7
Traction Module	MPa	2250
Thermal		
Vitreous Transition (T_g)	°C	150
Casting shrinkage	%	0.7-0.9
Thermal Conductivity	$W.m^{-1}.K^{-1}$	0.7-0.85
Advantages		Good mechanical properties, good dimensional stability, even in humid atmosphere, large range of using temperature (-80°C, +135°C)

THEORETICAL ANALYSIS: EQUATIONS AND RESOLUTION

We present a model of finite differences which includes the essential thermal transfers. This model is composed of a serial assembling of many elementary models. Each model is based on a nodal discretisation of a collector section and of the storage circuit.

Presentation of the Collector: Model and Assumptions

The simulation is done by dividing the transversal section collector into six isothermal regions: the glass cover (G), the air layer (A), the top half of the absorber (P1), the water layer (F), the bottom half of the absorber (P2) and the insulation (I); therefore, we propose to dividing the length collector into ten sections f_k in order to take into account the temperature distribution of the working fluid inside the collector.

This model also takes into account the following input physical parameters:

- diffuse and direct solar irradiances I_d and I_b
- temperature of the ambient outside air T_{ae}
- temperature of the inside air T_a of the house
- air velocity in front of the collector V
- sky temperature T_{sky} calculated from the following equation [18]:

$$T_{sky} = T_{ae}\left[0.8 + \frac{T_{dp} \cdot 273}{250}\right]^{1/4} \tag{1}$$

T_{dp}: the dew point temperature.

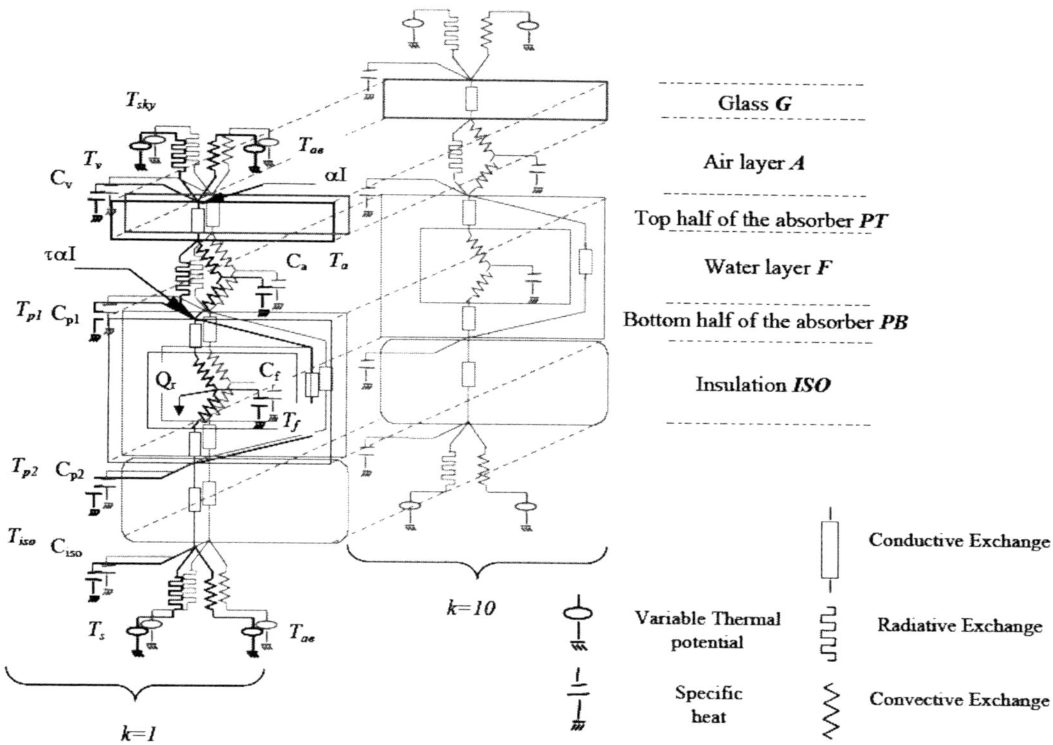

Figure 3. Topologic diagram of the polymer collector.

A detailed description of these regions is shown in *figure 3* and in *figure 4*. A heat balance is done for each region. These equations are elaborated from a model based on an electrical analogy where temperatures, flows, flow sources and imposed temperatures are respectively analogous to potentials, currents, current generators and voltage generators. The model is based on the following assumptions:

- the thermo-physical properties in solids and fluids are constant;

- the conductive transfers are neglected compared to the convective and radiative exchanges in the fluid layer;
- in the water layer of the collector, the conductive and radiative transfers are neglected with regard to the convective ones.
- the radiative transfers are neglected compared to the convective ones on the bottom face of the collector.

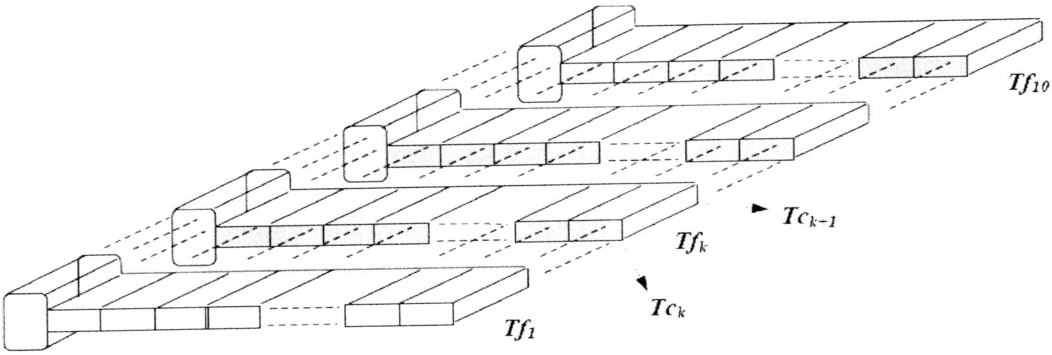

Figure 4. *Transversal section of the polymer absorber-exchanger.*

Global Equations

For each node, according to the Kirchoff's law:

$$CC_i \frac{dT_i}{dt} = \sum_j U_{ij}(T_j - T_i) + \phi \tag{2}$$

Where ϕ is the solar irradiance incident on the glass or on the top half of the absorber depending on the considered node.

The capacities CC_i come from the term of power variation $M_i\, C_i \frac{dT_i}{dt}$ from the balance equation for the mesh point I

$$CC_i = M_i\, C_i = \rho_i\, C_i\, \upsilon_i \tag{3}$$

The conductive term from the discretisation of the term $\lambda_i\, A_i \frac{dT_i}{dt}$ is expressed according to the Fourier's law by:

$$U_{ij} = \lambda_i\, A_i \frac{1}{\Delta_{ij}} \tag{4}$$

Where Δ_{ij} is the inter node spacing.

Concerning U_{ij}, if the material of the mesh point i is different from the material of the mesh point j, the conductivity is expressed like two serial electrical resistances. If we assume a good contact between these two materials, we have:

$$U_{ij} = \frac{U_i \times U_j}{U_i + U_j} \tag{5}$$

The convective term is given by the Newton's law
$h\, A_{ij}$
The radiative exchange term is given by

$$\sigma\, \varepsilon_i\, \alpha_j\, A_i\, R_{ij}\, (T_j^4 - T_i^4) \tag{6}$$

where the different surfaces are considered as grey.

When the temperature difference T_j-T_i is relatively small, the previous expression can be linearized to:

$$4\, \sigma\, \varepsilon_i\, \alpha_j\, A_i\, R_{ij}\, (T_j - T_i)\, \mathrm{T}^3_{\mathrm{m,ij}} \tag{7}$$

The radiative conductivity is then expressed by:

$$Ur_{ij} = \sigma\, \varepsilon_i\, \alpha_j\, A_i\, R_{ij}\, (T_j^2 + T_i^2)\, (T_j + T_i) = 4\, \sigma\, \varepsilon_i\, \alpha_j\, A_i\, R_{ij}\, T^3_{m.ij} \tag{8}$$

Lastly, the fluid transport is introduced into equation (2) by the factor $U_f\, (T_i - T_j)$ where $U_f = \dot{m}_c\, C_f$.

More particularly, in the collector the fluid transport is introduced by $U_f\, (Tc_{k+1} - Tc_k)$. The different expressions of temperature difference are given by

$$Tc_{2k+1} - Tc_{2k} = 2\,Tf_{2k+1} - 4\sum_{j=1}^{k} Tf_{2j} + 4\sum_{j=1}^{k} Tf_{2j-1} - 2\,T_{1f} \tag{9}$$

$$Tc_{2k} - Tc_{2k-1} = 2\,Tf_{2k} - 4\sum_{j=1}^{k} Tf_{2j-1} + 4\sum_{j=1}^{k} Tf_{2j-2} + 2\,T_{1f} \tag{10}$$

with Tf_k the average temperature in each section k of the collector ($k \in [1,10]$):

$$Tf_k = \frac{(Tc_{k+1} + Tc_k)}{2} \tag{11}$$

The natural and forced convective coefficients and the configuration factors are estimated from classical correlations.

The following relation [19] describes the convective thermal exchanges inside the collector pipes, for the laminar flow regime:

$$h_f = \frac{\lambda_f}{D} 1.607 \left(\frac{D}{L_c} Re\, Pr\right)^{1/3} \quad \text{for } Re < 2300 \tag{12}$$

The convective transfer coefficient on the outside collector surface is related to the wind speed V by the relation [20]:

$$h_c = 7.0 + 2.1V \tag{13}$$

Among all the elements taken into account in the total coefficients of the thermal transfer, two terms allow to justify the choice of a collector with an absorber using polymer material. The first one depends on the conductivity of the layer crossed by the thermal flux, and the second one depends on the convective transfer coefficient between the surface which is in contact with the fluid and the fluid itself. But, the first is much greater than the second. Thus, we propose an 'absorber-exchanger' in copolymer material, poor heat conductor, but which can be used with a low thickness.

Meteorological Data and Solar Energy Absorption Assumptions

The results presented are determined from real meteorological data collected in our laboratory situated at Ajaccio, a seaside Mediterranean site (Latitude 41°55'N, Longitude 8°48'). In this site, we have a complete meteorological station, where the direct normal and the global horizontal and tilted (30°, 45° and 60°) irradiances are respectively measured by an Eppley NIP pyrheliometer and a Kipp and Zonen (CM5) pyranometer. The standardisation of such instruments is maintained by the French meteorological organisation Meteo France. Other parameters such as ambient temperature, pressure, relative humidity, wind speed and direction are also recorded. The data are collected every minute and recorded. The diffuse radiation is not measured, but can be obtained easily from the following:

$$I_d = I - I_b \tag{14}$$

where I_d, I and I_b are respectively the diffuse, total and beam solar irradiance. We supposed that the collector faced to the south and 45° tilted [17].

The calculation of the absorbed solar irradiance S is based on the ASHRAE convention [21] and used by Tsilingiris P.T. [22]:

$$S = (\tau\alpha)_n \left(R_b\, I_b\, K_b + I_d\, K_d\, \frac{1+\cos\beta}{2} + (I_b + I_d)\, \rho_g\, K_g\, \frac{1-\cos\beta}{2}\right) \tag{15}$$

the incidence angle modifier for the direct component of solar radiation is given by:

$$K_b = 1 + b_0 \left(\frac{1}{\cos\theta_{eb}} - 1\right) \tag{16}$$

θ_{eb} is the incidence angle of the beam solar radiation and the numerical constant b_0=-0.1 is for single glazed collectors.

For the sky and ground diffuse radiation, this coefficient is given as a function of the diffuse radiation equivalent sky and ground incidence angles:

$$K_d = 1 + b_0 \left(\frac{1}{\cos\theta_{ed}} - 1\right) \tag{17}$$

$$K_g = 1 + b_0 \left(\frac{1}{\cos\theta_{eg}} - 1\right) \tag{18}$$

The equivalent sky and ground incidence angles for diffuse radiation are given as a function of the collector field tilt angle and calculated according to Duffie and Beckman [23]:

$$\theta_{ed} = 59.68 - 0.1388\beta + 0.001497\beta^2 \tag{19}$$

$$\theta_{eg} = 90 - 0.5788\beta + 0.002693\beta^2 \tag{20}$$

with the inclination of the collector β=45° and the ground albedo ρ_g=0.2 for our site. The normal transmittance-absorbance is fixed at 0.80.

Presentation of the Tank: Model and Assumptions

Water tank may operate with significant degrees of stratification with the top of the tank hotter than the bottom. In our case, the tank will be modeled in i sections (i nodes), [23] with energy balances written for each section of the tank.

To formulate these equations, we make assumptions about how the water entering the tank is distributed to the various nodes.

The water from the collector and from the load enter respectly at a temperature T_{22f} and T_l . It can be assumed that all water finds its way down inside the tank from node 1 to node i, where its density is nearly equal to the density of the water in the tank corresponding to the node i. To obtain this configuration, we use a manifold diffuser (figure 1) [16].

A collector control function B_c^i can be defined to determine which node receives collector return water:

$$B_c^i = \begin{cases} 1 & if \;\; i = 1 \; and \;\; T_{22,f} > T_{s,i} \\ 1 & if \;\; T_{s,i-1} \geq T_{22,f} > T_{s,i} \\ 0 & if \;\; i = 0 \; or \; i = N + i \\ 0 & otherwise \end{cases} \tag{21}$$

The i branches of the collector return water are controlled by this collector return control function as shown in *figure 5*.

The liquid returning from the load can be controlled in a similar manner with a load return control function B_l^i.

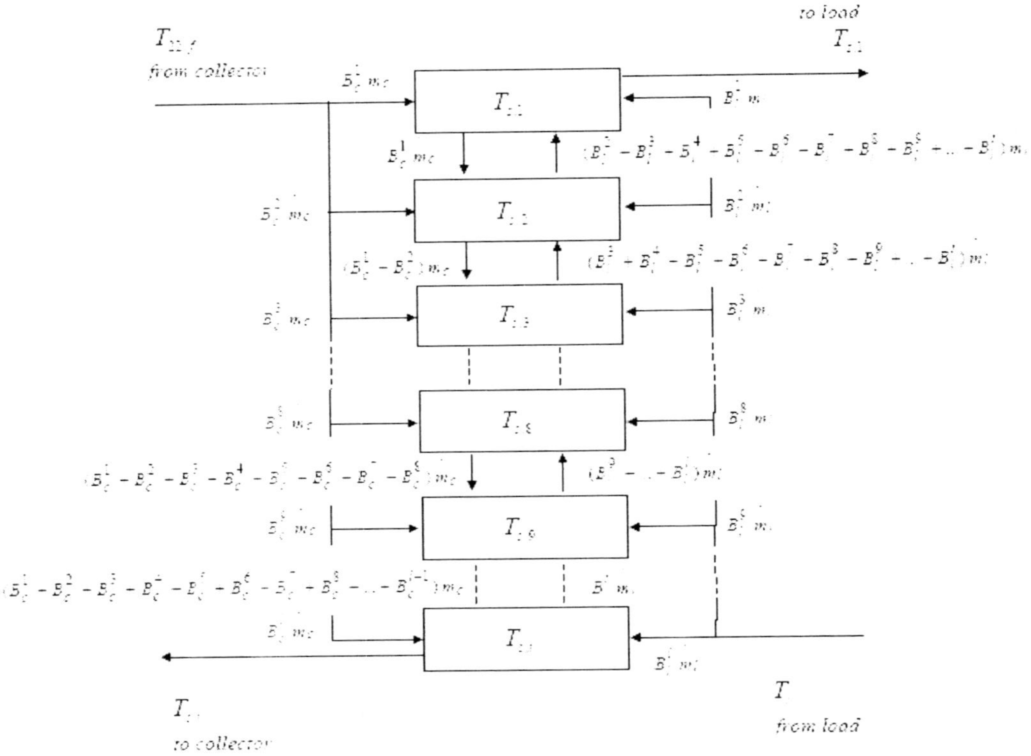

Figure 5. i nodes stratified liquid storage tank.

$$B_l^i = \begin{cases} 1 & if \;\; i = N \; and \;\; T_l < T_{s,N} \\ 1 & if \;\; T_{s,i-1} \geq T_l > T_{s,i} \\ 0 & if \;\; i = 0 \; or \; i = N + 1 \\ 0 & otherwise \end{cases} \tag{22}$$

The net flow between nodes can be either up or down depending upon the magnitudes of the collector and load flow rates and the values of the two control functions at any particular instant. It is convenient to define a mixed flow rate that represents the net flow into node i from node i-1, excluding the effects of flow, if any, directly into the node from load

$$\begin{cases} \dot{m}_{m,1} = 0 \\ \dot{m}_{m,i} = \dot{m}_c \sum_{j=1}^{i-1} B_c^j - \dot{m}_l \sum_{j=i+1}^{N} B_l^j \\ \dot{m}_{m,N+1} = 0 \end{cases} \quad (23)$$

With control functions defined above, an energy balance on node i can be expressed as:

$$\rho_i C_i V_i \frac{dT_{s,i}}{dt} = U_s (T_a - T_{s,i}) + B_c^i \dot{m}_c C_i (T_{22,f} - T_{s,i}) + B_l^i \dot{m}_l C_i (T_l - T_{s,i}) +$$

$$+ \begin{cases} \dot{m}_{m,i} C_i (T_{s,i-1} - T_{s,i}) & if \ \dot{m}_{m,i} > 0 \\ \dot{m}_{m,i+1} C_i (T_{s,i} - T_{s,i+1}) & if \ \dot{m}_{m,i+1} < 0 \end{cases} \quad (24)$$

where a term $U_s (T_a - T_{s,i})$ has been added to take into account the heat losses from node i to the environment (house) at the temperature T_a assumed constant and equal to $20°C$.

This model is based on the following assumption: the conductive transfers of the manifold diffuser is neglected compared to thermal exchange of mixed flow.

The various simulations have been developed for a normalised water consumption profile corresponding to an evening demand for which the hot water is totally used from 18 hours.

Connection Pipes

Modeling the connection pipes with insulation (e=3 cm) is simplified by considering the energy balance as a whole for the liquid returning from the load and returning from the collector. The temperature developments in the pipe connections are assumed linear. Thus, the energy balance can be described as:

$$L C_f \frac{dT_{w1}}{dt} = \dot{m}_c C_f (T_{22f} - T_{2f}) + U_p L_{wae} (T_{ae} - T_{w1}) + U_p L_{wa} (T_a - T_{w1}) \quad (25)$$

$$L C_f \frac{dT_{w2}}{dt} = \dot{m}_c C_f (T_{1f} - T_{s,i}) + U_p L_{wae} (T_{ae} - T_{w2}) + U_p L_{wa} (T_a - T_{w2}) \quad (26)$$

$$L = L_{wae} + L_{wa} \quad (27)$$

We supposed a pipe length $L_{wae} = L_{wa} = 1.5\ m$ corresponding respectively to the part of pipe outside and inside the house.

Method of Solution

All the energy balances can be written in the guise of a differential matrix equation:

$$[C]\frac{d\vec{T}(t)}{dt} = [M]\vec{T}(t) + [S]\vec{E}(t) \tag{28}$$

$\vec{T}(t)$ is a vector containing system temperatures at the 62+i nodes (Tv_k, Ta_k, Tpt_k, Tf_k, Tpb_k, $Tiso_k$, Tw_1, Tw_2, T_{si}) of the mesh, [C] is a diagonal matrix (dimension: 62+i) with all the values of the thermal capacities of the material, [M] is a squared matrix (dimension: 62+i, with i=1, 3, 6, 10, 15) with all the heat exchange coefficients between the elements of the mesh, [S] is a matrix (dimension: (62+i)x5) which joins together the 5 input physical parameters (or excitations) expressed by the vector $\vec{E}(t)$ (I_b, I_d, T_{ae}, T_a, T_{sky}) and 62+i elements of the mesh. By solving the equations of this analogical model, we get directly the expression of the temperature for each nodes of the model in function to the input parameters. This equation system is solved by the Runge Kutta Merson's method (order 4).

Results and Discussion

The sizing must respond to the needs of hot water for 3 inhabitants in a Mediterranean site. Thus, we considered a collector with a surface of 2 m², a storage tank of 150 litres. All the temperatures of the incoming water in the tank are summarised in *table 2*. The simulations have been performed with one year of real weather data. It is assumed that, in the early morning, the water inside the collector and all the parts of the collector follow the ambient temperature and the temperature of the water inside the tank are given in *table 2*.

Table 2. Average temperature of the incoming water in the tank.

Ajaccio 41°55'					
January	*February*	*March*	*April*	*May*	*June*
10°C	10°C	11°C	14°C	16°C	18°C
July	*August*	*September*	*October*	*November*	*December*
19°C	20°C	18°C	16°C	14°C	11°C

To take into account in the modeling the degree of tank stratification, it is necessary to divide the tank in a given number of nodes. Obviously, if the tank is fully mixed, the temperature at these various i nodes will be the same one whatever the number orifices of the manifold diffuser is. More the degree of stratification is high, more the number i of nodes (and so orifices) will have to be high. Thus, we want to determine the optimal number of nodes needed to maximise correctly the performances of the low flow SDHW. In a second time, we will calculate the optimum flow rate allowing a maximum production of thermal energy.

Influence of the Number of Nodes Used to Model the Tank

The *figure 6* show the influence of the number of nodes on the gained energy calculated on one year of data.

We modeled the tank using respectively 1, 3, 6 10 and 15 nodes corresponding to the orifices number of the manifold diffuser. According to the number of nodes, the calculated performances are different. More the flow rate is low, more the degree of stratification in the tank is important. So, in order to take into account this stratification, the tank must be modeled with a sufficient number of nodes and consequently the manifold diffuser must have sufficient number of orifices. In fact, using only one node correspond to a homogenous temperature in the tank which is a wrong hypothesis for the low flow but becomes more realistic for high flow as seen in *figure 6* where the gained energy is identical whatever the number of nodes is.

As a matter of fact, even with only one orifice, using low flow rate results in a thermal stratification.

We note that the modeling with 10 and 15 nodes conduces to the same results: so it seems to be not necessary to use more than 10 nodes and so, no more than 10 orifices, to study the behaviour of this solar domestic hot water. In the following of this article all the presented results will be computed with a ten nodes model for the tank and so with a manifold diffuser with ten orifices.

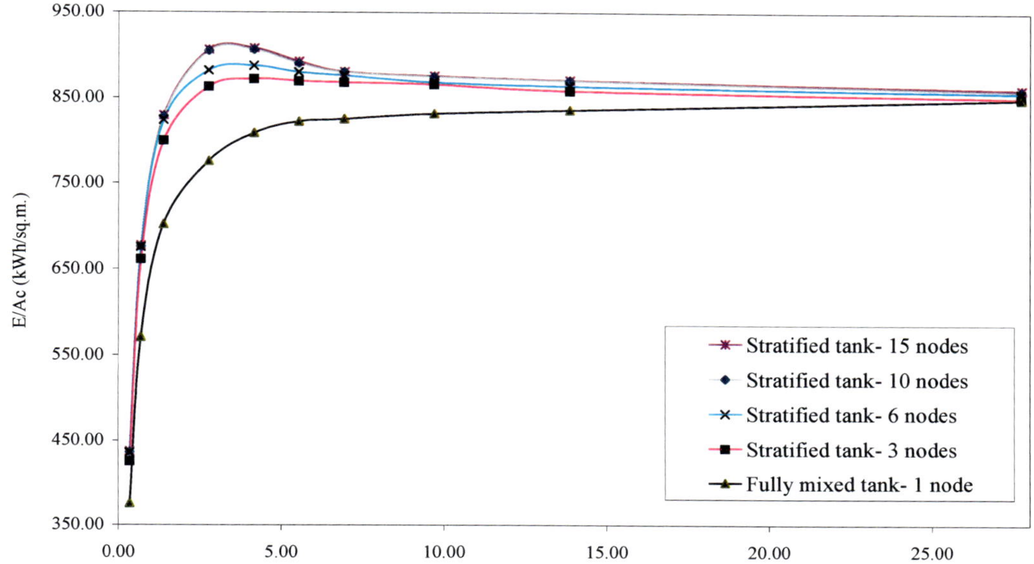

Figure 6. Gained energy versus the flow rate of the fluid (with a thickness of insulation = 2cm).

Influence of the Flow Rate

Using low flow operation results in an increased outlet temperature from the solar collector and a higher degree of thermal stratification inside the heat storage. We note that a stratified tank has much higher performance than a fully mixed tank.

The *figure 6* show that the mixing caused by a high flow rate begin for about 30 $kg.m^{-2}.s^{-1}$ because from this flow rate value the results obtained with 1 and 10 nodes are similar.

For a stratificated tank, it appears an optimal flow rate equal to $2.65.10^{-3}$ $kg.m^{-2}.s^{-1}$ (i.e. 0.12 tank volume per hour) which maximize energy saving.

With this flow rate conducing to a stratificated tank, the gained energy is 5.25% higher than a fully mixed tank. These results are close to results obtained by S. Furbo [14] in testing low flow solar heating.

Thus, the value of flow rate equal to $2.65.10^{-3}$ $kg.m^{-2}.s^{-1}$ seems to be adequate for our solar collector. This optimal value of the flow rate is used to determine the collected solar thermal energy and the daily mean of the efficiency over one year of use.

Daily Dynamical Evolution of Fluid Temperatures and Efficiency

For a given geometrical configuration, for a given running mode (insulating thickness e = 2 cm, flow rate $\dot{m}_c$ =2.65 .10^{-3} $kg.m^{-2}.s^{-1}$, fluid layer thickness equal to 1cm, [2]) and for a stratified tank (10 nodes), we show in *figures 7* and *8* the daily evolution of the efficiency, the temperature of the output collector fluid and the temperature of the storage tank fluid for a summer day (August 25) and for a winter day (February 28).

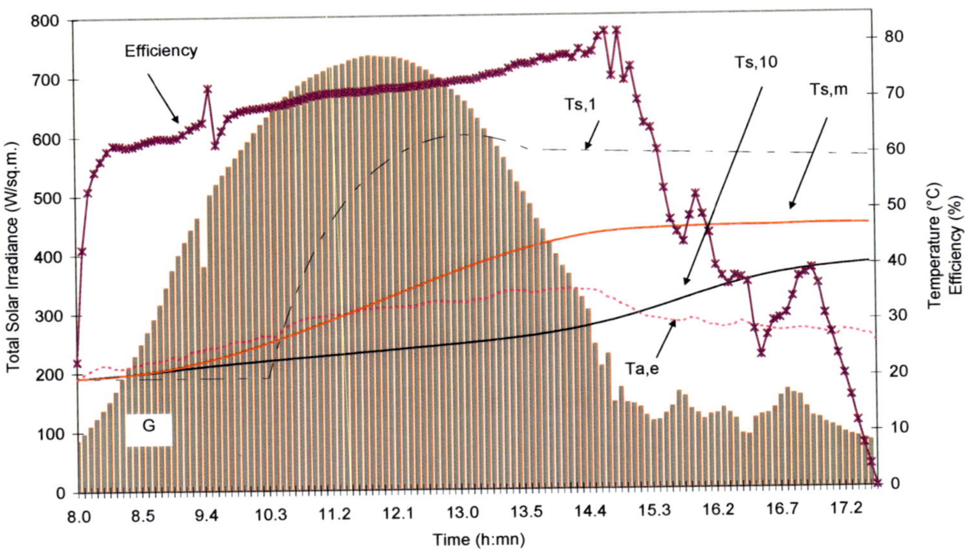

Figure 7. Evolution of the efficiency and storage fluid temperature (August 25).

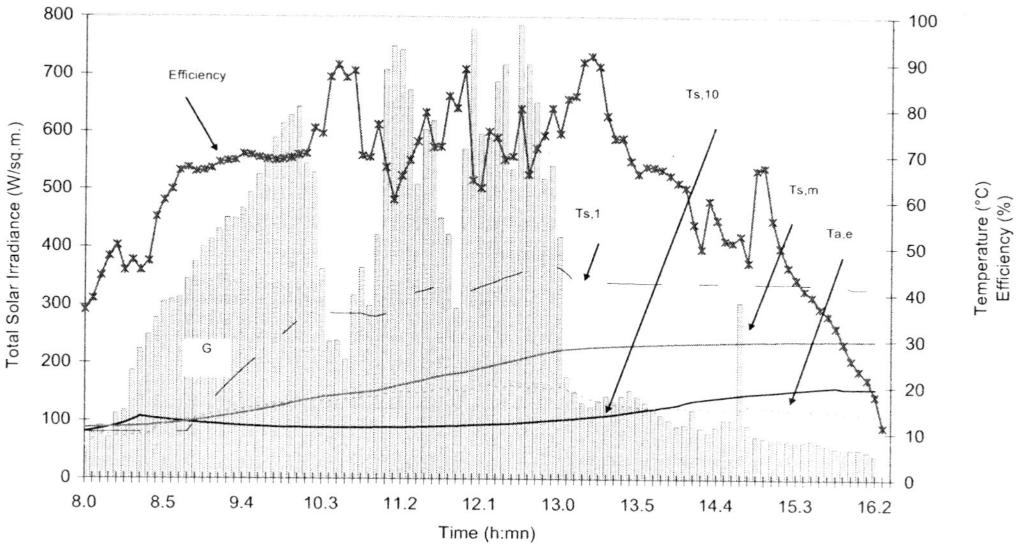

Figure 8. Evolution of the efficiency and storage fluid temperature (February 28). The daily mean efficiency is given by the following relation [23]:

$$\eta_{day} = \frac{\sum q_f}{A_c \int_{sunrise}^{sunset} I(t)\, dt} \tag{29}$$

The simulation based on the August 25, shows that the important radiation (daily peak greater than 700 W/m²) allows a constant energy supply during all the day to the storage tank from the collector. The *figure 8* presents a variable radiation due to cloud passages. We note a gain of energy in the storage fluid during the first half of the day. The water stays in the collector when the regulation is on, then the fluid temperature comes closer to the outdoor air temperature.

We note in *figures 9* and *10* the temperature profiles in the tank for a summer day and a winter day; the final average temperatures in the storage tank are respectively about 47.9 °C (degree of stratification max $(T_{s,1}-T_{s,10})=19.5°C$) and 29.6 °C (degree of stratification max $(T_{s,1}-T_{s,10})=21.0°C$), and with 71 % and 61.5 % respectively of mean collector efficiencies, with a wind speed taken equal to 0 $m.s^{-1}$.

We see in *figure 11* the temperature profile in the collector computed from simulations.

Figure 12 illustrates for a summer day (august 25) the temperature difference between outlet and inlet fluid of the collector. It varies in the range from 0°C to 7°C for fully mixed tank and in the range from 0°C to 40°C for high thermal stratified tank. Using low flow operation, we note a working period for the solar collector longer than a conventional flow operation where temperature difference is about 0°C during the last 2 hours and 30 minutes of the day.

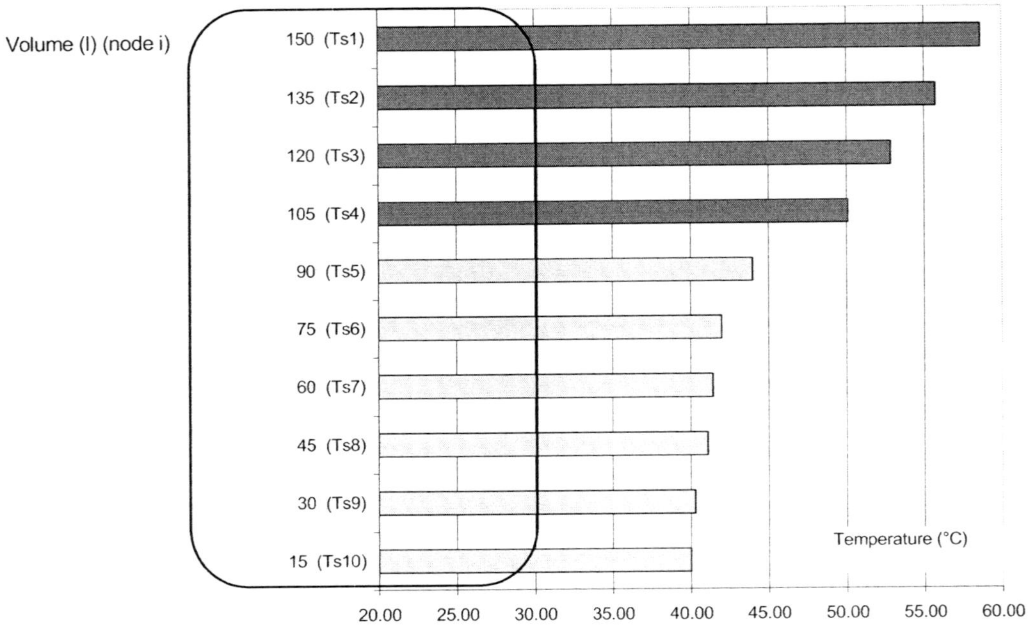

Figure 9. Temperature profile in tank (August 25).

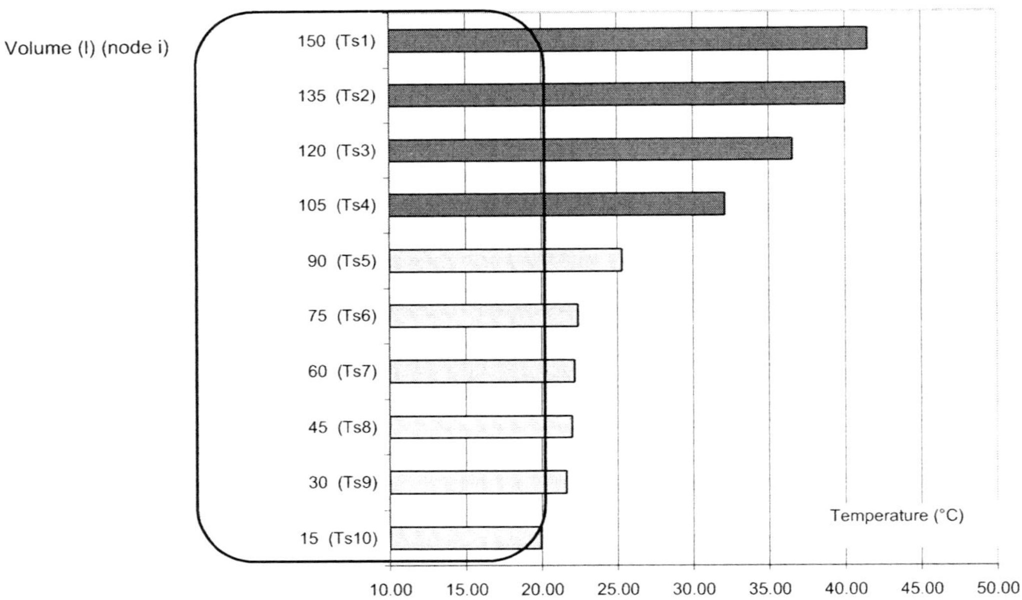

Figure 10. Temperature profile in tank (February 28).

Using low flow operation results in a decreased outlet temperature from the tank, a decreased of the annual mean of temperature solar collector (even if we note an increased outlet temperature from the solar collector) and so a decreased of DT and consequently a higher instantaneous efficiency of the system (figure 13).

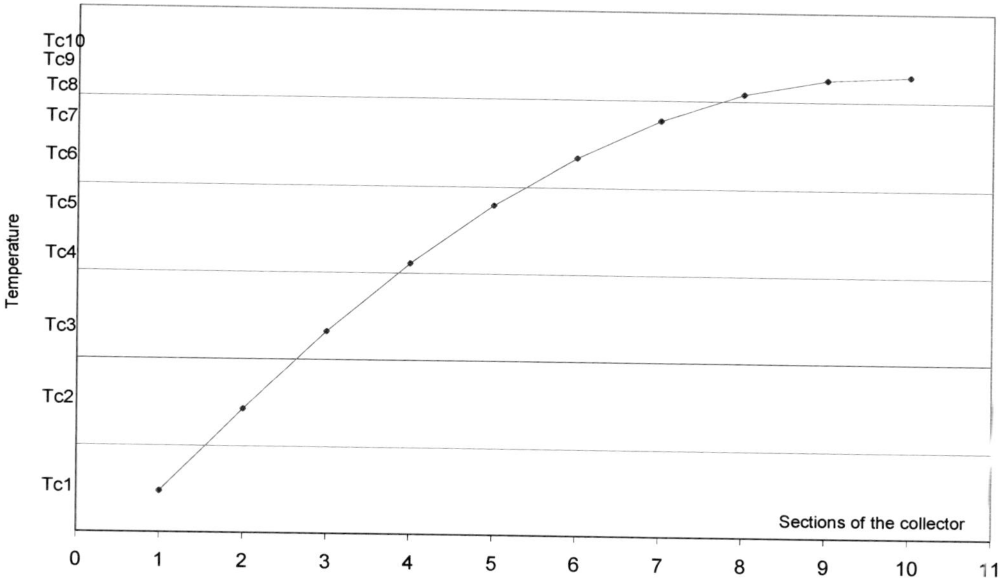

Figure 11. Temperature profile in the collector.

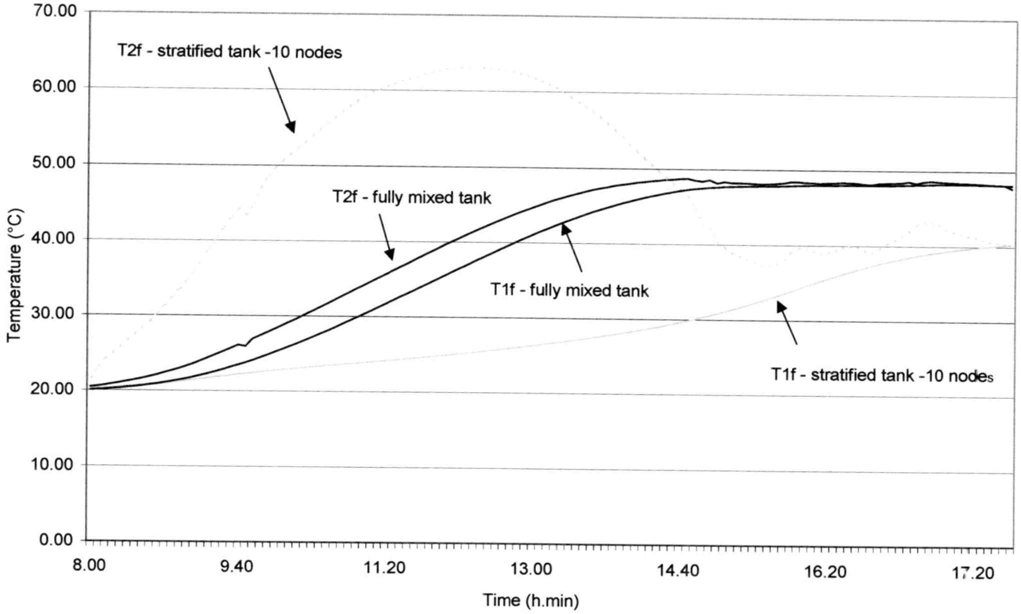

Figure 12. Temperature evolution of outlet and inlet fluid of the collector for stratified tank and fully mixed tank.

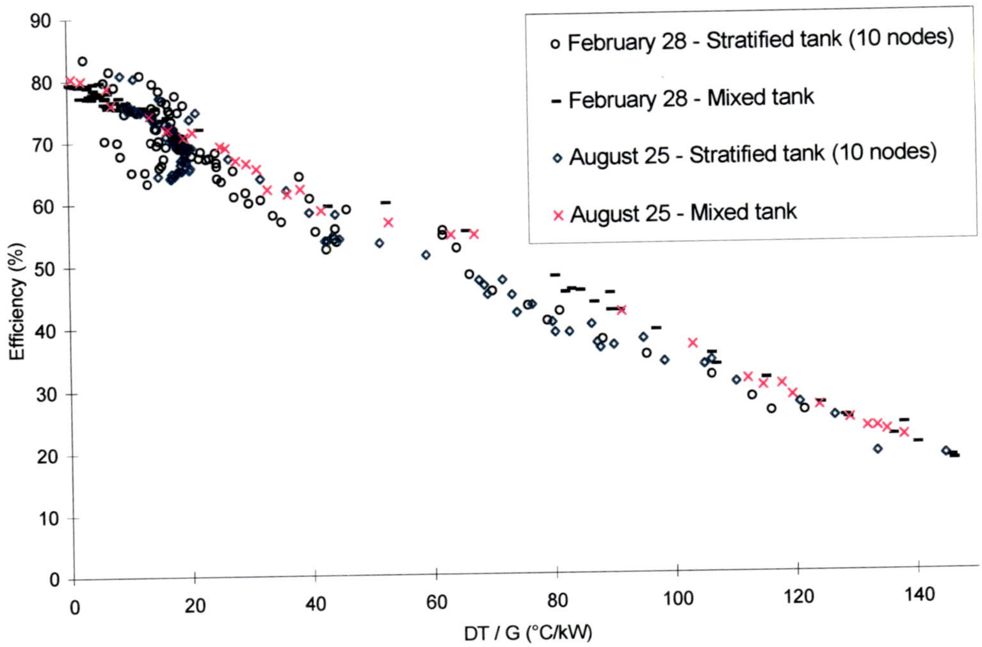

Figure 13. Instantaneous efficiency versus the ration "DT/G" for fully mixed tank and stratified tank.

Consequently, this low flow rate induces a thermal stratification of the tank which increase and maximise energy saving (figure 6).

Performance of the system

The flow rate has an important influence on the performances of the solar system. We studied this influence in calculating the average efficiency of the collector during five minutes (for a high thermal stratified tank $\dot{m}_c = 2.65 .10^{-3}$ kg.m^{-2}.s^{-1} and for a fully mixed tank $\dot{m}_c = 30 .10^{-3}$ kg.m^{-2}.s^{-1}).

For two particular days (winter and summer), the *figure 13* illustrates the effect of the flow rate on the response of the collector efficiency. This study has been performed on 365 days, we note a linear response of the efficiency [23] according to the flow operation:

when we used low flow operation, and so with high stratified tank, the working periods for the solar collector will be longer, which results in an increased nots in higher efficiency.

Equations of the straight lines (calculated by linear regression method on the basis of about 45000 points i.e. 365 days) are respectively for fully mixed tank and high stratified tank: $\eta = -0.40\ DT/G + 78.72$ and $\eta = -\ 0.43\ DT/G + 76.80$.

Daily Evolution of the Available Solar Energy

For a solar collector with a surface of 2 m², from *table 3*, the annual mean of daily efficiencies are 54.0 % and 52.0 % and the annual mean of daily productivities are 4.81 kWh and 4.66 kWh, respectively, for fully mixed tank and high stratified tank.

Figure 14 illustrates during a year the evolution of monthly mean temperature of the storage fluid for fully mixed tank and high thermal stratified tank. These monthly mean temperatures vary in a range from 19.2 °C to 69.6 °C for fully mixed tank and in a range from 20.0 °C to 71.1 °C for high thermal stratified tank.

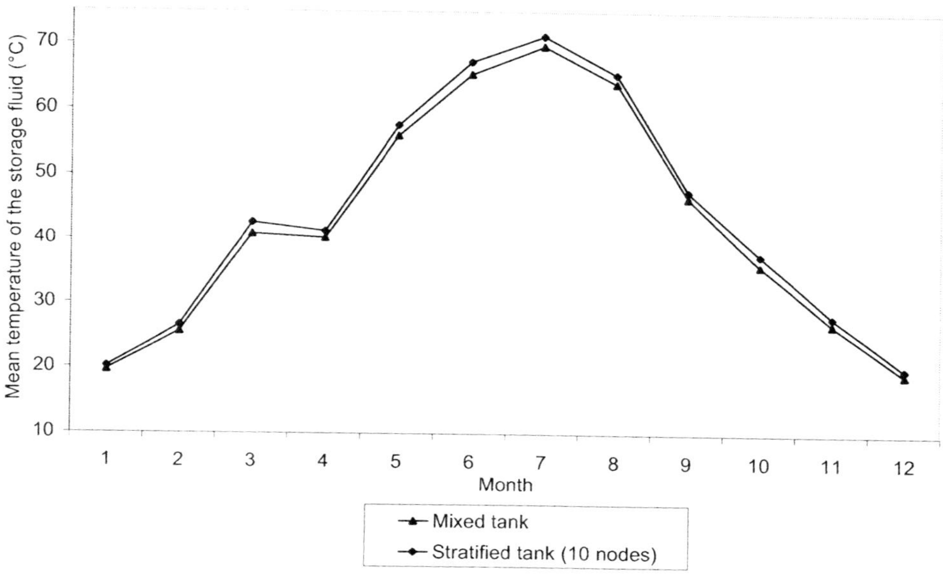

Figure 14. Monthly mean temperature of the storage fluid for fully mixed tank and stratified tank.

Table 3. Daily mean energy and efficiency for fully mixed tank and stratified tank for each season

Seasons	Mean Tempe-rature	Daily Mean Solar Energy (kWh) *(β=45°)*	Stratified Tank (10 nodes)		Mixed Tank	
			Daily Mean Energy collected (kWh)	Daily Mean Efficiency (%)	Daily Mean Energy collected (kWh)	Daily Mean Efficiency (%)
Winter	9.6	6	2.94	49	2.82	47
Spring	16.2	11.79	6.36	54	6.13	52
Summer	23.6	12.51	7.38	59	7.25	58
Autumn	13.8	4.78	2.58	54	2.44	51

CONCLUSION

In this paper, we illustrated the thermal behaviour of a solar collector using a copolymer material in regard to radiative conditions. We chose a copolymer material for the 'absorber-exchanger' satisfying chemical, thermal and mechanical constraints: the polycarbonat.

We developed a nodal model for the solar water heating installation and studied the influence the flow rate and thermal stratification tank. The optimised system has an insulation of 2 cm, a thickness of 1 cm for fluid layer and works with a collector flow rate of 2.65 $.10^{-3}$ $kg.m^{-2}.s^{-1}$. Such a system presents a yearly mean efficiency of about 54.0 % and 52.0 % and the annual mean of daily productivities are 4.81 kWh and 4.66 kWh respectively for fully mixed tank and high thermal stratified tank.

The results show that ten orifices for the manifold diffuser are enough to have a high stratified tank and a stratified tank has much higher performance than a fully mixed tank. With high degree of stratification saving energy is higher (5.25% over one year of use) than fully mixed tank.

The utilisation of a copolymer for the total design of the solar collector has the advantage of reducing the weight by more half in comparison with a traditional collector using essentially metals with similar performances.

REFERENCES

[1] Cristofari C., Notton G, Poggi P. and Louche A., Diagnostic de la situation énergétique de la région Corse – L'éolien peut-il être intégré dans une réelle politique énergétique de développement ?, La Revue de l'Energie Ed. *Techniques & Economiques* 521 (2000) 544-552.

[2] Cristofari C., Notton G, Poggi P. and Louche A., Modeling and performance of a copolymer solar water heating collector, *Solar Energy in press* (2001).

[3] Dickinson W.C., Clark A.F., Day J.A. and Wouters L.F., The shallow solar pond energy conversion system, *Solar Energy* 18 (1976) 3-10.

[4] Clark A. F., Casamajor A. B. and Hewett L. D., *Shalow solar ponds, Solar Energy Technology Handbook Part A* New York (1978) 377-402.

[5] Casamajor A. B., The application of shallow solar ponds for industrial process heat: case histories, *Proc. A.S. / ISES Silver Jubilee Congress* Atlanta GA,1979, 1029-1032.

[6] Tsilingiris P.T., Design, analysis and performance of low-cost plastic film large solar water heating systems, *Solar Energy* 60 (5) (1997) 245-256.

[7] Bliss R. W., The derivations of sereval plate efficiency factors useful in the design of flat plate solar heat collectors, *Solar Energy* 3 (1959) 55-63.

[8] Willier A., *Prediction of performance of solar collectors, in Applications of Solar Energy for Heating and Cooling of buildings*, ASHRAE New York, 1977.

[9] Hottel H. C. and Whillier A., Evaluation of flat plate collector performance, *Trans. of the Conference on the Use of Solar Energy* University of Arizona Press 2, 1958, 1-74.

[10] Klein S.A., Calculation of flat-plate collector loss coefficients, *Solar Energy* 17 (1975) 79-80.

[11] Hottel H. C. and Woertz B. B., Performance of Flat-plate Solar-Heat Collectors, *Transactions of the American Society Mechanical Engineers* 64 (1942) 91-103.

[12] Matrawy K. K. and Farkas I., Comparison study for three types of solar collectors for water heating, *Energy Conversion and Management*, 38 (1997) 861-869.

[13] Van Niekerk W. M. K., C. G. du Toit and T. B. Scheffler, Performance modeling of a parallel tube polymer absorber, *Solar Energy* 58 (1996) 39-44.

[14] Shah L. J., *Investigation and modeling of thermal conditions in low flow SDHW systems, Department of buildings and energy* Technical University of Denmark Report R-034, 1999.

[15] Furbo S., Optimum design of small DHW low flow solar heating systems, *ISES Solar World Congress Budapest Report* 93-24, 1993.

[16] Furbo S., Present and future SDHW System technology, *Proceedings EuroSun'98* Portorôz Slovenia, 1998.

[17] Chateauminois M., Mandineau D., Roux D., *Tableau des températures du réseau eau froide Calcul d'installations solaire à eau*, Collection de l'ESIM Edition EDISUD, 1979.

[18] Bliss R. W., Atmospheric radiation near the surface of the ground, *Solar Energy* 5 (3) (1961) 103-120.

[19] Lévêque M.A., Les lois de transmission de chaleur par convection, *Annal. Mines Mem.* 13 (1928) 201-205.

[20] Schott T., Operation temperatures of PV modules, *paper presented to 6th PVSEC Conference London*, 1985, 392-396.

[21] ASHRAE, *Methods of testing to determine the thermal performance of solar collectors*, ASHRAE New York, 1977.

[22] Tsilingiris P.T., Solar water heating design-A new simplified approach, *Solar Energy* 57 (1) (1996) 19-28.

[23] Duffie J.A. and Beckman W.A., *Solar Energy Thermal Processes*, Wiley-interscience publication New York 2nd Edition, ISBN 0-471-51056-4, 1991.

[24] Rongere F.X., Bonneau D., Covalet D. and Gautier B., Un logiciel modulaire pour l'élaboration de modèles thermiques pour l'habitat, *Revue Générale de thermique* 28 (335-336) (1989) 644-649.

[25] Touchais M., *Le chauffage solaire- Les insolateurs à fonction thermique*, 2ème édition Estrel, 1981.

In: Solar Energy: New Research
Editor: Tom P. Hough, pp.97-148

ISBN 1-59454-630-4

Chapter 4

SOLAR ENERGY FOR DOMESTIC HEATING AND COOLING AND HOT WATER PRODUCTION

Soteris Kalogirou*
Higher Technical Institute, Nicosia, Cyprus

ABSTRACT

Heating and cooling of buildings require a large amount of energy. For countries with good solar potential, solar energy can be considered a viable option for these applications. Traditionally solar energy is used for hot water production. With the increase in the price of fuels solar space heating became a viable option, however, the potential problem of such a system is that a lot of solar energy collectors need to be disconnected or shaded during summertime. Solar cooling offers possibilities of using the same solar collectors for space cooling as well through an absorption chiller. To be viable solar space heating systems need to utilize a low temperature supply system such as an air system. Absorption units on the other hand require temperature of the order of 90°C but the system during summertime is more effective and the available solar radiation is much more than in the wintertime. Additionally any effort to use solar energy for space heating and cooling must be done after the building is insulated well in order to reduce the thermal losses. A system suitable for a typical 196 m^2 house is modelled and simulated with TRNSYS program for two locations at different latitudes; Nicosia, Cyprus (35°) and Madison, Wisconsin (43°). These two locations offer two different climates; i.e. mild winters and severe summers for Nicosia and severe winters and mild summers for Madison. Three different types of solar collectors are considered; a flat plate, advance flat plate and evacuated tube collectors. The results show that the optimum system for Nicosia consists of 1.5 m^3 storage tank and 30 m^2 evacuated tube collectors installed at an inclination equal to latitude minus 5° and operated at a water flow rate of 36 kg/hr-m^2. For Madison the optimum system consists of the same size of storage tank (1.5 m^3) and

* Correspondence: Higher Technical Institute, P. O. Box 20423, Nicosia 2152, Cyprus; Tel. +357-22-406466, Fax. +357-22-406480, Email: skalogir@spidernet.com.cy

50 m^2 advance flat plate collectors installed at an inclination equal to the latitude and operated at a water flow rate of 27 kg/hr-m^2. The annual solar contribution is about 55% for the system in Nicosia and about 29% for the system in Madison. The economic analysis performed showed that both systems are viable as positive life cycle savings are obtained (5100 Euro for the system in Nicosia and 1340 Euro for the system in Madison). Therefore it can be concluded that solar energy should be used whenever possible.

Key words: Solar heating, solar cooling, solar hot water production, modelling, simulation, economic analysis

NOMENCLATURE

A	collector area (m^2)
AFP	advance flat plate collector
a_o	intercept efficiency
a_1	negative of the first-order coefficient of the efficiency (W/m^2 °C)
a_2	negative of the second -order coefficient of the efficiency (W/m^2 °C^2)
b_o, b_1	incidence angle modifier equation constants
C	investment cost (€)
C_a	area dependent cost (€/m^2)
C_{aux}	cost of auxiliary (€)
C_{FA}	cost rate of auxiliary energy (€/kJ)
C_{FL}	cost rate of conventional fuel (€/kJ)
C_{load}	cost of fuel to cover load (€)
C_o	operation cost (€)
C_s	investment cost of solar system (€)
C_v	volume dependent cost (€/m^3)
d	market discount rate (%)
ETC	evacuated tube collector
f	solar contribution (%)
$f_{(H)}$	annual solar contribution for the heating system (%)
FPC	flat plate collector
F_R	heat removal factor
$f_{(HW)}$	annual solar contribution for the hot water system (%)
$f_{(total)}$	total annual solar contribution of complete system (%)
FYFS	first year fuel savings (€)
G	total global solar radiation (kJ/m^2)
i	interest rate (%)
$k_{\alpha\tau}$	incidence angle modifier
LCS	life cycle savings (€)
n	thermal efficiency
N	number of years
PW	present worth (€)
Q_u	useful energy extracted (kJ)
$Q_{u,a}$	total annual Q_u (kJ)
Q_{aux}	auxiliary energy (kJ)

Q_{aux-C}	total cooling system auxiliary (kJ)
Q_{aux-H}	total heating system auxiliary (kJ)
Q_{aux-HW}	total hot water system auxiliary (kJ)
Q_{load}	energy required to cover load (kJ)
Q_{loss}	energy lost from storage tank and pipes (kJ)
T_a	ambient temperature (°C)
T_{load}	total annual system load (kJ)
T_{aux}	total annual auxiliary energy (kJ)
T_i	inlet temperature to the collector (°C)
U	heat loss coefficient ($W/m^2°C$)
U_L	overall heat loss coefficient of the collector ($kJ/m^2°C$)
V	storage tank volume (m^3)

Greek

ΔT	Temperature difference (°C) [$=T_i-T_a$]
θ	angle of incidence (deg.)
$\tau\alpha$	transmittance-absorptance product

INTRODUCTION

The sun is a sphere of intensely hot gaseous matter with a diameter of $1.39x10^9$ m. The solar energy strikes our planet a mere 8 minutes and 20 seconds after leaving the giant furnace, the sun which is $1.5x10^{11}$m away. The sun has an effective blackbody temperature of 5762K (Kreith and Kreider, 1978). The sun's total energy output is $3.8x10^{20}$ MW which is equal to 63 MW/m^2 of the sun's surface. This energy radiates outwards in all directions. From the total radiation emitted by the sun only a tiny fraction, of 45 billionth of a percent is intercepted by the earth. However, even with this small fraction it is estimated that 30 minutes of solar radiation falling on earth is equal to the world energy demand for one year.

Man realised that a good use of solar energy is in his benefit, from the prehistoric times. The first findings were related to the domestic life. The Greek historian Xenophon in his "memorabilia" records some of the teachings of the Greek Philosopher Socrates (470 - 399 BC) regarding the correct orientation of dwellings in order to have houses which were cool in summer and warm in winter.

Since prehistory, the sun has dried and preserved man's food. It has also evaporated sea water to yield salt. Since man began to reason, he has recognised the sun as a motive power behind every natural phenomenon. This is why many of the pre-historic tribes considered Sun as "God". Many scripts of ancient Egypt say that the Great Pyramid, one of the man's greatest engineering achievements, was built as a stairway to the sun (Anderson, 1977).

Basically all the forms of energy in the world as we know it are solar in origin. Oil, coal, natural gas and woods were originally produced by photosynthetic processes, followed by complex chemical reactions in which decaying vegetation was subjected to very high temperatures and pressures over a long period of time (Kreith and Kreider, 1978). Even the wind and tide energy have a solar origin since they are caused by differences in temperature in various regions of the earth.

The greatest advantage of solar energy as compared with other forms of energy is that it is clean and can be supplied without any environmental pollution. Over the past century fossil fuels have provided most of our energy because these are much cheaper and more convenient than energy from alternative energy sources, and until recently environmental pollution has been of little concern.

The reason for the rapid increase in oil demand occurred mainly because increasing quantities of oil, produced at very low cost, became available during the 50's and 60's from the Middle East and North Africa. For the consuming countries imported oil was cheap compared with indigenously produced energy from solid fuels.

But the main problem is that proved reserves of oil and gas, at current rates of consumption, would be adequate to meet demand for another 40 and 60 years respectively. The reserves for coal are in better situation as they would be adequate for at least the next 250 years.

If we try to see the implications of these limited reserves we will be faced with a situation in which the price of fuels will be accelerating as the reserves are decreased. Considering that the price of oil has become firmly established as the price leader for all fuel prices then the conclusion is that energy prices will increase over the next decades at something greater than the rate of inflation or even more. Additional to this is also the concern about the environmental pollution caused by the burning of the fossil fuels.

There are many alternative energy sources which can be used instead of fossil fuels. The decision as to what type of energy source should be utilised must, in each case, be made on the basis of economic, environmental and safety considerations. Because of the desirable environmental and safety aspects it is widely believed that solar energy should be utilised instead of other alternative energy forms, even when the costs involved are slightly higher.

In addition to the thousands of ways in which the sun's energy has been used by both nature and man through time, to grow food or dry clothes, it has also been deliberately harnessed to perform a number of other jobs. Solar energy is used to heat and cool buildings (both active and passive), to heat water for domestic and industrial uses, to heat swimming pools, to power refrigerators, to operate engines and pumps, to desalinate water for drinking purposes, to generate electricity, to produce various chemicals such as hydrogen, and many more. The objective of this paper is to present the design of a complete solar system for space heating, cooling and hot water production. The various types of collectors that can be used for this purpose are examined with respect to the economic viability of such systems located in two different climates, Nicosia, Cyprus having hot summers and mild winters and Madison, Wisconsin having mild summers and severe winters. The best collector type as well as the optimum slope, flow rate and area of the solar system and the optimum storage tank size are determined for each site.

Solar Collectors

Solar energy collectors are special kind of heat exchangers that transform solar radiation energy to internal energy of the transport medium. The major component of any solar system is the solar collector. This is a device which absorbs the incoming solar radiation, converts it into heat, and transfers this heat to a fluid (usually air, water, or heat transfer oil) flowing through the collector. The solar energy thus collected is carried from the circulating fluid either directly to the hot water or space conditioning equipment, or to a thermal energy storage tank from which can be drawn for use at night and/or cloudy days.

Solar energy collectors are basically distinguished by their motion as stationary, single axis tracking and two-exes tracking, and their operating temperature. Due to the increased maintenance requirements for domestic applications only stationary solar collectors are used. These collectors are permanently fixed in position and do not track the sun. Three types of collectors fall in this category:

1. Flat plate collectors (FPC)
2. Stationary compound parabolic collectors (CPC)
3. Evacuated tube collectors (ETC)

In the present application only the flat-plate and evacuated tube collectors are considered which are industrially available.

Flat-Plate Collectors (FPC)

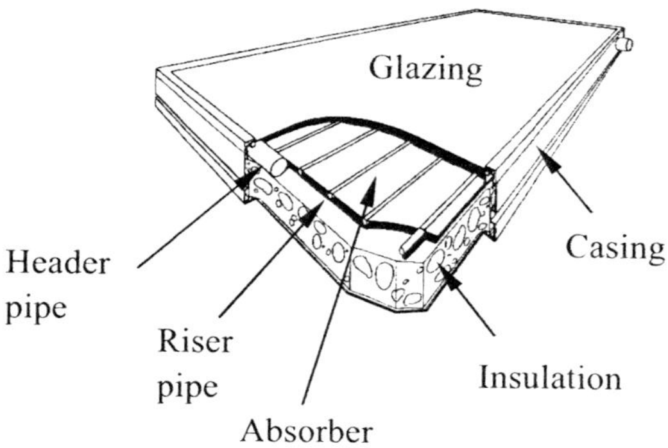

Figure 1. Pictorial view of a flat-plate collector

A typical flat-plate solar collector is shown in Figure 1. When solar radiation passes through a transparent cover and strikes the blackened absorber surface of high absorptivity, a large portion of this energy is absorbed by the plate and then transferred to the transport medium in the fluid tubes to be carried away for storage or use. The underside of the absorber plate and the side of casing are well insulated to reduce conduction losses. The liquid tubes can be welded to the absorbing plate, or they can be an integral part of the plate. The liquid tubes are connected at both ends by large diameter header tubes.

The transparent cover is used to reduce convection losses from the absorber plate through the restraint of the stagnant air layer between the absorber plate and the glass. It also reduces radiation losses from the collector as the glass is transparent to the short wave radiation received by the sun but it is nearly opaque to long-wave thermal radiation emitted by the absorber plate (greenhouse effect).

Flat plate collectors are usually permanently fixed in position and require no tracking of the sun. The collectors should be oriented directly towards the equator, facing south in the northern hemisphere and north in the southern.

A flat-plate collector generally consists of the following components as shown in Figure 2:

Glazing - One or more sheets of glass or other transparent material.

Tubes, fins, or passages - Conduct the heat transfer fluid from the inlet to the outlet.

Absorber plate - Flat, corrugated, or grooved plate, to which the tubes, fins, or passages are attached. The plate may be integral with the tubes.

Headers - Admit and discharge the fluid from the collector.

Insulation - Minimise the heat loss from the back and sides of the collector.

Casing - Surround the above components and keep them free from dust and moisture.

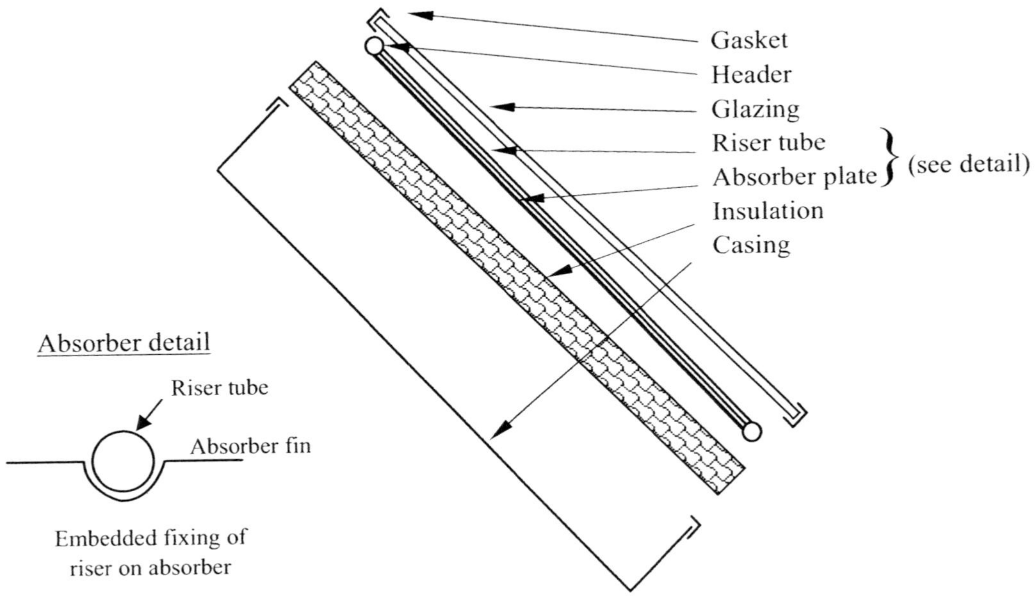

Figure 2. Exploded view of a flat-plate collector and absorber detail.

Flat-plate collectors are built in a large variety of designs and from a number of different materials. They are used to heat fluids such as water, water with antifreeze additive, or air. The purpose of the collector is to collect as much solar energy as possible at the lower possible cost. Despite the adverse effects of the sun's ultraviolet radiation, corrosion and clogging because of acidity, alkalinity or hardness of the heat transfer fluid, freezing of water, or deposition of dust or moisture on the glazing, the collector should have a long life. The glazing must also be able to withstand thermal expansion and avoid breakage due to hail, vandalism or other causes. These can be minimised by the use of tempered glass.

Flat plate collectors (FPC) are by far the most used type of collector. Flat–plate collectors are usually employed for low temperature applications up to 100°C, although some new types of collectors employing vacuum insulation and/or transparent insulation (TI) can achieve slightly higher values (Benz, *et al.*, 1998). Due to the introduction of highly selective coatings actual standard flat plate collectors can reach stagnation temperatures of more than 200°C. With these collectors good efficiencies can be obtained up to temperatures of about 100°C.

The characteristics of a typical water flat plate collector considered in this work are shown in Table 1.

Table 1. Characteristics of a typical flat plate collector (FPC) and advance flat plate collector (AFP) systems

Parameter	Simple flat plate collector	Advanced flat plate collector
Fixing of risers on the absorber plate	Embedded	Ultrasonically welded
Absorber coating	Black mat paint	Chromium selective coating
Glazing	Low-iron glass	Low-iron glass
Efficiency mode	n v_s $(T_i - T_a)/G$	n v_s $(T_i - T_a)/G$
G_{test} – flow rate per unit area at test conditions (kg/s-m^2)	0.015	0.015
a_o – intercept efficiency	0.79	0.80
a_1 – negative of the first-order coefficient of the efficiency (W/m$^{2\,\circ}$C)	6.67	4.78
b_o – incidence angle modifier constant	0.1	0.1

The performance of flat-plate collectors depend greatly on the way fins are attached to the riser tubes. Best results are obtained by welding the fins on the pipes thus improve heat conduction. Lately some modern manufacturing techniques have been introduced by the industry like the use of ultrasonic welding machines, which improve both the speed and the quality of welds. The greatest advantage of this method is that the welding is performed at room temperature therefore deformation of the welded parts is avoided. These collectors with selective coating are called advance flat plate collectors and the characteristics of a typical type are also shown in Table 1.

Two of the most important parts of these collectors that play an important role in their performance are the glazing materials and the absorber plates. These are examined in the following sections.

Glazing Materials

The glazing should admit as much solar irradiation as possible and reduce the upward loss of heat as much as possible. Although glass is almost opaque to the longwave radiation emitted by the hot collector plates, absorption of that radiation causes an increase in the glass temperature and a loss of heat to the surrounding atmosphere by radiation and convection.

Glass has been widely used to glaze solar collectors because it can transmit as much as 90% of the incoming shortwave solar irradiation while is not transmitting any of the longwave radiation emitted outward by the absorber plate. Glass with low iron content has a relatively high transmittance for solar radiation (approximately 0.85 to 0.90 at normal incidence), but its transmittance is essentially zero for the longwave thermal radiation (5.0 to 50 μm) emitted by sun-heated surfaces (ASHRAE, 1995).

The commercially available grades of window and green-house glass have normal incidence transmittances of about 0.87 and 0.85 respectively. For direct radiation, the transmittance varies considerably with the angle of incidence with values approaching zero at 90° incidence (ASHRAE, 1995).

Antireflective coatings and surface texture can also improve transmission significantly. The effect of dirt and dust on collector glazing may be quite small, and the cleansing effect of an occasional rainfall is usually adequate to maintain the transmittance within 2 to 4% of its maximum value.

Various prototypes of transparently insulated flat plate collectors have been built and tested in the last decade (Spate *et al.*, 1999; Schweiger, 1997). Low cost and high temperature resistant transparent insulating (TI) materials have been developed so that the commercialisation of these collectors becomes feasible. A prototype flat-plate collector covered by transparent insulation was developed by Benz *et al.* (1998). It was experimentally proved that the efficiency of the collector was comparable with that of evacuated tube collectors. However at the time of writing, no commercial collectors of this type are available in the market.

Collector Absorbing Plates

The collector plate absorbs as much of the irradiation passing through the glazing as possible, while loosing as little heat as possible upward to ambient and downward through the insulation. The collector plates must also transfer easily the retained heat to the transport fluid. The absorptance of the collector surface for shortwave solar radiation depends mainly on the colour of the coating and on the incident angle. Usually black colour is used, however various colour coatings have been proposed (Tripanagnostopoulos *et al.*, 2000; Wazwaz *et al.*, 2002; Orel *et al.*, 2002) mainly for aesthetic reasons.

By suitable electrolytic or chemical treatment, surfaces can be produced which have high solar radiation absorptance (α) value and low longwave emittance (ε) value. Essentially, typical selective surfaces consist of a thin upper layer, which is highly absorbent to shortwave solar radiation but relatively transparent to longwave thermal radiation, deposited on a surface that has a high reflectance and a low emittance for longwave radiation (ASHRAE, 1995). Selective surfaces are particularly important when the collector surface temperature is much higher than the ambient air temperature. Lately a low-cost mechanically manufactured selective solar absorber surface method is proposed (Konttinen *et al.*, 2003).

An energy efficient solar collector should absorb incident solar radiation, convert it to thermal energy and deliver the thermal energy to a heat transfer medium with minimum losses at each step. It is possible to use several different design principles and physical mechanisms in order to create a selective solar absorbing surface. Solar absorbers are based on two layers with different optical properties which are referred as tandem absorbers. A semiconducting or dielectric coating with high solar absorptance and high infrared transmittance on top of a non-selective highly reflecting material such as metal constitutes one type of tandem absorber. Another alternative is to coat a non-selective highly absorbing material with a heat mirror having a high solar transmittance and high infrared reflectance (Wackelgard *et al.*, 2001).

Today, commercial solar absorbers are made by electroplating, anodization, evaporation, sputtering and by applying solar selective paints. Much of the progress during recent years has been based on the implementation of vacuum techniques for the production of fin type absorbers used in low temperature applications. The chemical and electrochemical processes used for their commercialization were readily taken over from the metal finishing industry. The requirements of solar absorbers used in high temperature applications, however, namely extremely low thermal emittance and high temperature stability, were difficult to fulfill with conventional wet processes. Therefore, large-scale sputter deposition was developed in the late 70's. The vacuum techniques are nowadays mature, characterized by low cost and have the advantage of being less environmentally polluting than the wet processes.

For fluid-heating collectors, passages must be integral with or firmly bonded to the absorber plate. A major problem is obtaining a good thermal bond between tubes and absorber plates without incurring excessive costs for labour or materials. Material most frequently used for collector plates are copper, aluminium, and stainless steel. UV-resistant plastic extrusions are used for low temperature applications. It should be noted that if the entire collector area is in contact with the heat transfer fluid, the thermal conductance of the material is not important (ASHRAE, 1995).

Evacuated Tube Collectors (ETC)

Conventional simple flat-plate solar collectors were developed for use in sunny and warm climates. Their benefits however are greatly reduced when conditions become unfavourable during cold, cloudy and windy days. Furthermore, weathering influences such as condensation and moisture will cause early deterioration of internal materials resulting in reduced performance and system failure. Evacuated heat pipe solar collectors (tubes) operate differently than the other collectors available on the market. These solar collectors consist of a heat pipe inside a vacuum-sealed tube, as shown in Figure 3.

Evacuated tube collectors have demonstrated that the combination of a selective surface and an effective convection suppressor can result in good performance at high temperatures (ASHRAE, 1995). The vacuum envelope reduces convection and conduction losses, so the collectors can operate at higher temperatures than flat-plate collectors. Like flat-plate collectors, they collect both direct and diffuse radiation. However, their efficiency is higher at low incidence angles. This effect tends to give evacuated tube collectors an advantage over flat-plate collectors in day-long performance.

Evacuated tube collectors use liquid-vapour phase change materials to transfer heat at high efficiency. These collectors feature a heat pipe (a highly efficient thermal conductor) placed inside a vacuum-sealed tube. The pipe, which is a sealed copper pipe, is then attached to a black copper fin that fills the tube (absorber plate). Protruding from the top of each tube is a metal tip attached to the sealed pipe (condenser). The heat pipe contains a small amount of fluid (e.g. methanol) that undergoes an evaporating-condensing cycle. In this cycle, solar heat evaporates the liquid, and the vapour travels to the heat sink region where it condenses and releases its latent heat. The condensed fluid return back to the solar collector and the process is repeated. When these tubes are mounted, the metal tips up, into a heat exchanger (manifold) as shown in Figure 3. Water, or glycol, flows through the manifold and picks up the heat from the tubes. The heated liquid circulates through another heat exchanger and gives off its heat to a process or to water that is stored in a solar storage tank.

Because no evaporation or condensation above the phase-change temperature is possible, the heat pipe offers inherent protection from freezing and overheating. This self-limiting temperature control is a unique feature of the evacuated heat pipe collector. The characteristics of a typical evacuated tube collector considered in this work are shown in Table 2.

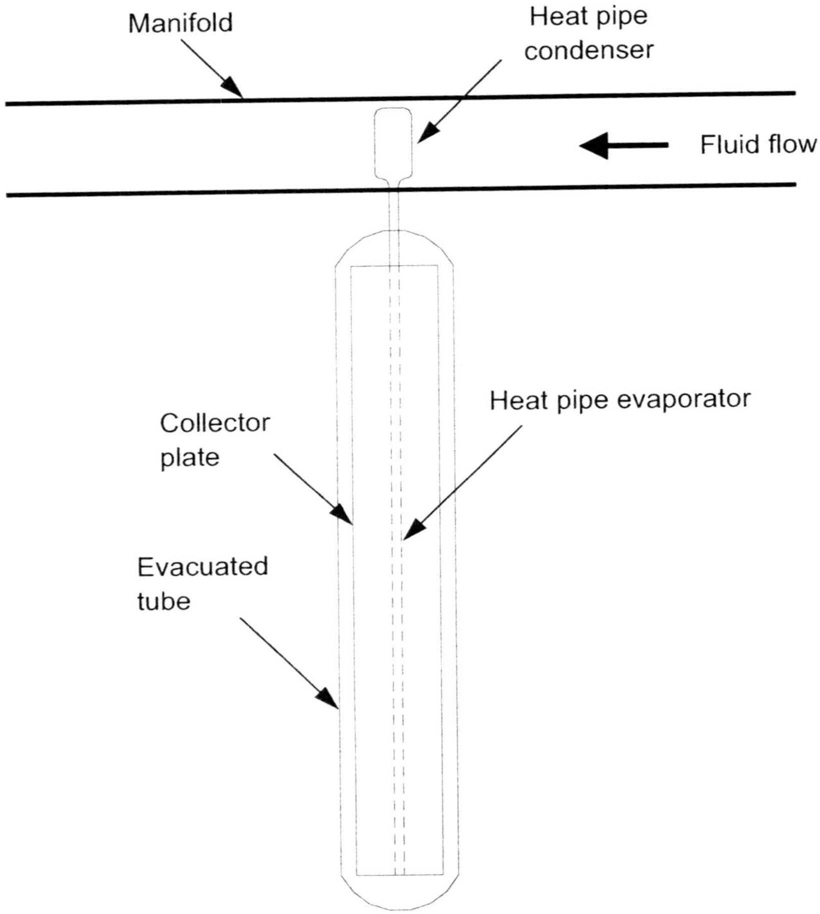

Figure 3. Schematic diagram of an evacuated tube collector

Evacuated tube collectors (ETC) basically consist of a heat pipe inside a vacuum-sealed tube. A large number of variations of the absorber shape of evacuated tube collectors are on the market (Lin and Fuerbo, 1998). Evacuated tubes with compound parabolic reflectors are also commercialised by several manufacturers. One manufacturer recently presented an all-glass evacuated tube collector, which may be an important step to cost reduction and increase of lifetime. Another variation of this type of collector is what is called Dewar tubes. In this two concentric glass tubes are used and the space in between the tubes is evacuated (vacuum jacket). The advantage of this design is that it is made entirely of glass and it is not necessary to penetrate the glass envelope in order to extract heat from the tube thus leakage losses are not present and it is also less expensive than the single envelope system (Morrison, 2001).

Table 2. Characteristics of a typical evacuated tube collector (ETC) system

Parameter	Value
Glass tube diameter	65mm
Glass thickness	1.6mm
Collector length	1965mm
Absorber plate	Copper
Coating	Selective
Absorber area for each collector	$0.1 m^2$
Efficiency mode	n vs $(T_i - T_a)/G$
G_{test} – flow rate per unit area at test conditions ($kg/s\ m^2$)	0.014
a_o – intercept efficiency	0.82
a_1 - negative of the first-order coefficient of the efficiency ($W/m^2\ °C$)	2.19
b_o - incidence angle modifier constant	0.2

Another type of collector developed recently is the integrated compound parabolic collector (ICPC). This is an evacuated tube collector in which at the bottom part of the glass tube a reflective material is fixed (Winston *et al.*, 1999). The collector combines the vacuum insulation and non-imaging stationary concentration into a single unit. In another design a tracking ICPC is developed which is suitable for high temperature applications (Grass *et al.*, 2000). These models however are experimental prototypes and as they are not commercialised yet they will not be considered in the present analysis.

Solar Water Heating Systems

In this section the solar collector subsystem is described. The main part of a solar water heating system is the solar collector panels that absorbs solar radiation and converts it to heat. This heat is then absorbed by a heat transfer fluid (water, non-freezing liquid, or air) that passes through the collector tubes and can be stored or used directly. As portions of the solar energy system are exposed to the weather conditions, they must be protected from freezing and from overheating caused by high solar irradiation levels during periods of low energy demand.

In solar water heating systems, potable water can be heated directly in the collector (direct systems) or indirectly by a heat transfer fluid that is heated in the collector and passes through a heat exchanger to transfer its heat to the service or potable water (indirect systems). The heat transfer fluid is transported either naturally (passive systems) or by forced circulation (active systems). Natural circulation occurs by density gradients (thermosyphoning), whereas for the forced circulation systems pumps or fans are used. Except for thermosyphon and integrated collector storage systems, which need no control, solar domestic and service hot water systems are controlled using differential thermostats.

Five types of solar energy systems can be used to heat domestic and service hot water: thermosyphon, integrated collector storage, direct circulation, indirect, and air. The first two are called passive systems as no pump is employed whereas the others are called active systems because a pump or fan is employed in order to circulate the fluid. For the complete space heating, cooling and hot water system considered in this chapter active systems are usually employed. Passive systems are mostly suitable for domestic hot water production only and thus they will not be considered in this work. For freeze protection, recirculation and drain-down are used for direct solar water heating systems and drain-back is used for indirect water heating systems.

All these systems offer significant economic benefits with payback times, depending on the type of fuel they replace, of a few years. The payback times depend on the economic indices, like the inflation rates and fuel price applied in a country. A wide range of collectors have been used for solar water heating systems. A review of the systems manufactured in the last twenty years is given in (Morrison and Wood, 1999; Kalogirou, 2004).

Direct Circulation Systems

In direct circulation systems, shown schematically in Figure 4, a pump is used to circulate potable water from storage to the collectors when there is enough available solar energy to increase its temperature and then return the heated water to the storage tank until it is needed. As a pump circulates the water, the collectors can be mounted either above or below the storage tank. The usual optimum flow rate for such units, suitable for domestic hot water systems, is about 54 kg/hr per m^2 of collector area. This value however will be optimized for the complete solar system. Direct circulation systems can be used in areas where freezing is not frequent. For extreme weather conditions, freeze protection is usually provided by recirculating warm water from the storage tank. Direct circulation systems often use a single storage tank equipped with an auxiliary water heater, but two-tank storage systems can also be used.

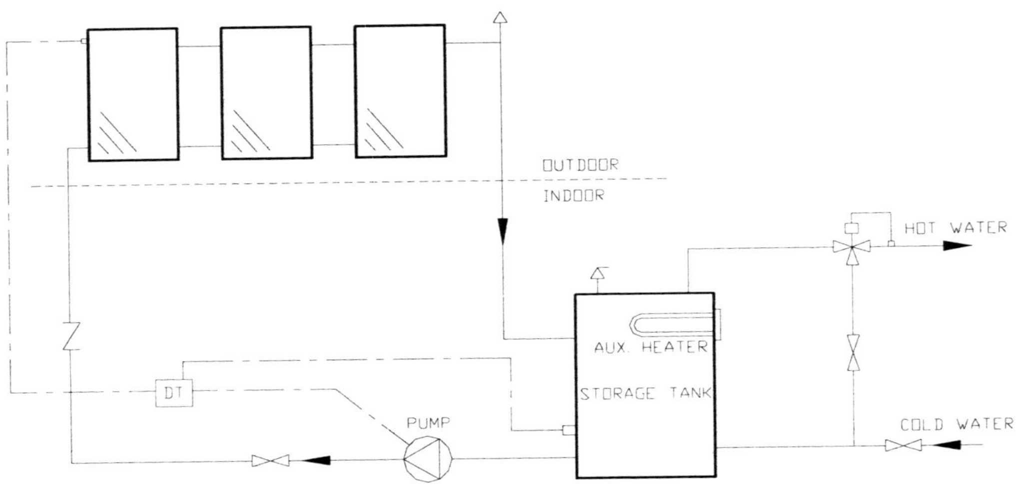

Figure 4. Direct circulation system

Direct circulation systems can be used with water supplied from a cold water storage tank or connected directly to city water mains. Pressure-reducing valves and pressure relief valves are required however when the city water pressure is greater than the working pressure of the collectors. Direct water heating systems should not be used in areas where the water is extremely hard or acidic because scale deposits may clog or corrode the collectors (ASHRAE, 1995).

A variation of the direct circulation system is the drain-down systems shown in Figure 5. In this case also potable water is pumped from storage to the collector panels where it is heated. When a freezing condition or a power failure occurs, the system drains automatically by isolating the collector array and exterior piping from the make-up water supply and draining is done by using the two normally open (NO) valves, shown in Figure 5. It should be noted that the solar collectors and associated piping must be carefully sloped so as to drain completely when circulation stops. The same comments about pressure and scale deposits apply here as for the direct circulation systems.

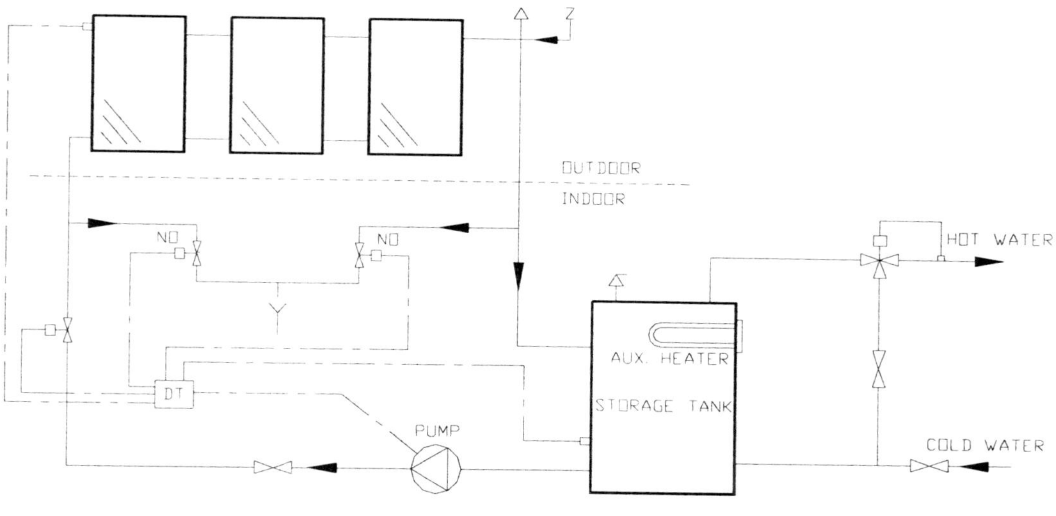

Figure 5. Drain-down system

Indirect Water Heating Systems

Indirect water heating systems, shown schematically in Figure 6, circulate a heat transfer fluid through the closed collector loop to a heat exchanger, where its heat is transferred to the potable water. The most commonly used heat transfer fluid is water/ethylene glycol solution, although other heat transfer fluids such as silicone oils and refrigerants can also be used. When fluids that are non-potable or toxic are used double-wall heat exchangers should be employed in case where no separate tank is used for domestic water production. The heat exchanger can be located inside the storage tank, around the storage tank (tank mantle) or can be external. It should be noted that the collector loop is closed and therefore an expansion tank and a pressure relief valve are required. Additional over-temperature protection is usually required to prevent the collector heat transfer fluid from decomposing or becoming corrosive (ASHRAE, 1995).

A variation of indirect water heating systems is the drain-back system. Drain-back systems are generally indirect water heating systems that circulate water through the closed collector loop to a heat exchanger, where its heat is transferred to the potable water. Circulation continues as long as usable energy is available. When the circulation pump stops the collector fluid drains by gravity to a drain-back tank. If the system is pressurised the tank serves also as an expansion tank when the system is operating and in this case it must be protected with a temperature and pressure relief valves. In the case of an unpressurised system (Figure 7), the tank is open and vented to the atmosphere (ASHRAE, 1995).

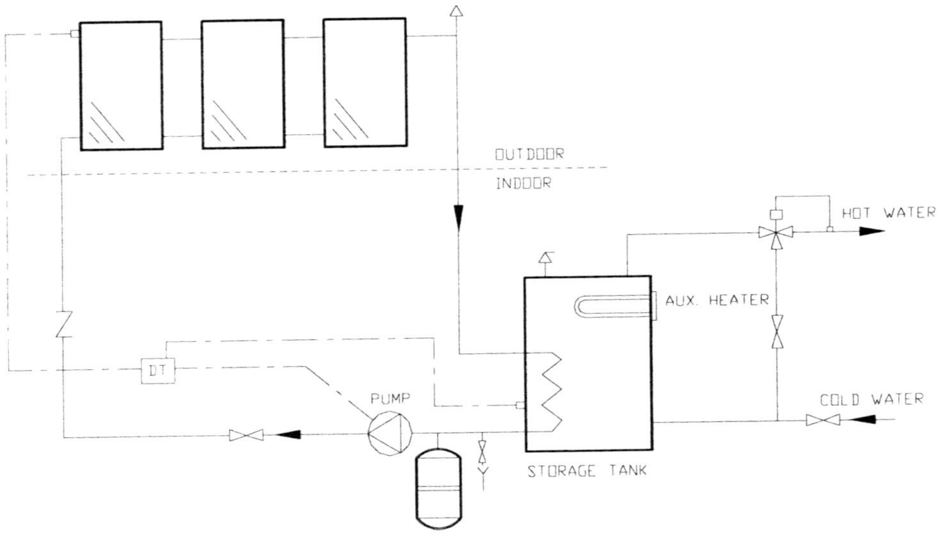

Figure 6. Indirect water heating system

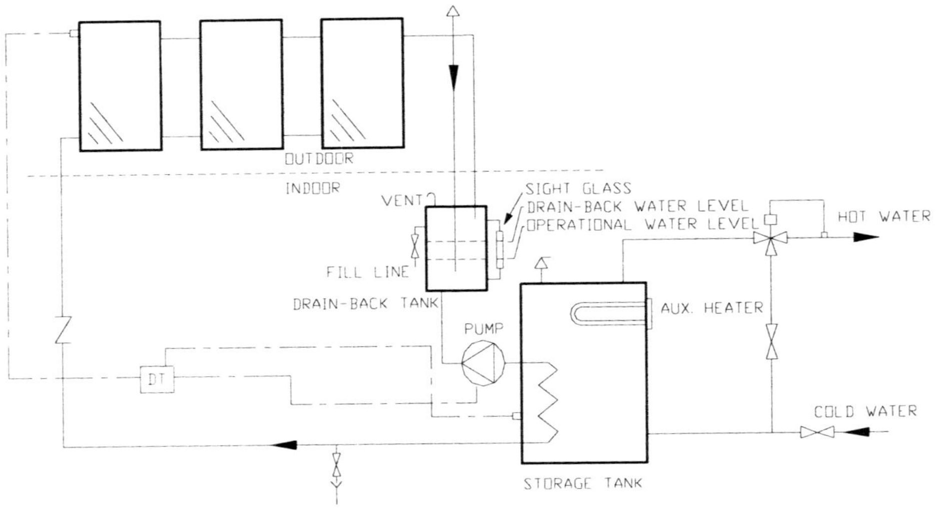

Figure 7. Drain-back system

The advantage of this system is that the collector loop is isolated from the potable water, so no valves are needed to actuate draining, and scaling is not a problem, however, the collector array and exterior piping must be adequately sloped to drain completely.

Air Systems

Air systems are indirect water heating systems that circulate air via ducts through the collectors to an air-to-liquid heat exchanger. In the heat exchanger, heat is transferred to the potable water, which is also circulated through the heat exchanger and returned to the storage tank. Figure 8 shows a double storage tank system. This type of system is used most often, because air systems are generally used for preheating domestic hot water and thus auxiliary is used only in one tank as shown to avoid heating the whole quantity of stored water with auxiliary energy.

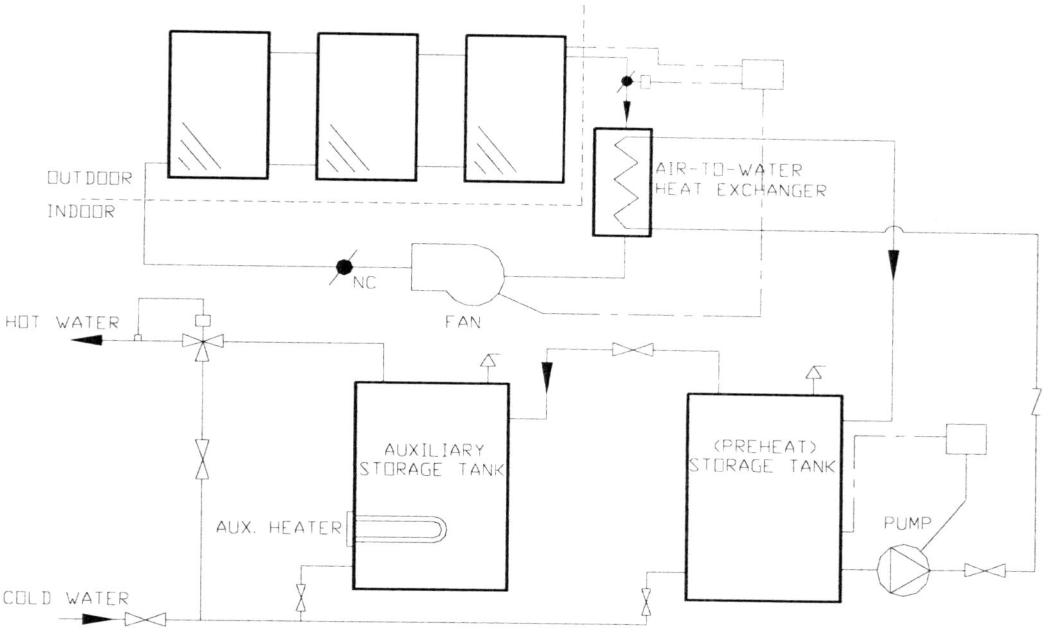

Figure 8. Air system

The main advantage of the system is that air does not need to be protected from freezing or boiling, is non-corrosive, and is free. The disadvantages are that air handling equipment (ducts and fans) need more space than piping and pumps, air leaks are difficult to detect, and parasitic power consumption is generally higher than that of liquid systems (ASHRAE, 1995).

The relative merits of solar water-based and air-based heating systems are numerous. A water system is always associated with problems and hazards such as collector freeze-up, boiling in the collector, corrosive effects of water, and accidental leakage. None of these is a problem in air systems. However, as the specific heat of water is much larger than that of air, the piping and storage of a water system is more compact than that for an air system. To have the same storage capacity, the storage volume needed in an air system can be as much as three times bigger than that in a water system. As to year-round air conditioning, water systems can be more easily adapted to supply energy to absorption refrigeration systems for summer cooling.

SOLAR SPACE HEATING AND COOLING

The components and subsystems discussed in previous section may be combined to create a wide variety of building solar heating and cooling systems. There are again two principal categories of such systems, passive and active. The term passive system is applied to buildings that include as integral part of the building, elements that admit, absorb, store and release solar energy and thus reduce the needs for auxiliary energy for comfort heating. In a passive system, solar radiation is admitted directly into the building through large transparent openings and massive building structures. In an active system, solar radiation is collected by solar collectors, thermal energy is stored in a storage tank, and transportation of fluid media is done mostly by mechanical means. In this chapter only active systems are considered.

Systems for space heating are very similar to those for water heating, described in the previous section, and as the same considerations for combination with an auxiliary source, boiling and freezing, controls, etc., apply to both these may not be repeated again. The most common heat transfer fluids are water, water and antifreeze mixtures and air. The load is the building to be heated or cooled. Although it is technically possible to construct a solar heating or cooling system which can satisfy 100% the design load, such a system would be non-viable since it would be oversized for most of the time. The size of the solar system may be determined by a life-cycle cost analysis.

Active solar space systems use collectors to heat a fluid, a storage unit to store solar energy until needed, and distribution equipment to provide the solar energy to the heated spaces in a controlled manner. A complete system includes additionally pumps or fans for transferring the energy to storage or to the load which require a continuous availability of non-renewable energy, usually electricity.

The load can be space cooling, heating, or a combination of these two with hot water supply as in the present application. In combination with conventional heating equipment solar heating and cooling provides the same levels of comfort, temperature stability, and reliability as conventional systems.

During daytime the solar system absorbs solar radiation with collectors and conveys it to storage using a suitable fluid. As the building requires heat this is obtained from storage. Control of the solar system is exercised by differential temperature controllers, i.e., the controller compares the temperature of the collectors and storage and whenever the temperature difference is more than a certain value (7-10°C) the solar pump is switched ON. In locations where freezing conditions are possible to occur, a low-temperature sensor is installed on the collector which controls the solar pump when a pre-set temperature is reached. This process wastes some stored heat, but it prevents costly damages to the solar collectors.

Solar cooling of buildings is an attractive idea as the cooling load and availability of solar radiation are in phase. Additionally, the combination of solar cooling and heating greatly improves the use factors of collectors compared to heating alone. Solar air conditioning can be accomplished by three types of systems; absorption cycles, adsorption (desiccant) cycles and solar mechanical processes. Some of these cycles are also used in solar refrigeration systems. It should be noted that the same solar collectors are used for both space heating and cooling systems when both are present.

A review of the various solar heating and cooling systems is presented by Hahne (1996) and a review of solar and low energy cooling technologies is presented by Florides *et al.* (2002b).

Space Heating and Service Hot Water

It is useful to consider solar systems as having five basic modes of operation, depending on the conditions that exist in the system at a particular time (Duffie and Beckman, 1991):

1. If solar energy is available and heat is not needed in the building, energy gain from the collector is added to storage.
2. If solar energy is available and heat is needed in the building, energy gain from the collector is used to supply the building need.
3. If solar energy is not available, heat is needed in the building, and the storage unit has stored energy in it, the stored energy is used to supply the building need.
4. If solar energy is not available, heat is needed in the building, and the storage unit has been depleted, auxiliary energy is used to supply the building need.
5. The storage unit is fully heated, there are no loads to met, and the collector is absorbing heat.

When the last mode occurs, there is no way to use or store the collected energy, and this energy must be discarded. This can be achieved through the operation of pressure relief valves or if the stagnant temperature will not be detrimental to the collector materials, the flow of fluids is turned off, thus the collector temperature will rise until the absorbed energy is dissipated by thermal losses. This is more suitable to solar air collectors.

Additional operational modes can also be employed such as to provide service hot water. It is also possible to combine modes, i.e., to operate in more than one mode at a time. Moreover, many systems do not allow direct heating from solar collector to building but always transfer heat from collector to storage whenever this is available and from storage to load whenever needed. In Europe solar heating systems for combined space and water heating are known as combisystems. In the rest of this section the design of residential-scale water systems is described.

There are many variations of systems used for both solar space heating and service hot water production. The basic configuration is similar to the solar water heating systems outlined in previous sections. When used for both space and hot water production this system allows independent control of the solar collector-storage and storage-auxiliary-load loops as solar-heated water can be added to storage at the same time that hot water is removed from storage to meet building loads. Usually, a bypass is provided around the storage tank to avoid heating the storage tank, which can be of considerable size, with auxiliary energy.

A detailed schematic of a liquid-based system is shown in Figure 9 (Duffie and Beckman, 1991). In this case a collector heat exchanger is shown between the collector and the storage tank, which allows the use of antifreeze solutions to the collector circuit. Relief valves are also required for dumping excess energy if the collector temperature reaches saturation. Means of extracting energy for service hot water are indicated. Auxiliary energy for heating is added so as to "top off" that available from solar energy system.

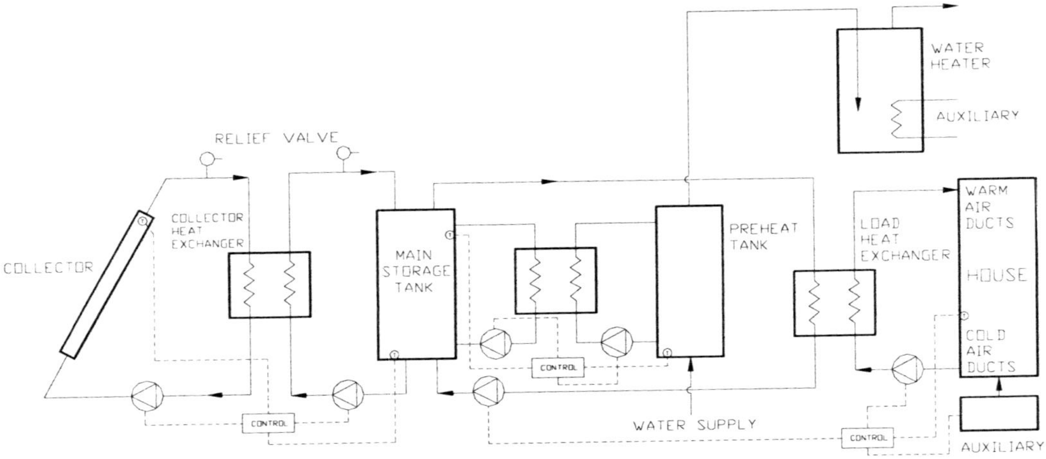

Figure 9. Detail schematic of a solar water heating system

A load heat exchanger is shown in Figure 9 to transfer energy from the tank to the air in the heated spaces. The load heat exchanger must be adequately designed to avoid excessive temperature drop and corresponding increase in the tank and collector temperatures.

Advantages of liquid heating systems include high collector F_R, smaller storage volume, and relatively easy adaptation to supply energy to absorption air conditioners (see next section).

Absorption Units

Absorption is the process of attracting and holding moisture by substances called desiccants. Desiccants are sorbents, i.e., materials that have an ability to attract and hold other gases or liquids, which have a particular affinity for water. During absorption the desiccant undergoes a chemical change as it takes on moisture, for example the table salt, which changes from a solid to a liquid as it absorbs moisture. The characteristic of the binding of desiccants to moisture, makes the desiccants very useful in chemical separation processes (ASHRAE, 1989).

Absorption systems are similar to vapour-compression air conditioning systems but differ in the pressurisation stage. In general an absorbent, on the low-pressure side, absorbs an evaporating refrigerant. The most usual combinations of fluids include lithium bromide-water

($LiBr-H_2O$) where water vapour is the refrigerant and ammonia-water (NH_3-H_2O) systems where ammonia is the refrigerant.

The pressurisation is achieved by dissolving the refrigerant in the absorbent, in the absorber section (Figure 10). Subsequently, the solution is pumped to a high pressure with an ordinary liquid pump. The addition of heat in the generator is used to separate the low-boiling refrigerant from the solution. In this way the refrigerant vapour is compressed without the need of large amounts of mechanical energy that the vapour-compression air conditioning systems demand.

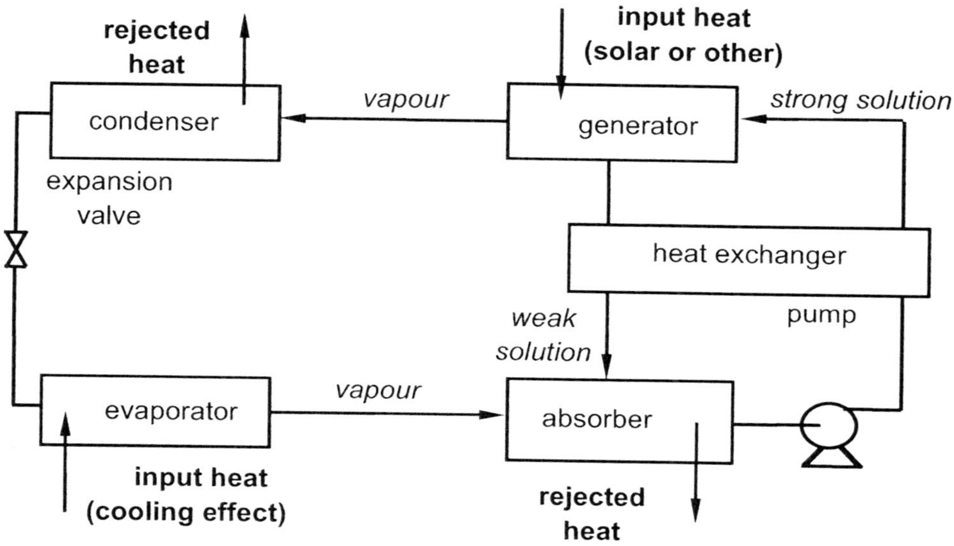

Figure 10. Basic principle of the absorption air conditioning system

The remainder of the system consists of a condenser, expansion valve and evaporator, which function in a similar way as in a vapour-compression air conditioning system.

The NH_3-H_2O system is more complicated than the $LiBr-H_2O$ system, since it needs a rectifying column that assures that no water vapour enters the evaporator where it could freeze. The NH_3-H_2O system requires generator temperatures in the range of 125°C to 170°C with air-cooled absorber and condenser and 95°C to 120°C when water-cooling is used. These temperatures cannot be obtained with flat-plate collectors. The coefficient of performance (COP), which is defined as the ratio of the cooling effect to the heat input, is between 0.6 to 0.7.

The $LiBr-H_2O$ system operates at a generator temperature in the range of 70°C to 95°C with water used as a coolant in the absorber and condenser and has COP slightly higher than the NH_3-H_2O systems. The COP of this system is between 0.6 and 0.8 (Duffie and Beckman, 1991). A disadvantage of the $LiBr-H_2O$ systems is that their evaporator cannot operate at temperatures much below 5°C since the refrigerant is water vapour. Commercially available absorption chillers for air conditioning applications usually operate with a solution of lithium bromide in water and use steam or hot water as the heat source. In the market two types of chillers are available, the single and the double effect.

The single effect absorption chiller is mainly used for building cooling loads, where chilled water is required at 6-7°C. The COP will vary to a small extent with the heat source and the cooling water temperatures. Single effect chillers can operate with hot water temperature ranging from about 80°C to 150°C when water is pressurised (Florides et al., 2002a).

The double effect absorption chiller has two stages of generation to separate the refrigerant from the absorbent. Thus the temperature of the heat source needed to drive the high-stage generator is essentially higher than that needed for the single-effect machine and is in the range of 155 to 205°C. Double effect chillers have a higher COP of about 0.9-1.2 (Dorgan *et al.,* 1995). Although double effect chillers are more efficient than the single-effect machines they are obviously more expensive to purchase. However, every individual application must be considered on its merits since the resulting savings in capital cost of the single-effect units can largely offset the extra capital cost of the double effect chiller.

The Carrier Corporation pioneered lithium-bromide absorption chiller technology in the United States, with early single-effect machines introduced around 1945. Due to the success of the product soon other companies joined the production. The absorption business thrived until 1975. Then the generally held belief that natural gas supplies were lessening, let to U.S. government regulations prohibiting the use of gas in new constructions and together with the low cost of electricity let to the declination of the absorption refrigeration market (Keith, 1995). Today the major factor on the decision on the type of system to install for a particular application is the economic trade-off between the different cooling technologies. Absorption chillers typically cost less to operate, but they cost more to purchase than vapour compression units. The payback period depends strongly on the relative cost of fuel and electricity assuming that the operating cost for the needed heat is less than the operating cost for electricity.

The technology was exported to Japan from the U.S. early in the sixties, and the Japanese manufacturers set a research and development program to improve further the absorption systems. The program led to the introduction of the direct-fired double-effect machines with improved thermal performance.

Today gas-fired absorption chillers deliver 50% of commercial space cooling load world-wide but less than 5% in the U.S., where electricity-driven vapour compression machines carry the majority of the load (Keith, 1995).

Many researchers have developed solar assisted absorption refrigeration systems. Most of them have been produced as experimental units and computer codes were written to simulate the systems. Some of these designs are presented here.

Hammad and Audi (1992) described the performance of a non-storage, continuous, solar operated absorption refrigeration cycle. The maximum ideal coefficient of performance of the system was determined to be equal to 1.6, while the peak actual coefficient of performance was determined to be equal to 0.55.

Haim *et al.* (1992) performed a simulation and analysis of two open-cycle absorption systems. Both systems comprise a closed absorber and evaporator as in conventional single stage chillers. The open part of the cycle is the regenerator, used to reconcentrate the absorber solution by means of solar energy. The analysis was performed with a computer code developed for modular simulation of absorption systems under varying cycle configurations (open- and closed-cycle systems) and with different working fluids. Based on the specified design features, the code calculates the operating parameters in each system. Results indicate a definite performance advantage of the direct-regeneration system over the indirect one.

Hawlader *et al.* (1993) developed a lithium bromide absorption cooling system employing an 11x11m collector/regenerator unit. They also have developed a computer model, which they validated against real experimental values with good agreement. The experimental results showed a regeneration efficiency varying between 38 and 67% and the corresponding cooling capacities ranged from 31 to 72 kW.

Ameel *et al.* (1995) give performance predictions of alternative low-cost absorbents for open cycle absorption using a number of absorbents. The most promising of the absorbents considered, was a mixture of two elements, lithium chloride and zinc chloride. The estimated capacities per unit absorber area were 50-70% less than those of lithium bromide systems.

Ghaddar *et al.* (1997) presented modelling and simulation of a solar absorption system for Beirut. The results showed that, for each ton of refrigeration, it is required to have a minimum collector area of 23.3m^2 with an optimum water storage capacity ranging from 1000 to 1500 liters, for the system to operate solely on solar energy for about 7 hours per day. The monthly solar fraction of total energy use in cooling is determined as a function of solar collector area and storage tank capacity. The economic analysis performed showed that the solar cooling system is marginally competitive only when it is combined with domestic water heating.

Erhard and Hahne (1997) simulated and tested a solar-powered absorption cooling machine. The main part of the device is an absorber/desorber unit, which is mounted inside a concentrating solar collector. Results obtained from field tests are discussed and compared with the results obtained from a simulation program developed for this purpose.

Hammad and Zurigat (1998) described the performance of a 1.5 ton solar cooling unit. The unit comprises a 14m^2 flat-plate solar collector system and five shell and tube heat exchangers. The unit was tested in April and May in Jordan. The maximum value obtained for actual coefficient of performance was 0.85.

Zinian and Ning (1999) describe a solar absorption air conditioning system which uses an array of 2160 evacuated tubular collectors of total aperture area of 540m^2 and a LiBr absorption chiller. Thermal efficiencies of the collector array are 40% for space cooling, 35% for space heating and 50% for domestic water heating. It was found that the cooling efficiency of the entire system is around 20%.

A method to design, construct and evaluate the performance of a single stage lithium bromide – water absorption machine is presented by Florides *et al.* (2003). In this paper the necessary heat and mass transfer relations and appropriate equations describing the properties of the working fluids are specified. Information on designing the heat exchangers of the LiBr-water absorption unit is also presented. Single-pass vertical-tube heat exchangers have been used for the absorber and for the evaporator. The solution heat exchanger was designed as a single-pass annulus heat exchanger. The condenser and the generator were designed using horizontal tube heat exchangers.

The Complete System

The objective of this work is to design a complete active solar space air conditioning (cooling and heating) system, together with a domestic water heating subsystem. Before giving the details of the complete system some design features of the individual systems when operated with solar energy are given.

Solar Heating and Hot Water System

A schematic diagram of a solar heating and hot water system is shown in Figure 11. Control of the solar heating system is based on two thermostats; the collector-storage temperature differential, and the room temperature. The differential thermostat compares the temperature between the collector outlet and the bottom of the storage tank. When this difference exceeds a preset number of degrees (6 to 10°C), the collector pump is activated and continues to run until the temperature differential drops to near zero (1-2°C); the pump is turned off. When the room thermostat calls for heat, the load pump is activated, drawing heated water from the main storage tank to meet the demand. If the energy in the storage tank cannot meet the load demand, a second stage of the thermostat activates the auxiliary heater to supply the balance of the heating requirements. Usually, the controller modifies also the valves so that the flow is entirely through the auxiliary heater whenever the solar energy storage is unable to meet the load demand.

The collector-storage design shown in Figure 11 is suitable for use in a non-freezing climate. To employ such a system in a freezing climate, provisions for complete and reliable drainage of the collector must be made. Collector drainage can be done through the installation of an automatic discharge valve, activated by the ambient temperature, and an air vent. This is called a drain-down system, where collector water is drained out of the system to waste. Alternatively, a drain-back system can be used (see previous section) in which the collector water is drained back to the storage whenever the solar pump stops. Air then enters the collector through a vent. It should be noted that pump power consumption is greater in a self-draining collector system than in a permanently filled one (Hsieh, 1986).

If the climate is characterized by frequent subfreezing temperatures, positive freeze protection through the use of an antifreeze solution in a closed collector loop is considered necessary. The most usual solution is water plus glycol. In this case, a heat exchanger is employed in between the collector loop and the water storage tank, as shown schematically in Figure 12. The involvement of a heat exchanger between the collector heat transfer fluid and the storage water introduces a temperature differential across the two sides, thereby lowering the storage tank temperature. This is a penalty for the system performance however this system design deserves favourable consideration to avoid the danger of malfunction in a self-draining system (Hsieh, 1986).

One important consideration in solar space heating systems is the decision for the best location of auxiliary energy. For maximum utilization of the energy supplied by an auxiliary source, the location of this energy input should be at the load, not at the storage. The supply of auxiliary energy at the storage will certainly increase the temperature of fluid entering the

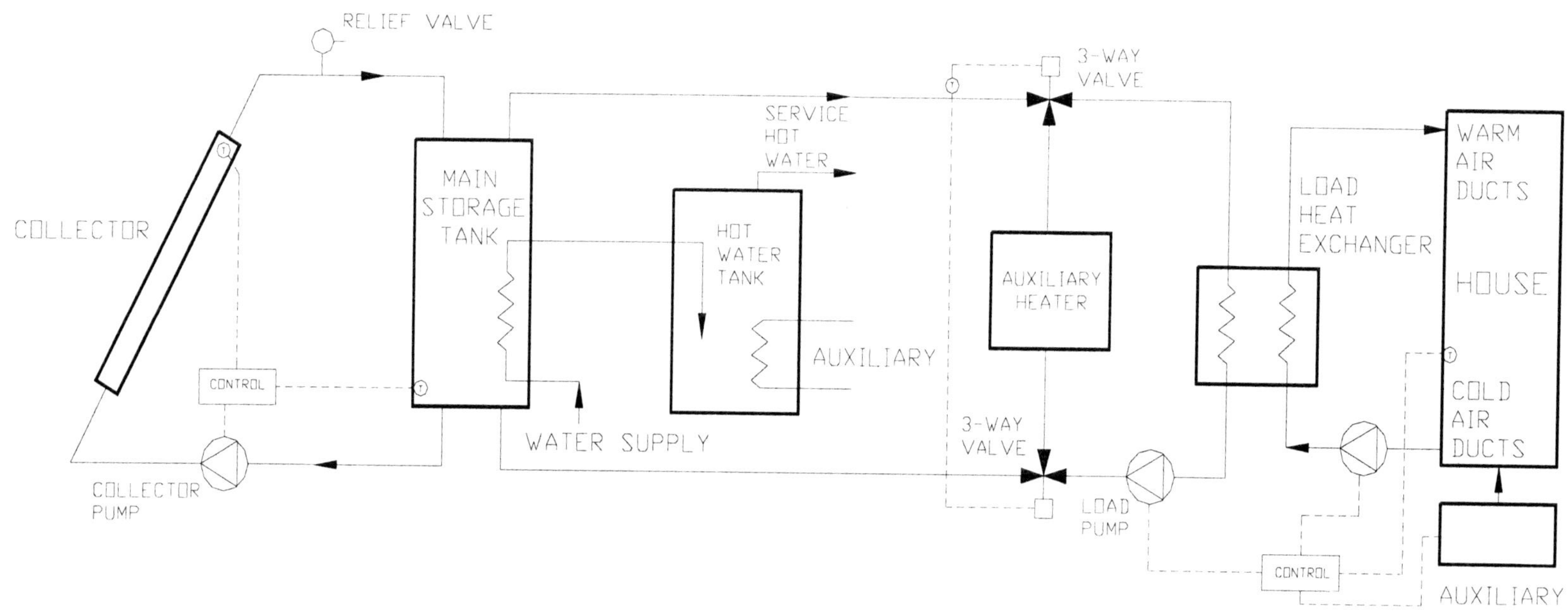

Figure 11. Schematic diagram of a solar space heating and hot water system

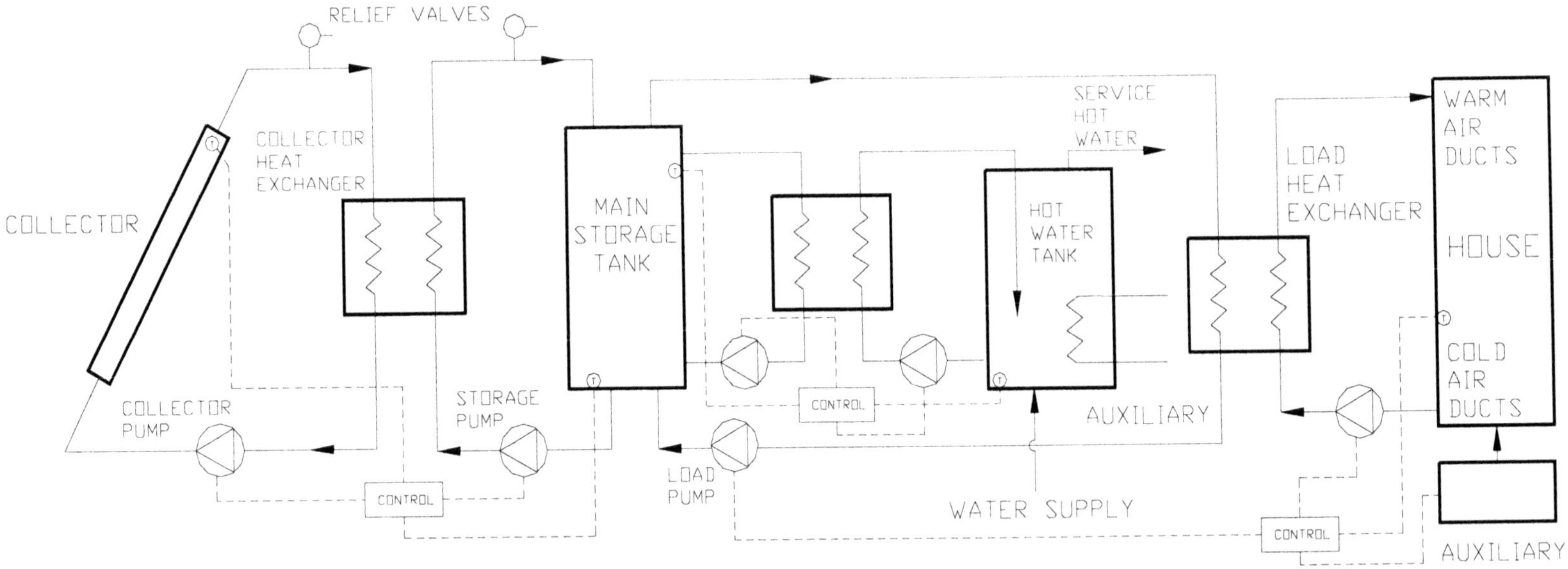

Figure 12. Schematic diagram of space heating and hot water system with antifreeze solution

collector, resulting in lower collector efficiency. When a water-based solar system is used in conjunction with a warm air space heating system, the most economical means of auxiliary energy supply is by the use of a fossil-fuel-fired boiler. In case of bad weather, the boiler can take over the entire heating load (Hsieh, 1986).

When a water-based solar system is used in conjunction with a water space heating system or to supply the heated water to an absorption air-conditioning unit, the auxiliary heater can be located in the storage-load loop, either in series or in parallel with the storage, as illustrated in Figure 13. When auxiliary energy is employed to boost the temperature of solar-heated water as shown in Figure 13a, maximum utilization of stored solar energy is achieved. This way of connecting the auxiliary supply, however, also has the tendency of boosting the storage tank temperature because water returning from the load may be at a higher temperature than the storage. Increasing the storage temperature by auxiliary energy has the undesirable effect of lowering the collector effectiveness, in addition to diverting the storage capacity to the storage of auxiliary energy instead of solar energy (Hsieh, 1986). This however depends on the operating temperature of the heating system. Therefore a low temperature system is required. This can be achieved with a water to air heat exchanger either centrally with an air handling unit or in a distributed way with individual fan coil units in each room to be heated. This system has the advantage of being able to be connected easily with space cooling system as for example with an absorption system as will be described in the next section. By using this type of systems solar energy can be used more effectively as a high temperature system would imply that the hot water storage remains at high temperature thus solar collectors will work at lower efficiency.

Another possibility is to use under-floor heating or an all water system employing the traditional heating radiators. In this case provisions need to be made during the design stage to operate the system at low temperature which implies the use of bigger-size radiators. Such a system is also suitable for retrofit applications.

Figure 13(b) illustrates an arrangement that makes it possible to isolate the auxiliary heating circuit from the storage. Solar-heated storage water is used exclusively to meet load demands when its temperature is adequate. When the storage temperature drops below the required level, circulation through the storage tank is discontinued and hot water from the auxiliary heater is used exclusively to meet space heating. This way of connecting the auxiliary supply avoids the undesirable increase of storage water temperature by auxiliary energy. However, it has the disadvantage that stored solar energy at lower temperatures is not fully utilized and this energy may be lost from the storage (through jacket losses). The same requirements for a low temperature system apply here so as to be able to extract as much energy as possible from the storage tank.

To emphasize the importance of the bypass operation mentioned in the preceding paragraphs, a numerical example is given. Suppose that the required water temperature in the heating system is 90°C. When the storage temperature is only 70°C, the auxiliary could be used to raise the temperature to 90°C. However, the water temperature drop in the heating radiators is only 10°C or less. Thus the returning water from the generator to the storage would be at 80°C. This is equivalent to the boosting of the storage with a 10°C temperature rise, thereby reducing the collector efficiency and useful solar energy storage. This example shows also the importance of using a low temperature heating system. However when the system is operating in cooling mode, where the operation temperature of the generation is of the order of 90°C, the solar system unavoidably works at higher temperature but this is not too much of a problem as during summertime a higher amount of solar radiation is available.

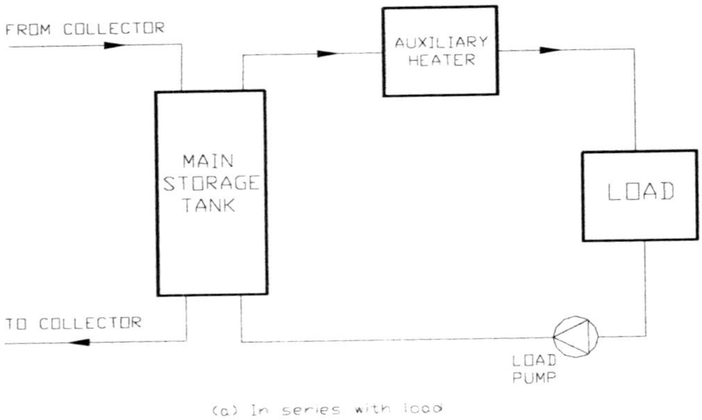

(a) in series with load

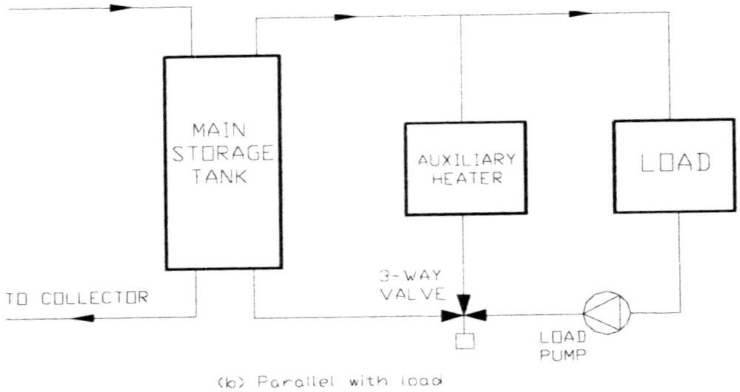

(b) parallel with load

Figure 13. Auxiliary energy supply in water-based systems

Solar Cooling System

The greatest disadvantage of solar heating system is that a large number of collectors need to be shaded or disconnected during summertime to reduce overheating. A way to avoid this problem and increase the viability of the solar system is to employ a combination of space heating and cooling.

The system is economically viable when the same collectors are used for both space heating and cooling. Flat-plate solar collectors are commonly used in solar space heating. However, because of the relatively low temperatures attainable by flat-plate collectors, only a few practical methods are available for flat-plate solar-operated cooling processes. One of the most promising designs is the utilization of an absorption refrigeration cycle with solar energy serving as the source of heat to the generator.

Figure 14 shows schematically a solar-operated absorption refrigeration system. The refrigeration cycle is the same as the one discussed in the previous section. The difference between this system and the customary fossil-fuel-fired units in that the energy supplied to the generator is from the solar collector-storage-auxiliary heater combination (see left side of Figure 14). Due to the intermittent nature of available solar energy a hot water storage tank is needed, thus the collected energy is first stored in the tank and used as energy source in the generator to heat the strong solution when needed. For this purpose the same storage tank of the solar heating system is used. When the storage tank temperature is too low to be effective in the generator, increase of the water temperature takes place in the auxiliary heater. Again here the same auxiliary heater of the space heating system can be used, at different set temperature. If the storage temperature is very low due to prolonged lack of sunshine (rarely happens during summertime), the storage is bypassed, as in space heating system, to avoid boosting of storage temperature by auxiliary energy, and the heating load of the generator is then entirely covered by the auxiliary heater.

It should be noted that the operating temperature range of the hot water supplied to the generator of a lithium bromide-water absorption refrigeration system is from 80 to 100°C. The lower temperature limit is imposed from the fact that hot water must be at a temperature sufficiently high (at least 80°C) to be effective for boiling the water from the solution in the generator. Also, the temperature of the concentrated lithium bromide solution on its return to the absorber must be kept high enough to prevent crystallisation of the lithium bromide. Usually an unpressurized water storage tank system is used and thus an upper limit of about 100°C is used to prevent water from boiling. In the present system a generator temperature of 93°C is considered which was found to be optimum for this type of systems (Florides et al., 2002c).

Since in an absorption-refrigeration cycle heat must be rejected from the absorber and the condenser an adequate cooling water system must be incorporated into the overall cycle. One way of providing cooling water is the use of a cooling tower as shown in Figure 14. Cooled water emerging from the cooling tower is first fed to the absorber and then to the condenser, because the former generally requires a lower temperature than the latter (Hsieh, 1986). It should be noted that the use of a cooling tower in a small residential system is problematic both with respect to space and maintenance requirements, thus whenever possible water drawn from a well can be used.

Just as in the case of space heating, the auxiliary heater can be arranged in parallel with the storage tank instead of in-series and the solar collection can be made through a heat exchanger that keeps the collector liquid separate from the storage tank water (indirect system).

A variation of the basic system shown in Figure 14 is the complete elimination of the hot storage tank and auxiliary heater and the supply of solar-heated fluid directly to the generator of the absorption system. This arrangement has the advantage of making higher temperature available to the generator on sunny days. However, it has the disadvantages of poor cooling control due to variations in solar input and the lack of stored energy to produce cooling during evenings or on cloudy days. This system can be more effective by using cool storage and in doing so, the intermittent effects of this arrangement, due to the absence of any hot storage, can be minimized. One way of achieving this is by using the absorption machine to produce chilled water, which is stored for cooling purposes (Hsieh, 1986). While hot storage tanks generally suffer from heavy heat losses, a chilled water storage has a relatively low rate of heat gain because of the smaller temperature difference between the chilled water and the

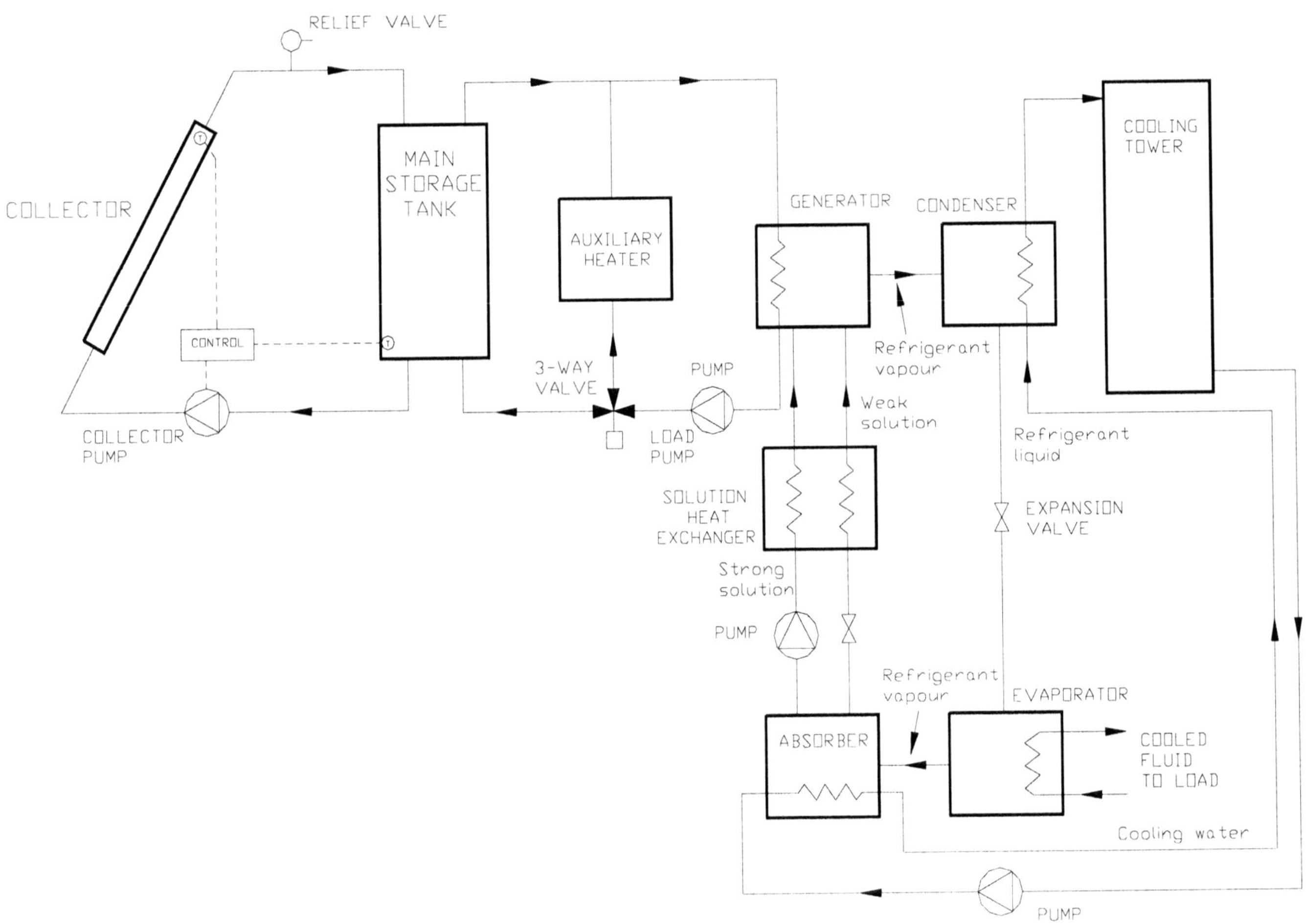

Figure 14. Schematic diagram of a solar-operated absorption refrigeration system

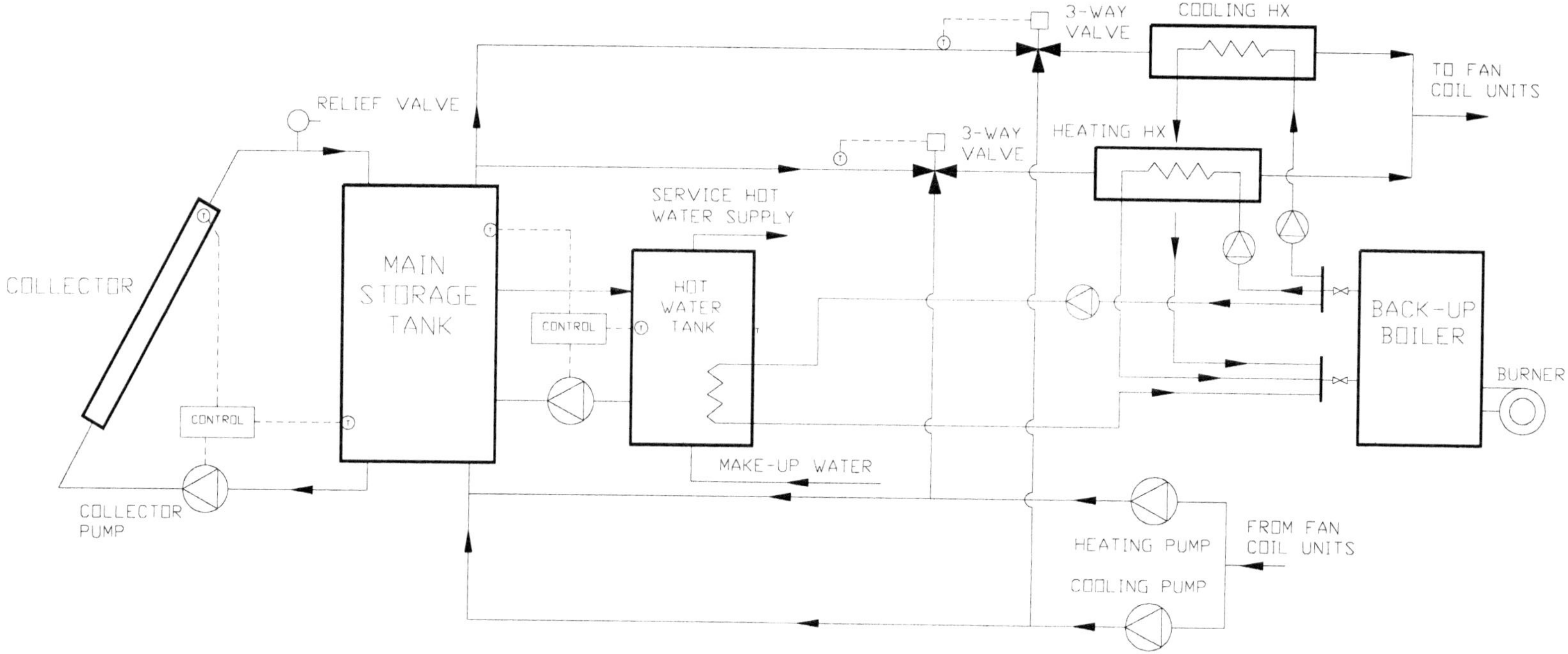

Figure 15. Schematic diagram of the complete space heating, cooling and hot water system

surroundings. However, as the temperature range of a cool storage is small in comparison with that of a hot storage, for the same amount of energy a larger storage volume is needed in cool storage than in hot storage. As solar heating systems always employ a storage tank, this arrangement is not considered in this work as the storage tank is readily available.

The Complete Solar System

A typical active solar space air conditioning (cooling and heating) system, together with a domestic water heating subsystem, is shown schematically in Figure 15. Solar liquid-heating collectors are used to absorb and transform solar radiation into thermal energy of the transport medium. Water is used as both the heat transfer and the storage medium. The water heated by solar energy is pumped into the storage tank waiting to supply space and domestic water loads. A fossil-fuel-fired conventional heater is integrated with the solar system to provide an auxiliary energy supply should the solar system fail to meet the building energy requirement in case of cloudy weather.

System Model

The complete system is simulated with TRNSYS program for two locations at different latitudes; Nicosia, Cyprus (35°) and Madison, Wisconsin (43°).

The proper sizing of the components of a solar system is a complex problem, which includes both predictable (collector and other performance characteristics) and unpredictable (weather data) components. In this section it is indicated how all the necessary components of the solar thermal system can be formulated into a single model with the aid and use of TRNSYS simulation program (TRNSYS, 2004).

Simplified analysis methods have advantages of computational speed, low cost, rapid turnaround, which is especially important during iterative design phases, and ease of use by persons with little technical experience. Disadvantages include limited flexibility for design optimization, lack of control over assumptions and a limited selection of systems that can be analyzed. Thus, if the system application, configuration or load characteristics under consideration are significantly non-standard, a detailed computer simulation may be required to achieve accurate results. Program TRNSYS offers such a detailed modelling capability.

The initial step in modelling a system is the derivation of a structure to be used to represent the system. It will become apparent that there is no unique way of representing a given system. Since the way the system is represented often strongly suggests specific modelling approaches, the possibility of using alternative system structures should be left open while the modelling approach selection is being made. The structure that represents the system should not be confused with the real system. The structure will always be an imperfect copy of reality. However, the act of developing a system structure and the structure itself will foster an understanding of the real system. In developing a structure to represent a system, system boundaries consistent with the problem being analyzed are first established. This is accomplished by specifying the items, processes, and effects that are internal to the system and those which are external. By simulating the system on an annual basis the long-term integrated system performance can be evaluated and the dynamic system's behaviour can be obtained.

TRNSYS Program Description

TRNSYS is an acronym for a "transient simulation program" and is a quasi-steady simulation model. The program consists of many subroutines that model subsystem components. The mathematical models for the subsystem components are given in terms of their ordinary differential or algebraic equations. With a program such as TRNSYS which has the capability of interconnecting system components in any desired manner, solving differential equations and facilitating information output, the entire problem of system simulation reduces to a problem of identifying all the components that comprise the particular system and connecting them together to represent the actual system.

Once all the components of the system have been identified it is necessary to construct an information flow diagram for the system. The purpose of the information flow diagram is to facilitate identification of the components and the flow of information between them. Each component is represented as a box, which requires a number of constant PARAMETERS and time dependent INPUTS and produces a time dependent OUTPUTS. An information flow diagram shows the manner in which all system components are interconnected. A given OUTPUT may be used as an INPUT to any number of other components. A simplified information flow diagram for the solar system under investigation is shown in Figure 16.

From the flow diagram shown in Figure 16 a deck file has to be constructed containing information on all the components of the system, weather data file, and the output format.

TRNSYS Type 1 employing the mode that considers second order collector performance equation (Eq. 1) is used to model the collectors. For the stationary collectors however, the standard single-order collector performance equation is used ($a_2=0$) in which only the terms (a_o) and (a_1), shown in Eq. 1, are used to model the collector.

$$n = a_o K_{\alpha\tau} - a_1\left(\frac{\Delta T}{G}\right) - a_2\left(\frac{\Delta T^2}{G}\right) \tag{1}$$

It should be noted that for the single-order performance equation, if n is plotted against $\Delta T/G$ a straight line is obtained with intercept (a_o) and slope (a_1). The intercept depends on the optical properties of the collector whereas the slope depends on the collector heat loss coefficient. A high performance collector has a high a_0 value (close to unity) and a low a_1 value, i.e., low heat losses. The values of a_0 and a_1 of the collectors considered in this work are shown in Tables 1 and 2.

The following model of incidence angle modifier is used in TRNSYS Type 1:

$$k_{\alpha\tau} = 1 - b_o\left(\frac{1}{\cos(\theta)} - 1\right) - b_1\left(\frac{1}{\cos(\theta)} - 1\right)^2 \tag{2}$$

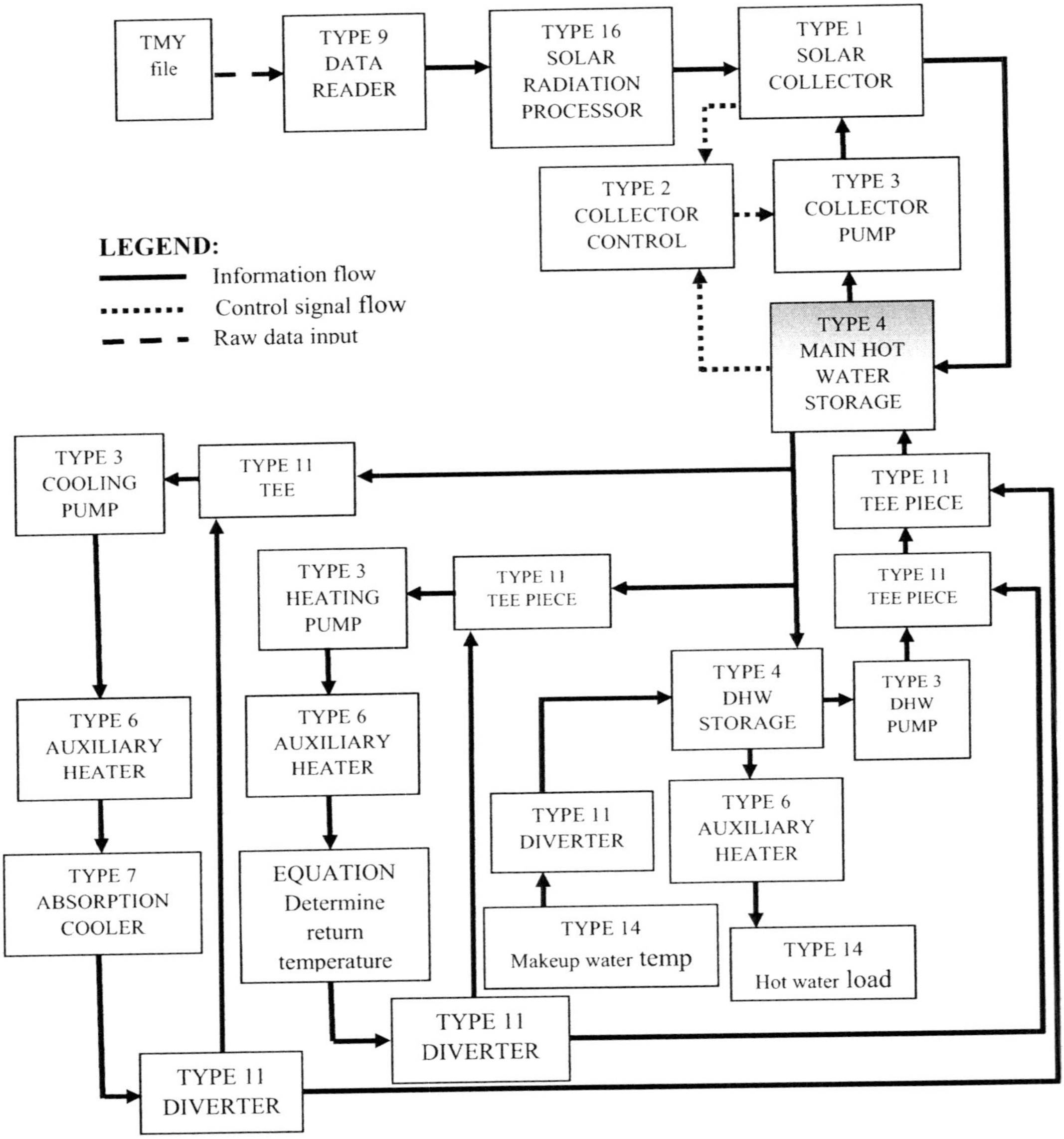

Figure 16. Information flow diagram for the solar system model

For the stationary collectors considered in this work only factor b_o is required (the values of b_o are also shown in Tables 1 and 2). The useful energy extracted from the collectors is given by:

$$Q_u = F_R A[k_{\alpha\tau}(\tau\alpha)G - U_L(T_i - T_a)] \tag{3}$$

The total useful energy for the whole year is obtained from:

$$Q_{u,a} = \sum_{d=1}^{365} \sum_{h=1}^{24} Q_u \tag{4}$$

and the annual auxiliary energy required, Q_{aux} is:

$$Q_{aux} = Q_{load} - \left(Q_{u,a} - Q_{loss}\right) \tag{5}$$

Where Q_{loss} is the thermal losses from the storage tank, pipes and the relief valve.

As can be seen from the above equations the energy obtained from the solar collector field depends on the inlet temperature to the collector T_i, which depends on the thermal load and the thermal losses (Q_{loss}).

The specifications of the complete system are shown in Table 3. Water heated from solar energy is stored in a TRNSYS Type 4 stratified fluid storage tank. The vertical cylinder construction is made of copper and is thermally insulated with polyurethane. Also the tank is protected by a galvanised outer shell 0.6mm thick. A second smaller size tank is used as part of the domestic hot water subsystem as shown in Figure 16. The size of this tank is 160 lt and is determined from the hot water consumption.

Table 3. Specifications of the complete system

Parameter	Nicosia system	Madison system	Units
Collector area	Variable	Variable	m^2
Main storage tank volume	Variable	Variable	m^3
DHW subsystem storage tank volume	0.16	0.16	m^3
Solar differential thermostat ON	10	10	°C
Solar differential thermostat OFF	2	2	°C
DHW subsystem auxiliary heater	12,000	12,000	kJ/hr
DHW subsystem heater set temperature	50	50	°C
Heating subsystem auxiliary heater	40,000	90,000	kJ/hr
Heating subsystem heater set temperature	46	46	°C
Cooling subsystem auxiliary heater	86,000	53,000	kJ/hr
Cooling subsystem heater set temperature	93	93	°C
DWH subsystem pump capacity	200	200	kg/hr
Heating subsystem pump capacity	1,195	2,685	kg/hr
Cooling subsystem pump capacity	3,300	2,010	kg/hr
Absorption cooler capacity	52,000	32,000	kJ/hr

Although in practice one auxiliary boiler is used for satisfying the needs for auxiliary energy of all three subsystems (as shown in Figure 15), in the model three auxiliary heaters (TRNSYS Type 6) are used one for the DHW subsystem, one for the heating subsystem and one for the cooling subsystem set at an upper temperature of 50°C, 46°C and 93°C for the three subsystems respectively.

The LiBr-water absorption air conditioner (TRNSYS Type 7) is a single-effect unit, based on Arkla model WF-36. Its nominal capacity is taken as 65,000 kJ/hr, assuming no internal auxiliary heater and working in energy rate control.

Although the hot water demand is subject to a high degree of variation from day to day and from consumer to consumer it is impractical to use anything but a repetitive load profile. This is not quite correct during the summer period, where the consumption pattern is somewhat higher. However, during this period, the temperature requirement for hot water is not as high as during winter. Consequently, the total thermal energy requirement is reasonably constant throughout the year. For the present simulation, the hot water consumption profile illustrated in Figure 17 is employed, which assumes a daily hot water consumption of 120 litres at 50°C for a family of four (30 litres/person).

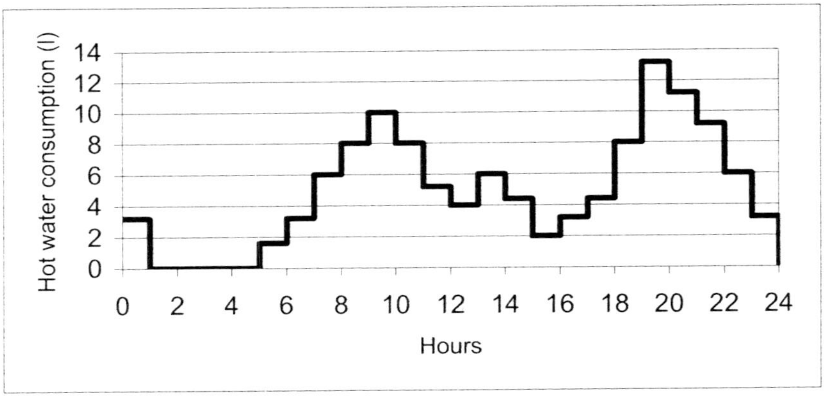

Figure 17. Hot Water daily consumption profile

Typical Meteorological Year

The systems were simulated with TRNSYS using Typical Meteorological Year (TMY) data for Nicosia, Cyprus and Madison, Wisconsin. TMY is defined as a year which sums up all the climatic information characterizing a period as long as the mean life of the system. The selection of typical weather conditions for a given location is very crucial in computer simulations for performance predictions and has led various investigators either to run long periods of observational data or to select a particular year, which appears to be typical from several years of data. Klein *et al.* (1976) have constructed the "average year" by selecting the monthly data from an 8-year period, which corresponded most closely to the average monthly insolation and ambient temperature. The TMY for Cyprus was generated from hourly measurements, of solar irradiance (global and diffuse on horizontal surface), ambient temperature, wind speed and direction, and humidity ratio, for a seven-year period, from 1986 to 1992 using the Filkenstein – Schafer statistical method (Kalogirou, 2003). The measurements were recorded by the Cyprus Meteorological Service at the Athalassa region, an area at the suburbs of the town of Nicosia. The TMY is considered as a representative year for the Cypriot and Mediterranean environment. The TMY for Madison, Wisconsin is readily available with TRNSYS program.

The adequacy of using an average or typical year of meteorological data with a simulation model to provide an estimate of the long-term system performance depends on the sensitivity of system performance to the hourly and daily weather sequences. Regardless of how it is selected, an "average" year cannot be expected to have the same weather sequences as those occurring in the long term. However, the simulated performance of a system for an "average year" may provide a good estimate of the long-term system performance if the weather sequences occurring in the average year are representative of those occurring in the long term or if the system performance is independent of the weather sequences (Klein *et al.*, 1976).

The annual values of global radiation on a horizontal surface and direct normal radiation as well as the annual mean air temperature of the two locations as obtained from the data contained in the two TMYs are shown in Table 4. As can be seen the weather of Nicosia is warm and sunny whereas both the mean air temperature and annual solar radiation are much lower in Madison.

Table 4. Annual values of radiation and mean annual temperature for the two locations considered

Location	Total annual global radiation on a horizontal surface (kWh/m^2)	Total annual direct normal radiation (kWh/m^2)	Mean annual air temperature (°C)
Nicosia, Cyprus	1691.2	1749.1	18.5
Madison, Wisconsin	1369.5	1317.2	7.9

Economic Analysis Method Description

A life cycle analysis is performed in order to obtain the total cost (or life cycle cost) and the life cycle savings of the systems. The economic scenario used in this work is that 30% of the initial cost of the solar system is paid at the beginning and the rest is paid in equal instalments in 10 years. The period of economic analysis is taken as 20 years (life of the system), whereas the inflation rates of fuel and electricity, used for pumps, are mean values of the last 10 years in Cyprus. Maintenance and parasitic costs are also considered. Light fuel oil (LFO) is assumed to be used for a fuel-only system. From the addition of fuel savings incurred because of the use of the system and the tax savings the mortgage, maintenance and parasitic costs are subtracted and thus the annual solar savings of the system are estimated which are converted into present worth values of the system. These are added up to obtain the life cycle savings. The economic analysis can be performed either within the TRNSYS environment or in a spreadsheet program. For the present work, the spreadsheet application is used. A detailed description of the method of economic analysis of solar systems using spreadsheets is given in (Kalogirou, 1996).

In general, the present worth (or discounted cost) of an investment or cost C at the end of year N, at a discount rate of d and interest rate of i is obtained by:

$$PW_N = \frac{C(1+i)^{N-1}}{(1+d)^N} \tag{6}$$

In the case of this work, the various costs and savings are estimated annually. The life cycle savings are obtained according to the following equation:

$$PW_{LCS} = \sum_{N=1}^{N} \frac{Solar\ Savings}{(1+d)^N} \tag{7}$$

The fuel savings are obtained by subtracting the annual cost of the conventional fuel used for the auxiliary energy from the fuel needs of a fuel only system. The integrated cost of the auxiliary energy use for the first year, i.e. solar back up, is given by the formula:

$$C_{aux} = \int_0^t C_{FA} Q_{aux}\, dt \tag{8}$$

The integrated cost of the total load for the first year i.e. cost of conventional fuel without solar, is:

$$C_{load} = \int_0^t C_{FL} Q_{load}\, dt \tag{9}$$

where C_{FA} and C_{FL} are the cost rates for auxiliary energy and conventional fuel respectively. In case that the same fuel is used for both $C_{FA}=C_{FL}$.

Table 5 shows the estimated costs per square meter of the collectors and other economic parameters considered in the present analysis.

Table 5. Investment cost of the collectors and other economic parameters considered in this study

Collector type	Collector price (Ca)
FPC	187 €/m^2
AFP	204 €/m^2
ETC	425 €/m^2
Economic parameters	**Value**
Market discount rate	6.5%
Interest rate	5.2%
General inflation rate	4.5%
Conventional fuel price	0.6€/lt
Storage tank price (C_v)	1700 €/m^3

Collector prices include collector mountings and field piping
Storage tank price includes also the cost of the solar pump

The investment cost of the stationary solar systems is estimated by:

$$C_s = C_a A + C_v V \tag{10}$$

Where C_a is the area dependent cost (€/m^2) and C_v is the volume dependent cost (€/m^3).

For the operation cost (C_o), maintenance and parasitic cost are considered. The former are estimated to be 1% of the initial investment and are assumed to increase at a rate of 0.5% per year of the system operation. The latter accounts for the energy required (electricity) to drive the solar pump. The total annual cost is given by:

$$C = C_s + C_o \quad (11)$$

It should be noted that the cost of the small domestic hot water system storage tank is not considered in the economic analysis as such a tank is present in the installation irrespective if the system is solar or not, thus only extra equipment is considered against the benefits obtained from the solar system.

Description of the Building

For the estimations carried out in this chapter a model house is considered, as illustrated in Figure 18. This is a typical house of 196m^2 floor area. The typical house consists of four identical external walls. Every wall is 14m in length by 3m in height and has a double glazing window of 5.2m^2. The typical house is constructed with double-walls made of 10cm hollow brick and 2cm plaster on each side and a layer of 5cm insulation in between. The roof is constructed from fair-face 15cm heavy concrete, 5cm polystyrene insulation, 7cm screed and 0.4cm asphalt, covered with aluminium paint of 0.55 solar absorptivity. The model house is further divided into four identical zones and the partition walls are considered as walls separating the four zones.

For the present study TRNSYS Type 19 model is used to estimate the heating and cooling loads for a typical house by utilizing the transfer function method (TRNSYS, 2004). The transfer function method estimates the heating and cooling loads arising from walls, windows, flat roofs and floors. For every zone a separate TRNSYS Type 19 unit, is necessary. Details of the input parameters required to model the typical house are given in Florides *et al.* (2000).

Type 19 model has two basic modes of operation, i.e., the energy rate and the temperature level control modes. In the present study the energy rate control mode is selected, since with this mode the temperature of the building can be maintained between specified limits and the energy required to maintain the zone in the specified temperature is given as output along with the limit temperature. The zone humidity ratio is also allowed to float between a maximum and a minimum limit specified by the user and the humidification or dehumidification energy is considered. A controller is used in conjunction with this mode to control the heating or cooling equipment.

It is important to emphasize here that any attempt to apply solar heating or cooling in a house should be done only after the building is insulated well to reduce the thermal loads. The above construction results in the annual and hourly maximum thermal loads shown in Table 6. These are estimated by keeping the room temperature during wintertime equal to 21°C and during summertime equal to 25°C and the humidity ratio between 0.005 and 0.008 kg of water per kg of dry air.

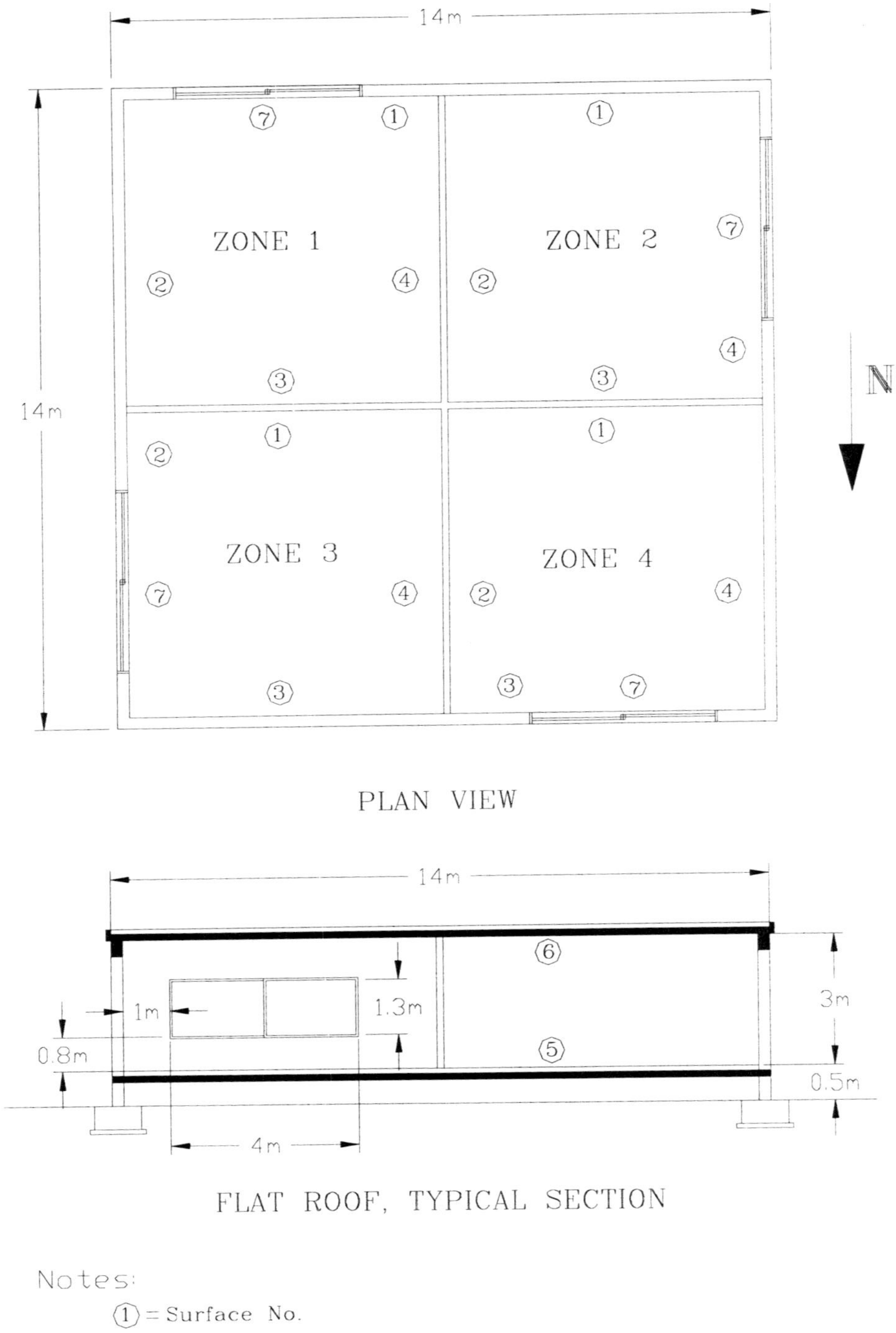

Figure 18. Model house drawing

As can be seen for the model house located in Nicosia the maximum thermal load is the cooling load whereas in Madison the maximum thermal load is the heating load. These are expected results which agree with the environmental conditions prevailing at each location as indicated in Table 4.

Table 6. Annual and hourly maximum thermal loads

Location	Heating load		Cooling load	
	Annual (GJ)	Maximum hourly (MJ/hr)	Annual (GJ)	Maximum hourly (MJ/hr)
Nicosia, Cyprus	12.03	39.61	79.38	87.37
Madison, Wisconsin	135.2	51.44	25.5	31.60

Winter room temperature = 21°C

Summer room temperature = 25°C

Humidity ratio range = 0.005-0.008 kg of water / kg of dry air

RESULTS

What is the ultimate objective of this kind of analysis is to determine the size and characteristics of the system which make the system viable and give the highest life cycle savings. In order to achieve this, the collector inclination, mass flow rate and area as well as the storage tank size needs to be optimized. Initially the optimization of the collector inclination and mass flow rate is presented.

Table 7. Optimisation of the collector inclination

Location	Collector	Slope (°)	Total auxiliary (GJ)	System f (%)
Nicosia	FPC	Lat-10°	82.57	14.81
		Lat-5°	*82.08*	*15.32*
		Lat	82.34	15.05
		Lat+5°	83.08	14.29
Nicosia	AFP	Lat-10°	61.38	36.68
		Lat-5°	*59.72*	*38.39*
		Lat	60.49	37.59
		Lat+5°	60.61	37.47
Nicosia	ETC	Lat-10°	24.97	74.24
		Lat-5°	*24.96*	*74.25*
		Lat	26.63	72.53
Madison	FPC	Lat-5°	147.9	11.03
		Lat	147.5	11.24
		Lat+5°	*147.1*	*11.52*
		Lat+10°	147.5	11.28
Madison	AFP	Lat-5°	130.2	19.03
		Lat	*129.8*	*19.51*
		Lat+5°	130.3	19.47
Madison	ETC	Lat-5°	106.3	36.12
		Lat	*105.9*	*36.29*
		Lat+5°	106.0	36.22

For all cases A=50m^2, V=2.5m^3, Flow=54 kg/hr-m^2

f = solar contribution, shows percent of load covered by solar

Numbers in italic indicate maximum (optimum) values

Table 8. Collector flow rate optimisation

Nicosia, Cyprus					Madison, Wisconsin				
Collector	Area (m^2)	Tank (m^3)	Flow rate ($kg/hr\text{-}m^2$)	T_{aux} (GJ)	Collector	Area (m^2)	Tank (m^3)	Flow rate ($kg/hr\text{-}m^2$)	T_{aux} (GJ)
FPC	20	1.5	*9*	91.71	FPC	20	1.5	9	160.5
	20	1.5	*18*	*89.81*		20	1.5	*18*	*159.2*
	20	1.5	27	92.01		20	1.5	27	159.5
	40	2.0	9	80.79		40	2.0	9	147.6
	40	2.0	*18*	*77.90*		40	2.0	*18*	*145.2*
	40	2.0	27	78.77		40	2.0	27	146.2
	60	2.5	9	72.82		60	2.5	9	137.1
	60	2.5	*18*	*67.20*		60	2.5	*18*	*133.3*
	60	2.5	27	68.79		60	2.5	27	133.4
AFP	20	1.5	18	80.78	AFP	20	1.5	18	151.1
	20	1.5	*27*	*80.42*		20	1.5	*27*	*150.4*
	20	1.5	36	82.64		20	1.5	36	152.0
	40	2.0	18	59.16		40	2.0	18	129.2
	40	2.0	*27*	*59.13*		40	2.0	*27*	*129.1*
	40	2.0	36	61.63		40	2.0	36	130.6
	60	2.5	18	45.45		60	2.5	18	111.6
	60	2.5	*27*	*41.56*		60	2.5	*27*	*111.0*
	60	2.5	36	42.89		60	2.5	36	114.4
ETC	20	1.5	27	65.22	ETC	20	1.5	18	137.6
	20	1.5	*36*	*61.77*		20	1.5	*27*	*136.5*
	20	1.5	45	62.34		20	1.5	36	137.2
	40	2.0	27	33.30		40	2.0	18	106.8
	40	2.0	*36*	*31.06*		40	2.0	*27*	*104.2*
	40	2.0	45	32.00		40	2.0	36	104.8
	60	2.5	27	20.09		60	2.5	18	84.48
	60	2.5	*36*	*16.85*		60	2.5	*27*	*83.99*
	60	2.5	45	17.11		60	2.5	36	87.59

Table 9. Optimum flow rates

Location	Collector	Optimum flow rate (kg/hr-m^2)
Nicosia, Cyprus	FPC	18
	AFP	27
	ETC	36
Madison, Wisconsin	FPC	18
	AFP	27
	ETC	27

Table 10. Performance parameters of optimum systems

Location	Collector	Area (m^2)	Storage tank (m^3)	Slope (°)	Flow rate (kg/h-m^2)	Qu (GJ)	Q_{aux-HW} (GJ)	$f_{(HW)}$ (%)	Q_{aux-H} (GJ)	$f_{(H)}$ (%)	Q_{aux-C} (GJ)	$f_{(total)}$ (%)
Nicosia	FPC	20	1.0	30	18	30.28	0.169	96.94	7.92	34.14	81.43	7.64
	AFP	40	1.5	30	27	78.90	0.032	99.41	2.74	77.22	56.34	39.02
	ETC	30	1.5	30	36	95.20	0.025	99.55	2.23	81.43	40.97	55.40
Madison	FPC	20	1.0	48	18	24.47	0.918	83.39	127.0	6.084	31.09	4.36
	AFP	50	1.5	43	27	77.39	0.504	90.88	101.4	25.02	16.45	28.82
	ETC	30	1.5	43	27	78.33	0.557	89.93	104.5	22.70	14.09	177.0

Optimisation of the Collector Inclination

When a hot water only system is designed the usual inclination of the collector is latitude plus 5 degrees. However as the sun during summertime is higher on the sky than in wintertime and as a considerable amount of load needs to be covered during summertime as well for the solar cooling system an investigation needs to be carried out to determine the collectors' inclination which gives the smaller amount of total auxiliary energy. The total auxiliary energy is the addition of backup energy over the year required to cover the heating, cooling and hot water loads. The results for the three types of collectors for the two locations considered are shown in Table 7.

As can be seen for Nicosia, where the prevailing load is the cooling load, the optimum collector inclination obtained for all types of collectors is latitude minus 5 degrees, i.e., 30°, so as to have the collector more perpendicular to the sun during summertime. For Madison, where the prevailing load is the heating load, the optimum collector inclination is found to be equal to latitude plus 5 degrees (48°) for the flat-plate collector and equal to the latitude (43°) for the AFP and ETC so as the collectors would receive more sunshine during wintertime.

Optimisation of the Collector Flow Rate

To find the optimum flow rate three combinations of collector and storage tank sizes were examined as follows:

1. Collector area=20m^2, storage tank size=1.5m^3
2. Collector area=40m^2, storage tank size=2.0m^3
3. Collector area=60m^2, storage tank size=2.5m^3

These combinations cover the whole range of collector area and storage tank volume that can be applied in the present systems. The optimisation parameter is again the minimum total auxiliary energy (T_{aux}) required by the systems to cover the total load (heating, cooling and hot water). The results are shown in Table 8. For these runs the optimum slope of the collectors obtained in previous section were used. Also to avoid having long lists of data only three values are presented for each case, the optimum (marked in italic) and the values immediately before and after the optimum. The optimum values found for each collector and for each location are shown in Table 9. As can be seen a general guideline is that the poorer the collector characteristics the lower the flow rate required. It should be noted that all values obtained are much lower than the generally accepted value of 54 kg/hr-m^2, usually considered in solar systems. The obtained values are also more cost effective as lower flow rate would require lower electrical power input.

Optimisation of the Collector Area and Storage Tank

The optimum sizes of collector area and storage tank are determined from the economic analysis. The optimisation parameter is the life cycle savings (LCS), i.e., the optimum combination of these two parameters is required to be determined which maximises the LCS.

The results of this exercise are shown in Tables 10 and 11, namely Table 10 shows the performance parameters and Table 11 the economic parameters of optimum systems. As can be seen the advanced flat-plate collector (AFP) which has a good performance and a comparatively low cost requires a greater amount of collector area compared to the other collector types considered. The standard flat plate collector (FPC) is comparatively expensive for the energy it supplies as indicated by the low values of LCS. The evacuated tube collectors (ETC) although they are very expensive can find applications in environments with good solar potential, like Cyprus. In fact for Cyprus this type of collector gives the highest LCS. The corresponding collector for Madison is the AFP. The contribution of the solar systems to cover the load of the various subsystems is represented by the solar contribution (f value). This contribution indicated is the percent of load (Q_{load}) covered with solar given by:

$$f = \frac{Q_{load} - Q_{aux}}{Q_{load}} \tag{12}$$

Table 11. Economic parameters of optimum systems

Location	Collector	Area (m^2)	Storage tank (m^3)	T_{aux} (GJ)	FYFS (€)	LCS (€)
Nicosia	FPC	20	1.0	89.52	438.9	1757
	AFP	40	1.5	59.11	924.7	3868
	ETC	30	1.5	43.23	1178.3	5107
Madison	FPC	20	1.0	195.0	300.3	-412
	AFP	50	1.5	118.3	950.4	2093
	ETC	30	1.5	119.1	937.6	1340

Two such contributions are $f_{(HW)}$, the solar contribution for the hot water subsystem and $f_{(H)}$, the solar contribution for the heating subsystem. It is not possible to give a solar contribution value for the cooling subsystem as due to the COP of the single-effect absorption cooler, which is of the order of 0.6, more thermal energy is required than the cooling load. Alternatively the total (annual) solar contribution $f_{(total)}$ is used, given by:

$$f_{total} = \frac{T_{load} - T_{aux}}{T_{load}} \tag{13}$$

Where T_{load} is the total annual load (heating + cooling + hot water load) and T_{aux} is the total auxiliary needed in one year.

To find the optimum systems shown in Tables 10 and 11 require the investigation of a number combinations of collector areas and storage tank volumes. To show all the values obtained for the various collectors would require a lot of space and will not add more value to this chapter. Therefore the results of only the best systems for the two locations are presented as shown in Tables 12 and 13 for Nicosia and Madison respectively.

Table 12. Optimisation of evacuated tube collectors (ETC) for Nicosia

Area (m^2)	Storage tank (m^3)	Qu (GJ)	$f_{(HW)}$	$f_{(H)}$	T_{aux} (GJ)	$f_{(total)}$	FYFS (€)	LCS (€)
10	1.0	34.14	96.49	31.16	84.60	12.72	517.50	3025
20	1.0	65.06	99.13	67.71	61.61	36.44	884.67	4955
30	1.0	94.89	99.59	81.49	44.26	54.34	1161.83	5008
40	1.0	125.30	99.64	83.59	32.56	66.41	1348.83	4588
20	1.5	65.57	98.82	63.56	62.34	35.69	873.00	4143
30	*1.5*	*95.20*	*99.55*	*81.43*	*43.23*	*55.40*	*1178.33*	*5107*
40	1.5	125.30	99.63	85.54	31.15	67.87	1371.33	4312
20	2.0	65.60	98.68	63.95	62.74	35.27	866.67	3415
30	2.0	95.71	99.53	80.76	44.37	54.23	1160.17	4192
40	2.0	125.40	99.66	87.77	31.06	67.96	1372.67	3705
50	2.0	153.10	99.77	93.01	21.78	77.53	1521.00	2210
30	2.5	95.89	99.47	81.15	44.41	54.18	1159.50	3553
40	2.5	125.40	99.69	89.06	31.14	67.87	1371.50	3055
50	2.5	153.60	99.76	93.00	21.80	77.51	1520.67	1575
60	2.5	182.00	99.76	93.17	16.85	82.62	1599.67	-1002
30	3.0	96.15	99.46	82.29	46.10	52.44	1132.50	2502
40	3.0	125.60	99.71	89.90	31.50	67.51	1365.67	2337
50	3.0	153.90	99.77	93.20	21.71	77.60	1522.00	968
60	3.0	182.30	99.77	93.21	16.62	82.85	1603.33	-1573
70	3.0	210.90	99.78	93.23	12.84	86.75	1663.83	-4442
50	3.5	154.10	99.77	93.19	21.57	77.75	1524.33	375
60	3.5	182.80	99.77	93.23	16.58	82.89	1604.00	-2192
70	3.5	211.70	99.78	93.25	12.91	86.68	1662.67	-5088

Table 13. Optimisation of advanced flat-plate collectors (AFP) for Madison

Area (m^2)	Storage tank (m^3)	Qu (GJ)	$f_{(HW)}$	$f_{(H)}$	T_{aux} (GJ)	$f_{(total)}$	FYFS (€)	LCS (€)
10	1.0	19.44	82.02	3.75	163.00	1.95	370.50	725
20	1.0	35.66	84.96	9.32	150.50	9.45	661.33	1460
20	1.5	36.04	85.48	9.31	151.10	9.11	659.67	805
30	1.5	51.12	87.38	14.57	140.10	15.72	937.67	1340
40	1.5	64.86	89.34	20.18	128.30	22.79	1175.67	1250
50	*1.5*	*77.39*	*90.88*	*25.02*	*118.30*	*28.82*	*950.33*	*2093*
60	1.5	88.61	92.10	29.32	109.80	33.95	1086.17	2040
20	2.0	36.16	85.64	9.33	151.30	8.99	651.67	52
30	2.0	51.43	87.47	14.42	140.30	15.59	929.67	587
40	2.0	65.35	89.19	19.79	129.20	22.29	1175.67	620
50	2.0	78.08	90.69	24.75	118.90	28.49	1362.50	-270
30	2.5	51.82	87.44	14.14	141.00	15.14	913.67	-293
40	2.5	66.05	89.01	19.32	130.60	21.45	1156.50	-308
50	2.5	78.89	90.54	24.61	119.40	28.18	1362.00	-923
60	2.5	91.30	91.66	28.57	111.00	33.22	1498.50	-2585

Table 13. Optimisation of advanced flat-plate collectors (AFP) for Madison - Continued

Area (m^2)	Storage tank (m^3)	Qu (GJ)	$f_{(HW)}$	$f_{(H)}$	T_{aux} (GJ)	$f_{(total)}$	FYFS (€)	LCS (€)
30	3.0	52.00	87.44	14.15	141.30	15.00	897.67	-1172
40	3.0	66.39	88.91	18.83	131.50	20.88	1140.50	-1188
50	3.0	79.76	90.37	24.07	120.40	27.54	1356.50	-1620
60	3.0	92.05	91.45	28.45	111.50	32.94	1499.00	-3205
70	3.0	104.40	92.29	31.97	104.50	37.11	1600.33	-5435
50	3.5	80.36	90.21	23.84	121.30	27.05	1348.00	-2385
60	3.5	93.06	91.22	28.23	112.10	32.55	1494.00	-3913
70	3.5	105.70	92.08	31.67	105.30	36.65	1602.17	-6033

As can be seen bigger systems with greater number of collector area and larger storage tank, although satisfy a large amount of the load are not viable, represented by the negative LCS. In this case, negative LCS represent the money lost by installing and operating the solar system instead of satisfying the same demand with fuel. It seems that the cost of the storage tank plays also a very important role in this optimisation. This can be seen by comparing the performance of systems with the same collector but different storage tank volume. In the case of Nicosia the solar system satisfies almost all the hot water needs of the house, irrespective of the collector area, whereas in the case of Madison slightly lower values are obtained.

Generally the greater the collector area the higher the useful energy collected (Q_u) and the lower the total annual auxiliary (T_{aux}) required to cover the load. Except for the case of hot water production, as more collectors are added for Nicosia more heating load is covered whereas the effect on cooling load is less pronounced as the cooling load is much bigger than the heating load. The reverse is true for Madison.

Performance of the Optimum Systems

TRNSYS can give results on an hourly, daily, monthly and annual basis. So far the annual results were analysed. In this section the monthly performance of the two optimum systems is presented. This analysis can reveal the actual contribution of the solar collectors on a monthly basis.

Optimum System for Nicosia

The performance of the evacuated tube collector system (optimum system) for Nicosia is presented in Figures 19 and 20. Figure 19 shows the energy flows and Figure 20 the solar contributions. As can be seen the useful energy supplied from the collectors is maximum during the summer months and the maximum value is during July and August with a value of about 11.3 GJ. Cooling load is also maximised during summertime and reaches a maximum value of 16.9 GJ during July. The cooling auxiliary energy is also maximised during summertime and the maximum value is reached during July and September and is about 7.9

GJ. During wintertime the energy supplied from the collectors satisfy a large amount of the heating needs of the house. The maximum heating auxiliary is required during January and is equal to 1.1 GJ. The only other months requiring heating auxiliary is February (0.61 GJ) and December (0.54 GJ). In all other months requiring heating load (March and November) this is satisfied completely by the solar system. The annual heating solar contribution as shown in Figure 20 is 0.81 and the lowest value is in January and is equal to 0.75. Hot water is almost completely covered from solar energy. As can be seen from Figure 20 both the monthly and the annual values are close to 1.

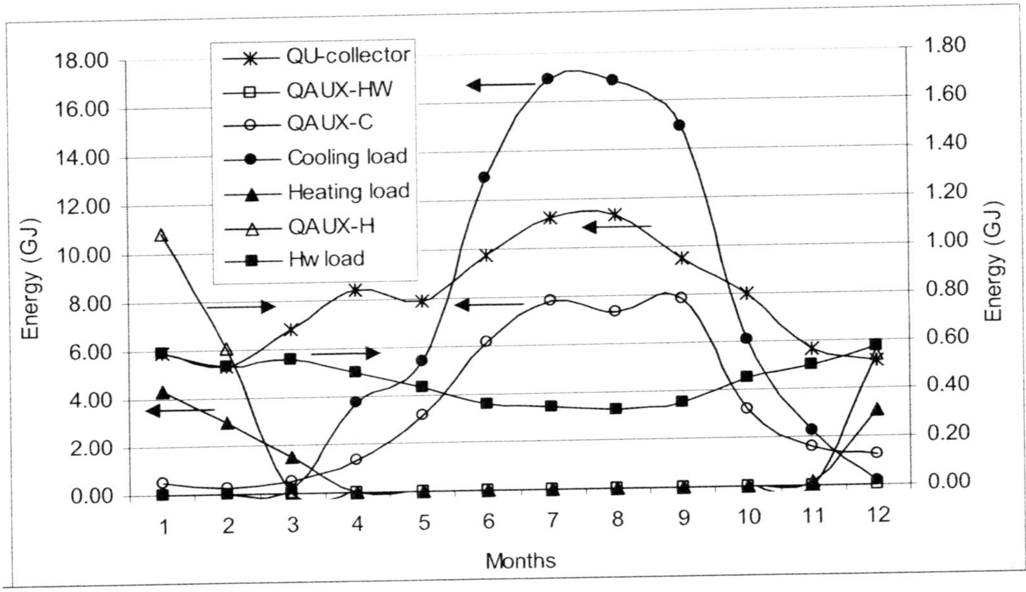

Figure 19. Energy flows of the ETC system installed in Nicosia

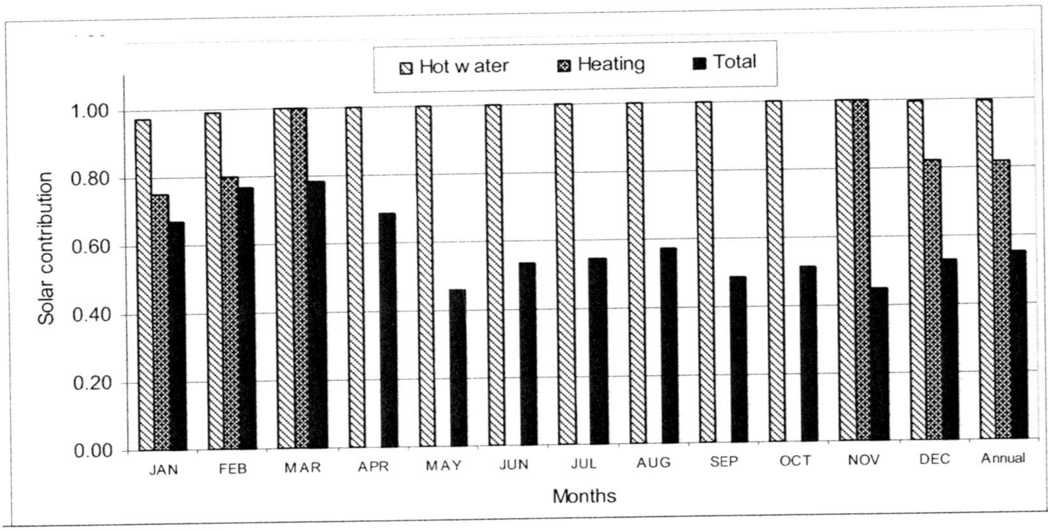

Figure 20. Solar contribution of the ETC system installed in Nicosia

The annual total contribution is equal to 0.55, the maximum value is in March and is equal to 0.78 and the minimum is in May and is equal to 0.46.

Optimum System for Madison

The performance of the advanced flat-plate collector system (optimum system) for Madison is presented in Figures 21 and 22. Figure 21 shows the energy flows and Figure 22 the solar contributions. As can be seen the useful energy supplied from the collectors is maximum during March and is equal to 10.2 GJ. In Madison the useful energy supplied form the collectors is more evenly distributed in the various months of the year than in Nicosia. As it is obvious, the predominant load in Madison is the heating load. This is maximised in January and reaches a value of 33.5 GJ. The auxiliary gets also its maximum value during winter time. The heating auxiliary is also maximised during January and is equal to 29.9 GJ. This high value of auxiliary leads to a low contribution of 0.11 as shown in Figure 22. The highest value of heating contribution is in September where the solar system covers completely the heating load. During September the solar system covers also completely the hot water requirements. For the hot water contribution this is also true for the months of June, July, August and September, whereas in April and May the contribution is almost 100%. The lowest value of solar contribution for hot water production is during January and is equal to 0.75 (maximum auxiliary equal to 0.6 GJ) whereas the annual value is equal to 0.91, i.e., 91% of the requirements for hot water are satisfied with solar energy. Cooling load is required in Madison only during the four summer months. The maximum value is during July and is equal to about 5.5 GJ. The cooling auxiliary also is maximised during July and is equal to 4.5 GJ.

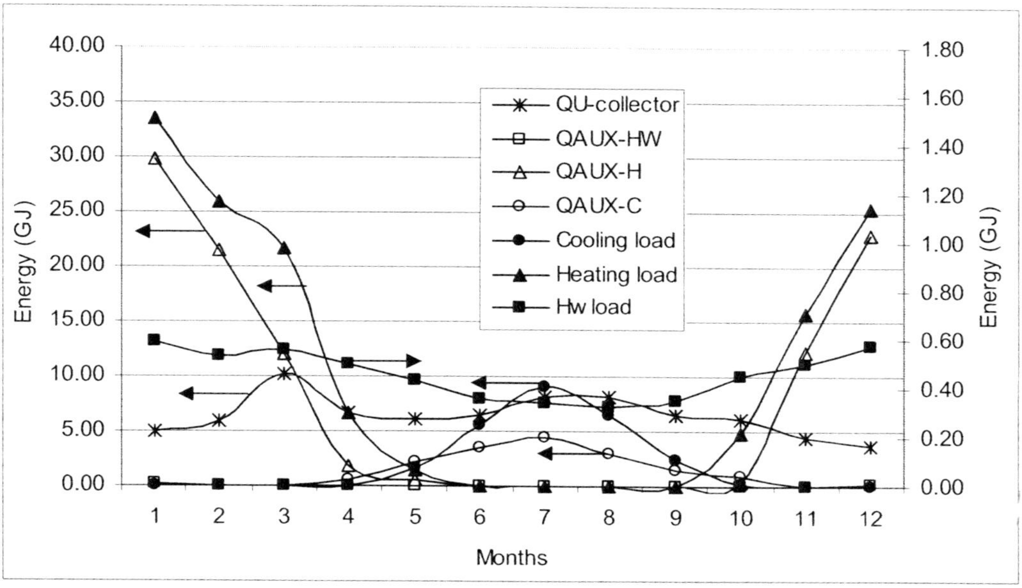

Figure 21. Energy flows of the AFP system installed in Madison

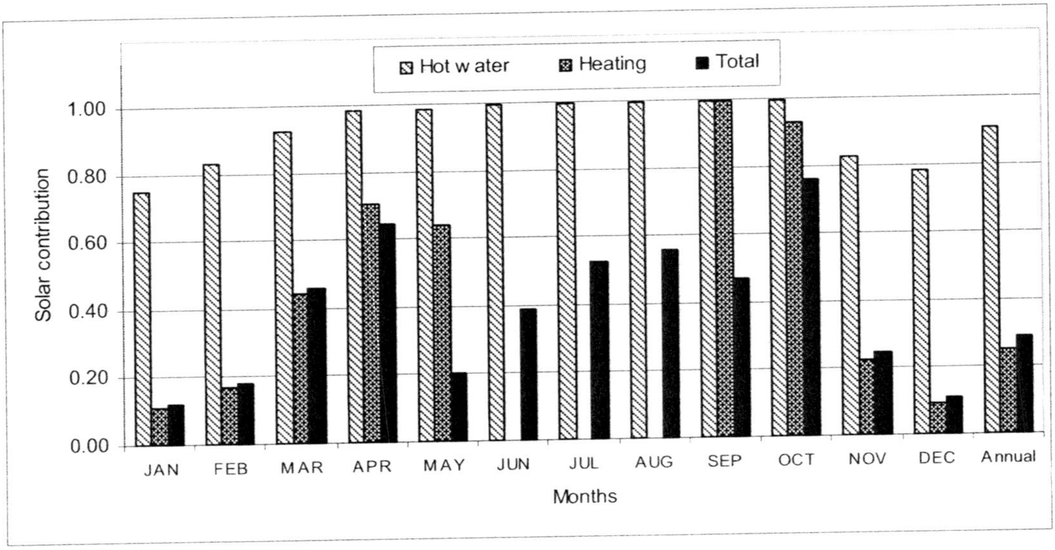

Figure 22. Solar contribution of the AFP system installed in Madison

The annual total contribution is equal to 0.29, the maximum value is in April and is equal to 0.65 and the minimum is in December and is equal to 0.11.

Effect of Heating System Supply Load Temperature

The importance of the heating system supply load temperature was emphasised in previous sections. In this section this is demonstrated by simulating the two optimum systems at higher load supply temperatures. The results are shown in Table 14. As can be seen the increase of the load supply temperature in both cases results in much higher need for auxiliary energy with a consequent drop of the solar contribution. For Nicosia, by increasing the heating system operating temperature from 46°C to 80°C, the heating system auxiliary increases from 2.23 GJ to 7.17 GJ, i.e., there is an increase of about 320% and the heating solar contribution dropped from 81.43% to about half (40.35%). However the effect of this increase on the total auxiliary energy required by the system and on the total solar contribution is not that much as the predominant load in Nicosia is the cooling load. For Madison, for the same operating temperature increase, the heating system auxiliary increased from 101.4 GJ to 128 GJ (increase 123%) and the heating solar contribution dropped from 25.02% to about a fifth (5.28%). In Madison the effect is less pronounced, compared to that expected, because the heating load is very large and most of it is supplied by auxiliary even at the low load operation temperature of 46°C. Both the total system auxiliary energy and the total system solar contribution are affected more than in the case of Nicosia as the heating load is the predominant load. Generally, the heating supply temperature has an insignificant effect on the cooling auxiliary energy. The examined temperature affects the mean storage temperature and thus the useful energy supplied from the collectors. When the storage temperature remains at a higher value the temperature entering the collector is higher thus the collector efficiency is lower. As can be seen the effect is more pronounced in the case of AFP collector which has poorer performance characteristics compared to those of the ETC.

Finally the economics of the systems operating at higher temperatures are examined. As can be seen from the results presented in Table 14, as the same size solar system is used in all cases (same initial cost), which replace a lower quantity of conventional fuel when the load temperature increases, the LCS drop considerably with the increase of load temperature and become even negative in the case of Madison.

Table 14. Effect of heating system load supply temperature

Location	Temperature (°C)	Qu (GJ)	Q_{aux-HW} (GJ)	$f_{(HW)}$ (%)	Q_{aux-H} (GJ)	$f_{(H)}$ (%)	Q_{aux-C} (GJ)	T_{aux} (GJ)	$f_{(total)}$ (%)	LCS (€)
Nicosia	46	95.20	0.025	99.55	2.23	81.43	40.97	43.23	55.40	5107
	60	94.72	0.0131	99.76	3.83	68.11	40.92	44.77	53.82	4722
	80	93.52	0.0131	99.76	7.17	40.35	40.70	47.88	50.60	3943
Madison	46	77.39	0.504	90.88	101.4	25.02	16.45	118.3	28.82	2093
	60	72.01	0.0455	99.18	113.5	16.06	16.37	129.9	21.85	-807
	80	64.59	0.0103	99.81	128.0	5.28	15.58	143.6	13.58	-4228

For Nicosia: A=30m^2, V=1.5m^3, Slope=30°, Flow rate=36 kg/hr-m^2, Collector=ETC
For Madison: A=50m^2, V=1.5m^3, Slope=43°, Flow rate=27 kg/hr-m^2, Collector=AFP

It can therefore be concluded that in order to be effective a solar space heating needs to be operated at low load supply temperature. For this purpose a water to air system is required as described in this chapter.

CONCLUSIONS

The design and performance of a complete solar space heating and cooling and hot water production system is presented in this chapter. Solar cooling offers possibilities of using the same solar collectors for space cooling as well through an absorption chiller. To be viable solar space heating systems need to utilize a low temperature supply system. Absorption units on the other hand require temperature of the order of 90°C but the system during summertime is more effective and the available solar radiation is much more than the amount available during the wintertime. Additionally any effort in using solar energy for space heating and cooling must be done after the building is well insulated in order to reduce the thermal losses. Such a system need to be optimized with respect to the area of solar collectors and their inclination, the size of the storage tank and the operating temperature during summer and winter. A system suitable for a typical 196 m^2 house is modelled and simulated with TRNSYS program for two locations at different latitudes; Nicosia, Cyprus (35°) and Madison, Wisconsin (43°). These two locations offer two different climates; i.e., mild winters and severe summers for Nicosia and severe winters and mild summers for Madison. The results of these simulations show that a solar system employing evacuated tube collectors is viable for Nicosia. For Madison the corresponding system employs advanced flat plate collectors. Both systems cover a considerable percentage of the heating, cooling and hot water loads of the building. The annual solar contribution is about 55% for the system in Nicosia and about 29% for the system in Madison. The system in Nicosia covers completely the hot water load and about 81% of the heating load. The system in Madison covers about 90% of the hot water load and about 25% of the heating load. The economic analysis performed showed that both systems are viable as positive life cycle savings are obtained (5100 Euro for the system in Nicosia and 1340 Euro for the system in Madison). It can therefore be concluded that solar energy should be used whenever possible as the environmental pollution caused by the burning of the fossil fuels should never be underestimated.

ACKNOWLEDGEMENTS

Figures 1 to 10 were reprinted from Kalogirou S., Solar Thermal Collectors and Applications, *Progress in Energy and Combustion Science*, Vol. 30, No. 3, pp. 231-295, copyright 2004, with permission from Elsevier.

REFERENCES

Ameel T.A., Gee K.G. and Wood B.D., 1995. Performance Predictions of Alternative, Low Cost Absorbents For Open-Cycle Absorption Solar Cooling, *Solar Energy*. Vol. 54, No. 2, pp. 65-73.

Anderson B., 1977. *Solar Energy: Fundamentals in Building Design*. McGraw-Hill Book Company.

ASHRAE, 1995. *Handbook of HVAC Applications*. Chapter 30. Atlanta.

ASHRAE, 1989. *Handbook of Fundamentals*. Atlanta.

Benz N., Hasler W., Hetfleish J., Tratzky S. and Klein B., 1998. Flat-Plate Solar Collector with Glass TI. *Proceedings of Eurosun'98 Conference on CD-ROM*, Portoroz, Slovenia.

Dorgan C.B., Leight S.P. and Dorgan C.E., 1995. Application Guide for Absorption Cooling/Refrigeration Using Recovered Heat. *American Society of Heating, Refrigerating and Air Conditioning Engineers, Inc.*

Duffie J.A. and Beckman W.A., 1991. *Solar Engineering of Thermal Processes*, John Willey & Sons, New York.

Erhard A. and Hahne E., 1997. Test and Simulation of a Solar-Powered Absorption Cooling Machine, *Solar Energy*, Vol. 59, No. 4-6, pp. 155-162.

Florides G., Kalogirou S., Tassou S. and Wrobel L., 2000. Modeling of the Modern Houses of Cyprus and Energy Consumption Analysis. *Energy-The International Journal,* Vol. 25, No. 10, pp. 915-937.

Florides G., Kalogirou S., Tassou S. and Wrobel L., 2002a. Modelling and Simulation of an Absorption Solar Cooling System for Cyprus, *Solar Energy*. Vol. 72, No. 1, pp. 43-51.

Florides G., Tassou S., Kalogirou S. and Wrobel L., 2002b. Review of Solar and Low Energy Cooling Technologies for Buildings, *Renewable and Sustainable Energy Reviews*. Vol. 6, No. 6, pp. 557-572.

Florides G., Kalogirou S., Tassou S. and Wrobel L., 2002c. Modelling, Simulation and Warming Impact Assessment of a Domestic-Size Absorption Solar Cooling System, *Applied Thermal Engineering*, Vol. 22, No. 12, pp. 1313-1326.

Florides G., Kalogirou S., Tassou S. and Wrobel L., 2003. Design and Construction of a Lithium Bromide-Water Absorption Machine, *Energy Conversion and Management*. Vol. 44, No. 15, pp. 2483-2508.

Ghaddar N.K., Shihab M. and Bdeir F., 1997. Modelling and Simulation of Solar Absorption System Performance in Beirut. *Renewable Energy*, Vol. 10, No. 4, pp. 539-558.

Grass C., Benz N., Hacker Z. and Timinger A., 2000. Tube Collector with Integrated Tracking Parabolic Concentrator, *Proceedings of the Eurosun'2000 Conference on CD-ROM*, Copenhagen, Denmark.

Hahne E., 1996. Solar Heating and Cooling, *Proceedings of Eurosun'96*, Freiburg, Germany. Vol. 1, pp. 3-19.

Hammad M. and Zurigat Y., 1998. Performance of a Second Generation Solar Cooling Unit, *Solar Energy*, Vol. 62, No. 2, pp. 79-84.

Hammad M.A. and Audi M.S., 1992. Performance of a Solar LiBr-Water Absorption Refrigeration System. *Renewable Energy*, Vol. 2, No. 3, pp. 275-282.

Haim I., Grossman G. and Shavit A., 1992. Simulation and Analysis of Open Cycle Absorption Systems for Solar Cooling, *Solar Energy*, Vol. 49, No. 6, pp. 515-534.

Hawlader M.N.A., Noval K.S. and Wood B.D., 1993. Unglazed Collector/Regenerator Performance for a Solar Assisted Open Cycle Absorption Cooling System, *Solar Energy*, Vol. 50, No. 1, pp. 59-73.

Hsieh J.S., 1986. *Solar Energy Engineering*, Prentice-Hall Inc. New Jersey.

Kalogirou S., 1996. Economic Analysis of Solar Energy Systems Using Spreadsheets, *Proceedings of the World Renewable Energy Congress IV*, Denver, Colorado, USA, Vol. 2, pp. 1303-1307.

Kalogirou S., 2003. Generation of Typical Meteorological Year (TMY-2) for Nicosia, Cyprus, *Renewable Energy*, Vol. 28, No. 15, pp. 2317-2334.

Kalogirou S., 2004. Solar Thermal Collectors and Applications, *Progress in Energy and Combustion Science*, Vol. 30, No. 3, pp. 231-295.

Keith E.H., 1995. Design Challenges in Absorption Chillers. *Mechanical Engineering – CIME*. Vol. 117, No. 10, pp. 80-84.

Klein S.A., Beckman W.A. and Duffie J.A., 1976. A Design Procedure for Solar Heating Systems, *Solar Energy*, Vol. 18, pp. 113-127.

Kreith F. and Kreider J.F., 1978. *Principles of Solar Engineering*. McGraw-Hill Book Company.

Konttinen P., Lund P.D. and Kilpi R.J., 2003. Mechanically Manufactured Selective Solar Absorber Surfaces, *Solar Energy Materials and Solar Cells*, Vol. 79, No. 3, pp. 273-283.

Lin Q. and Furbo S., 1998. Solar Heating Systems with Evacuated Tubular Solar Collector, *Proceedings of the Eurosun'98 Conference on CD-ROM*, Portoroz, Slovenia.

Morrison G. and Wood B., 1999. Packaged Solar Water Heating Technology: Twenty Years of Progress. *Proceedings of ISES Solar World Congress on CD-ROM*, Jerusalem, Israel.

Morrison G.L., 2001. Solar Collectors, In Gordon J. (Ed.) *Solar Energy: The state of the art*, pp. 145-221, ISES.

Orel Z.C., Gunde M.K. and Hutchins M.G., 2002. Spectrally Selective Solar Absorbers in Different Non-Black Colours. *Proceedings of WREC VII, Cologne on CD-ROM*.

Schweiger H., 1997. *Optimisation of Solar Thermal Absorber Elements with Transparent Insulation*. Thesis, Universitat Politecnica de Catalunya, Terrassa (Barcelona), Spain.

Spate F., Hafner B. and Schwarzer K., 1999. A System for Solar Process Heat for Decentralised Applications in Developing Countries, *Proceedings of ISES Solar World Congress,* Jerusalem, Israel on CD-ROM.

Tripanagnostopoulos Y., Souliotis M. and Nousia Th., 2000. Solar Collectors with Colored Absorbers. *Solar Energy*, Vol. 68, pp. 343-356.

TRNSYS program Manual, 2004. *Solar Energy Laboratory*, University of Wisconsin, Madison, USA.

Wackelgard E., Niklasson G.A. and Granqvist C.G., 2001. Selective Solar-Absorbing Coatings, In Gordon J (Ed.) *Solar Energy: The state of the art*, pp. 109-144, ISES.

Wazwaz J., Salmi H. and Hallak R., 2002. Solar Thermal Performance of a Nickel-Pigmented Aluminium Oxide Selective Absorber. *Renewable Energy*, Vol. 27, pp. 277-292.

Winston R., O'Gallagher J., Muschaweck J., Mahoney A. and Dudley V., 1999. Comparison of Predicted and Measured Performance of an Integrated Compound Parabolic Concentrator (ICPC), *Proceedings of ISES Solar World Congress on CD-ROM*, Jerusalem, Israel.

Zinian H.E. and Ning Z., 1999. A Solar Absorption Air-Conditioning Plant Using Heat-Pipe Evacuated Tubular Collectors, *Proceedings of ISES Solar World Congress on CD-ROM*, Jerusalem, Israel.

In: Solar Energy: New Research
Editor: Tom P. Hough, pp.149-198
ISBN 1-59454-630-4

Chapter 5

SOLAR THERMAL POWER GENERATION ON MARS

*Viorel Badescu**
Candida Oancea Institute, Faculty of Mechanical Engineering,
Polytechnic University of Bucharest, Buchares, Romania.

ABSTRACT

A "dynamic" solar power plant (which consists of a solar collector - thermal engine combination) is proposed as an alternative for the more usual photovoltaic cells. Upper bounds for the efficiency of solar thermal power plants operating in the Martian environment are first evaluated. A general thermodynamic approach, first presented here, clearly shows which of the three theories usually quoted in literature gives the exergy of thermal radiation. Recent works reporting accurate upper bounds for the efficiency of thermal radiation energy conversion into work are subsequently used in this chapter. The results refer to thermal engines powered by direct or diffuse solar radiation on Mars. Diffuse solar radiation is modeled as diluted or multiply scattered thermal radiation. A more elaborated model uses an endoreversible Carnot cycle to describe solar engine operation. Two strategies to collect solar radiation are analyzed: a solar horizontal collector and a solar collector whose tilt and orientation are continuously adjusted to keep the receiving surface perpendicular on Sun rays. Meteorological data measured at Viking Landers (VL) sites are used in computations. Results show that generally the influence of latitude on performance is important. In some situations the meteorological effects compensate the latitudinal effects and the output power is quite similar at both VL1 and VL2 sites. During a winter dust-storm day the maximum output power is much smaller than during autumn. High efficiency thermal engines should be used in combination with solar collectors kept perpendicular on Sun's rays. When a horizontal solar collector is considered, the dependence of the maximum output power on optimum solar efficiency seems to be quadratic at both VL1 and VL2 sites. When a collector perpendicular on Sun's rays is considered, this dependence is more complicate, but keeps the quadratic

---

* Candida Oancea Institute, Faculty of Mechanical Engineering, Polytechnic University of Bucharest, Spl. Independentei 313, Bucharest 79590, Romania. Phone: ++40.21.402.9428, Fax: ++40.21.410.4251, email: badescu@theta.termo.pub.ro

feature. No obvious difference exists between power plant performances in the two years of VL2 operation. A solar Stirling engine based on a horizontal selective flat-plate converter is analyzed in the last part of this chapter. All the computations were performed for a solar collection area similar in size with that of Mars Pathfinder's Sojourner. The solar efficiency at noon is as high as 18 %. The power provided by the engine is as high as 16 W during autumn and winter. These results suggest that under the Martian environment the performance of properly designed solar Stirling engines is comparable with that of PV cell power systems.

1. Introduction

Mars is a key objective of space exploration. Starting in the 1960s, different missions took place on the surface of the Red Planet with various scientific goals as the determination of the internal structure of the planet, the chemical and mineralogical analyses of Martian rocks and soils and the study of atmospheric circulation and weather patterns. Future missions include both robotic and manned expeditions. They require the availability of power generation units utilising chemical energy, radio-isotopes, solar radiation or nuclear fission as the primary energy source (Aldrich,1992; Angelino and Invernizzi,1993). The utilisation of systems, which use internally stored nuclear and chemical energy, implies in many cases limited life and excessive weight. Properly designed solar power systems can offer long life in space.

During Mars spacecraft exploration the photovoltaic (PV) cells were almost exclusively envisaged as the main source of energy. They were proposed to power small rovers (Hibbs, 1989), long - endurance, remotely piloted aircraft capable of flight within the Martian environment (Collozza, 1990) and to ensure the life support for a 40 day manned Mars surface scientific expedition (McKissock et al, 1990). More recently, the Sojourner of the Pathfinder mission was powered by PV cells (see e.g. Golombek et al, 1997).

However, for many years researchers realized that in some specific situations solar dynamic power systems with thermal energy storage can provide significant savings in life cycle costs and launch mass when compared with conventional photovoltaic power systems with battery storage (Menetrey (1963), Secunde et al (1989), Prisnjakov (1991), Prisnjakov et al (1991)). A standard solar dynamic power system uses a mirror to concentrate solar radiation onto an absorber structure. By conduction through a solid material or circulation of a working fluid, the absorber heat is transferred to a thermal engine (i.e. a turbogenerator, Stirling engine, termocouple or thermionic-emitter). Alternators coupled to these thermal engines may generate electrical energy. Previous practice proved that three different cycles can be used for thermodynamic conversion of solar radiation: Brayton, Rankine and Stirling (Menetrey (1963), Prisnjakov (1991), Prisnjakov et al (1991)). For continuos operation during dark periods the use of melted materials to store thermal energy is being considered. Among the advantages of dynamic power systems one could quote their ability to provide electrical energy and heat simultaneously, the fact that the power plant may be unified by using either solar or nuclear energy or their relative invulnerability to corpuscular particles and to electromagnetic radiation and the possibility of power control according to a given power consumption schedule (Prisnjakov, 1991).

Earlier studies reported solar dynamic systems based on thermodynamic cycles and thermionic converters, respectively. Two systems of the former type were described in Menetrey (1963). One of them is a 3 kW solar dynamic system of 10 % overall efficiency consisting of a parabolic concentrator and a turbo-generator unit which uses mercury as working fluid in a Rankine cycle. The other one is a 15 kW system for space vehicle applications. Its working fluid is rubidium while the overall electrical efficiency was 21.7 %. The solar - thermionic system was designed to provide 135 Watts at Mars distance from the Sun with a system weighting about 14 kg. More recent Russian studies are using permanent gases (mainly mixtures of xenon and helium) as working fluid in Brayton cycles. They have output power of 3 to 5 kW, specific weights up to 40-45 kg/kW and overall efficiency of 31.2 % (Prisnjakov (1991), Prisnjakov et al (1991), Prisnjakov et al (1994)). Significant efforts to develop solar dynamic power systems (25 kW) for application to a low Earth orbit space station have been conducted by NASA Lewis Research Center (LeRC) as an integral part of the US space station development program (McLallian et al, 1988). The culmination of the LeRC programme demonstration program was a full system test in a thermal/vacuum tank of a Space Station Freedom type solar dynamic system at the 2 kWe size in February 1995. Related reports can be found in (Secunde et al (1989), Weingartner et al (1994)). A comparison between the various power systems designed for the Space Station Freedom shows the sun-to-user efficiency is 5.7 to 6.8 % for the PV system and 18.7 to 19.8 % for the dynamic system. If, in addition, the storage subsystem is taken into consideration, the solar dynamic system becomes even more attractive, mainly because the efficiency of the latent heat storage is higher compared to the Ni-Cd batteries of the pV system (for details see Weingartner et al (1994)).

A natural question arising about power systems to be used on Mars surface is: are properly designed dynamic systems comparable in performance with the PV – based power systems? Some of the above references seem to indicate an affirmative answer. However, they all refer to systems operating in the interplanetary space. A solar power system placed on Mars surface has to take into consideration one very important feature of the Martian weather. The period between the areocentric longitude L_s=161 - 326° and variants thereof have been referred to as the "dust storm season". This period is nearly centered on perihelion, which is the time of maximum insolation on Mars (Martin and Zurek (1993), Hourdin et al (1995)). In the northern hemisphere, the Martian atmosphere varies greatly from year to year during the dust storm season while at the other seasons, most phenomena are remarkably similar from year to year. Zurek et al (1992) provide a comprehensive description of Martian meteorology. The dust storm periods are characterized by an almost vanishing direct solar radiation (see e.g. Badescu (1998a) and references therein). The main consequence is that isotropic diffuse solar radiation can not be concentrated and flat plate solar collectors have to be used in this case.

The structure of this chapter is as follows. In the next section one gives details on the meteorological data measured at Viking Landers (VL) sites that were subsequently used in computations. Upper bounds for the efficiency of solar thermal power plants operating in the Martian environment are evaluated in section 3. A model of endoreversible Carnot solar power plant is presented in section 4 while section 5 refers to solar Stirling engines based on horizontal selective flat-plate converters. These models provide lower upper limits for the performance of actual thermal solar engines. Section 6 contains the conclusions.

2. Meteorologic and Actinometric Data on Mars Surface

The atmospheric (vertical) optical depth τ is an indicator of solar radiation attenuation in the Martian atmosphere. One can shortly define τ as follows. Let us denote by I_{top} and I_{ground} the direct solar irradiance at the top of the atmosphere and at ground level, respectively. Then, usage of τ allows writing the Beer's law as $I_{top} / I_{ground} = e^{-\tau/\cos\theta_0}$ where θ_0 is the zenith angle (i.e. the angle between the direction of solar direct radiation and the normal to a horizontal surface). During the Viking mission to Mars, cameras on the two Viking Landers used a Sun diode to measure the direct solar irradiance I_{ground}. Since both I_{top} and the atmospheric path are known when the Sun is viewed, application of the Beer's law provides directly the vertical column visible dust opacity of the atmosphere above the Viking Lander sites (see Pollack et al (1990), Badescu (2001) and references therein). Mars may be considered "clear" when the dust content in the atmosphere is low, but when local or global storms occurs the optical depth increases and the direct beam solar radiation decreases drastically.

During the next sections we shall use as input data values of ambient pressure and temperature, wind speed and atmospheric optical depth measured at Viking Lander 1 (VL1) (22.3° N, 47.9° W) and Viking Lander 2 (VL2) (47.7° N, 225.7° W) sites. The data were taken from the Viking Lander Meteorology and Atmospheric Opacity Data Set Archive from the Planetary Data System database made publicly available on a compact disc (Lee, 1995). Three electronic files were used here, namely VL_MBIN.DAT and VL_TBIN.DAT (which contain values of atmospheric pressure and temperature and zonal and meridional wind speed, among others) and VL_OPAC.DAT (which contains atmospheric optical depth data).

The file VL_MBIN.DAT (8325 records) contains binned and splined data obtained from the Viking Meteorology Instrument System (VMIS) through portions of the Viking Lander 1 mission. The file VL_TBIN.DAT (25750 records) contains binned and splined data obtained from the VMIS through most of the Viking Lander 2 mission and the early days of the Lander 1 mission. In both cases the numbering of the Martian solar days (sols) started at day 0 when each Lander touched down. Both files VL_MBIN.DAT and VL_TBIN.DAT contain mean values of meteorological parameters in each bin. The bin number starts with 0 at local midnight and increments by 1 each 1/25 of a sol. The file VL_OPAC.DAT (1044 records) stores measurements of atmospheric optical depth and associated errors estimates. Each record contains the optical depth value and its associated local solar time (in Earth hours) beginning at midnight.

As the files VL_MBIN.DAT and VL_TBIN.DAT are more "dense" than the VL_OPAC.DAT file, a time interpolation procedure was used to obtain the values of the atmospheric pressure and temperature and wind velocity associated to a given record of atmospheric optical depth (for details see Badescu (1998b)). The atmospheric pressure values are missing for eleven records. In these cases the following procedure was adopted:

i) in case of the record VL2, year 1, areocentric longitude = 229.808 (autumn), sol = 200, hour = 11.41, we used the arithmetic mean of the pressure values corresponding to sol = 199, hour = 13.87 and sol = 201, hour = 11.42, respectively.

ii) in case of the record VL2, year 2, areocentric longitude = 19.228 (spring), sol = 454, hour = 9.62, we used the arithmetic mean of the pressure values corresponding to sol = 453, hour = 17.33 and sol = 454, hour = 17.25, respectively.
iii) in case of VL2, year 2, the first nine records during summer (areocentric longitudes between 93.363 and 100.510) we used the pressure value corresponding to the tenth record (areocentric longitude = 101.697, sol = 634, hour = 6.98).

A set of *complete* records was finally obtained. Each complete record associates to a given sol number and a local solar time value a set of four meteorological values, namely optical depth τ and atmospheric temperature T_a, atmospheric pressure p_a and wind speed w. The number of complete records is rather small (Table 2.1).

Table 2.1. The number of *complete* meteorological data records available at VL1 and VL2 sites

		VikingLander 1		VikingLander 2	
		Sol numbers	Number of complete Records	Sol numbers	Number of complete records
Year 1	Spring	-	-	-	-
	Summer	6 to 155	22	6 to 111	37
	Autumn	168 to 306	108	120 to 257	77
	Winter	307 to 350	56	269 to 415	55
Year 2	Spring	-	-	416 to 608	147
	Summer	-	-	610 to 784	58
	Autumn	-	-	792 to 872	25
	Winter	-	-	-	-

The optical depth values were used to evaluate the incoming fluxes of solar radiation. The resulting file of solar energy data was used as input during the simulation of solar heat engines operation. The three components of global radiation flux (namely direct, diffuse and ground reflected) were computed by using the procedure we previously proposed in Badescu (1998a). The incoming fluxes of solar radiation were evaluated for all the complete records of Table 2.1.

A few basic facts about Martian meteorology are now summarized for reader convenience. The near-surface atmospheric temperature at Viking Landers latitude ranges from about -17°C during summer to –107° C in winter. Also, the daily average atmospheric pressure ranges between about 7 and 10 hPa, with a minimum near areocentric longitude L_s=150° and a maximum around L_s=300° (Badescu, 2001). The solar constant at the level of Mars orbit is 590 W/m^2. The solar global irradiance on a horizontal surface at ground level could be as high as 400 W/m^2 at the noon of a clear summer day and as low as 80 W/m^2 at midday during a winter dust storm (Badescu, 1998a).

3. UPPER LIMITS FOR SOLAR ENERGY CONVERSION EFFICIENCY INTO WORK

The purpose of this section is to give information about the upper limits of the maximum efficiency of converting solar energy into mechanical work on the surface of Mars. This subject was already treated in a number of papers in case of solar energy conversion systems placed on Earth surface. Changing the conversion system on Mars does not require a different theoretical framework. However, the characteristics of Mars meteorology is expected to have considerable influence on the numerical results.

The upper bounds for the performance presented in this section are obtained by using very idealized models. They have however the advantage of giving a perspective for the results obtained with the more realistic models developed in the next sections.

Exergy of Solar Radiation and Simple Upper Bounds for Conversion Efficiency

The exergy of thermal radiation was studied in many papers, starting with the seminal work of Petela (1964). Controversial arguments were presented in literature (see e.g. Bejan (1988) for an early review). Recently, the discussion was re-opened in relation with the exergy of solar radiation (Petela (2003); Candau (2003)). Also, many authors studied the thermodynamics of converting radiation energy into mechanical work (for early reviews see Landsberg and Tonge (1979), Bejan (1988)). Three different theories were proposed by Jeter (1981), by Landsberg, Petela and Press (Petela (1961), Landsberg and Mallison (1976), Press (1976)) and by Spanner (1964), respectively. The three theories were unified by Bejan (1987) while a generalization of Bejan's approach was given in Badescu (1988a).

Computing the maximum work obtained by converting the radiation energy, on one hand, and the radiation exergy, on the other one hand, are two different things as also noted in Petela (2003). Usually, in the first case there is a given configuration of the conversion system and one try to improve the design and operation of that configuration. However, the exergy represents the maximum quantity of work that can be produced in some given environment. Therefore, computing the exergy implies a high degree of idealization and is usually associated with the mental construction of a "reversible conversion system". Two types of difficulties could occur in this last case. The first difficulty appears when treating systems producing power (i.e. work rate) and is finally related to the question whether a Carnot (reversible) engine is allowed to provide a non-null work rate. A second difficulty appears when treating particular irreversible processes whose reversible ideal limit is difficult to imagine as a non-null flux process. Absorption/emission of radiation is such a couple of processes as the associated entropy generation is strictly positive as far as the temperature of the absorber/emitter is different from the radiation source temperature (see e.g. Bejan (1988) p. 483-487). There are a few ways of approaching these difficulties and two are reminded here. The traditional approach assumes that (i) Carnot (reversible) engines are allowed to provide non-null power and (ii) any irreversible (non-null flux) process is allowed to have its own reversible ideal limit. A second, more recent approach assumes (i) but rejects (ii) above and chooses to include the irreversibilities in the treatment as they are, as far as they occur at the boundary of the conversion system. This is the so called endoreversible approach and its main scope is to minimize the effect of these *existing* irreversibilities (for example by the minimization of the entropy generation). Both approaches were used in the past when treating thermal radiation energy conversion into work.

Now, we derive the exergy of thermal radiation from a simple, general, original thermodynamic argument. Let us consider a system Σ containing a radiation absorber coupled to a heat engine. The temperature outside the system (i.e. the ambient temperature) is T_0. Thermal radiation (carrying properties, like energy and entropy) enters and leaves the system. The absorber converts part of the energy of the incident radiation into heat that is transferred to the engine producing work. The other part of the incident radiation is re-radiated by the absorber at its own temperature. Work and heat (transferred by other mechanisms than radiation, e.g. by conduction and/or convection) leave the system Σ. An appropriate temperature T_s and an energy flux φ_s are associated to the radiation entering the system. The work rate leaving the system is $\dot{W}$. One defines the efficiency of converting the energy of the radiation entering the system into work as $\eta \equiv \dot{W} / \varphi_s$. One asks about the exergy flux contained in the radiation energy flux φ_s and about the maximum estimated value of η.

Various (irreversible) processes take place inside Σ during the conversion of radiation energy into work. One denotes by $\dot{S}_{gen,i} (\geq 0)$ the entropy generation rate of process i. Of course, when the process i is treated (or, in other words, is modeled) as a reversible process the entropy generation vanishes (i.e. $\dot{S}_{gen,i} = 0$). The Gouy-Stodola theorem applied to the system Σ gives (Bejan (1988) p. 113-114)

$$\dot{W} = \dot{W}_{rev} - T_0 \sum_i \dot{S}_{gen,i} \tag{3.1}$$

where $\dot{W}_{rev}$ is the work rate leaving the system in case all processes inside Σ are treated as reversible. Use of Eq (3.1) and the efficiency definition yields:

$$\eta = \eta_{rev} - T_0 \frac{\sum_i \dot{S}_{gen,i}}{\varphi_s} \tag{3.2}$$

where obviously

$$\eta_{rev} \equiv \frac{\dot{W}_{rev}}{\varphi_s} = \eta_{Carnot} = 1 - \frac{T_0}{T_s} \tag{3.3}$$

When all the irreversibilities inside Σ are taken into account, from Eq (3.2) one could compute the real efficiency η of the thermal collector-heat engine combination. When part of the processes inside Σ are treated as reversible (i.e. their entropy generation rate $\dot{S}_{gen,i}$ is assumed to vanish), Eq (3.2) allows to obtain *upper bounds* for the real efficiency η. When all processes inside Σ are treated as reversible processes from Eqs (3.1) and (3.3) one finds the "highest" upper bound for the real efficiency η.

Thermal radiation exergy was evaluated in the past by both traditional and endoreversible approaches. This is well summarized in Figure 7 of Petela (2003) and briefly described below.

(a) When all processes are treated as reversible, there is no entropy generation inside Σ and Eqs (3.2) and (3.3) yield:

$$\eta_J = 1 - \frac{T_0}{T_s} > \eta \tag{3.4}$$

This is the efficiency proposed by Jeter (1981) and applies to Figure 7(1) of Petela (2003). Note that the above assumption implies using a reversible (i.e. Carnot) heat engine. Treating the absortion/emission in the limit case of a reversible process means of course that the absorber and the radiation source have strictly the same temperature. Then, the work rate $\dot{W}$ that leaves that system is given by:

$$\dot{W} = \dot{W}_{rev} = \varphi_s\left(1 - \frac{T_0}{T_s}\right) \tag{3.5}$$

and this is the exergy flux expression proposed by Jeter (1981).

(b) When one irreversibility, namely "filling" the system Σ with incident radiation, is taken into account, from Eq (3.2) one obtains (Bejan (1987), Badescu (1999), see also Section 4 of Petela (2003))

$$\eta_S = 1 - \frac{4}{3}\frac{T_0}{T_s} \tag{3.6}$$

This is the efficiency proposed by Spanner (1964) and keeping its value positive requires some constraints on temperatures (see Petela, 2003).

(c) When two irreversibilities are taken into account (namely "filling" and "emptying" the system Σ with/of radiation) from Eq (3.2) one finds (Bejan (1987), Badescu (1999), see also Section 4 of Petela (2003)):

$$\eta_{LPP} = 1 - \frac{4}{3}\frac{T_0}{T_s} + \frac{1}{3}\left(\frac{T_0}{T_s}\right)^4 > \eta \tag{3.7}$$

This is the Landsberg-Petela-Press efficiency and is related to Figure 7(3) of Petela (2003). The same result applies in case the absorber (or the cascade of absorbers, as a limit case adopted in Candau (2003)) operates at a lower temperature than the radiation source. This treats the absorption/emission as an irreversible process. If the remaining processes are treated as reversible, one shall find the "exergy" expression first proposed by Petela (1964), and later on by Badescu (1988b) (his equation (28)) and Candau (2003) (his equation (9)), among others.

The procedure above explains why Bejan (1987) concluded that the efficiencies η_J, η_S and η_{LPP} do not contradict each other. All these relationships are correct, as they are rigorously derived upper bounds for the real efficiency η (under some temperature limitations however in case of η_S). All of them are equally elegant, being expressed in a Carnot-like style, namely in terms of just the hot and cold heat reservoir temperatures. However:

i) One of the three efficiencies (i.e. η_J) is obtained in case all processes inside system Σ are treated as reversible. It predicts a performance similar to that of a Carnot engine and therefore, according to the traditional approach, it gives the exergy flux of thermal radiation (i.e. Eq (3.5)).

ii) One of the three efficiencies (i.e. η_{LPP}) is more accurate, because it considers a larger number of irreversibilities taken place inside Σ and, consequently, it estimates a *lower* upper limit for the maximum value of η.

Let us apply the above results in case of solar radiation on Mars. The temperature of solar direct radiation is $T_s = 5760\ K$ while the average temperature of Mars surface is $T_0 = 248\ K$. Then, the efficiency predicted by Jeter efficiency Eq (3.4) is 0.9569 and for a solar direct irradiance at ground level $\varphi_s = 400\ W/m^2$ the exergy flux given by Eq (3.5) is 382.76 W/m^2. Using Spanner efficiency Eq (3.6) and Eq (3.4) yield the efficiency 0.9426 and the work flux 377.04 W/m^2, respectively. Finally, using Landsberg-Petela-Press efficiency Eq (3.7) and Eq (3.4) yield again the efficiency 0.9426 and the work flux 377.04 W/m^2, respectively. Apart from the theoretical divergences these results are very close each other.

Accurate Upper Bounds for Solar Energy Conversion Efficiency

All theories in the previous subsection predict efficiencies of converting radiation energy into work too high to be of practical interest. Much more accurate rigorous upper bounds of η were reported in recent years (Badescu, 1998c, 1998d). They are consequences of taking account of additional processes and irreversibilities than those used to derive η_{LPP}. Some of these new results are listed below and applied to the special case of solar energy conversion on Mars surface.

(i) In case of a smaller than hemispherical blackbody radiation source and a blackbody receiver coupled to a reversible engine the upper bound of η is (Badescu, 1998c):

$$\eta_i = 1 - \frac{4}{3}\left(f\frac{T_0}{T_s}\right) + \frac{1}{3}\left(f\frac{T_0}{T_s}\right)^4 \quad > \eta \tag{3.8}$$

where the geometric factor

$$f \equiv \left(\frac{\Omega}{\pi}\right)^{-1/4}\left(1 - \frac{\Omega}{4\pi}\right)^{-1/4} \tag{3.9}$$

is a function of the solid angle Ω subtended by the radiation source. In case of a spherical source of radiation (e.g. the Sun) the following simple relation applies:

$$f = \sin^{-1/2}\delta \tag{3.10}$$

where δ is the half-angle of the cone subtended by the source. Of course, when a hemispherical source of radiation is considered, $\delta = \pi/2$ and η_i given by Eq (3.8) reduces to η_{LPP} given by Eq (3.7). However, generally $\eta_i < \eta_{LPP}$. One can see that η_i depends on just two temperatures (namely the temperatures of the heat reservoirs) as any Carnot-like efficiency. But it also takes account of the geometry of the radiation source and this additional information input increases its accuracy.

Let us use Eqs (3.8)-(3.10) to solar energy conversion on Mars. The case of direct solar radiation is considered. Then, δ is the half-angle of the cone subtended by the Sun when viewed from Mars. The mean distance Sun to Mars is $d_{Sun-Mars} = 227.94 \cdot 10^6\ km$ while Sun radius is $R_s = 696 \cdot 10^3\ km$. Consequently, $\delta = \sin^{-1}(R_s / d_{Sun-Mars}) = 0.003053$. Equation (3.10) yields $f = 18.096$. Use of Eq (3.8) gives the upper bound $\eta_i = 0.08393$. This is obviously a *more accurate* upper bound than those predicted in section 2.1.

(ii) In case of smaller than hemispherical blackbody radiation source and a blackbody receiver coupled to a Chambadal-Novikov-Curzon-Ahlborn (CNCA) engine the upper bound of η is (Badescu, 1998c)

$$\eta_{ii} = 1 - \frac{4}{3}\left(f\frac{T_0}{T_s}\right)^{1/2} + \frac{1}{3}\left(f\frac{T_0}{T_s}\right)^2 \quad > \eta \tag{3.11}$$

This is of course a more accurate estimate for η that η_i is, because part of the entropy generation related to heat engine operation is taken into account (one reminds that the CNCA engine is an endoreversible engine). As an application, using Eq (3.11) in case of solar direct radiation on Mars yields $\eta_{ii} = 0.02542$ which is lower than the value derived previously for η_i, as expected.

Note that a formula corresponding to η_{LPP}, in case the system Σ contains a CNCA engine, can be obtained from Eq (3.11) by making $f = 1$. One finds

$$\eta'_{ii} \equiv 1 - \frac{4}{3}\left(\frac{T_0}{T_s}\right)^{1/2} + \frac{1}{3}\left(\frac{T_0}{T_s}\right)^2 \tag{3.12}$$

Equation (3.12) is a more accurate estimate for η than η_{LPP} is, but a less accurate estimate than η_{ii} is. Using Eq (3.12) in case of solar direct radiation on Mars yields $\eta'_{ii} = 0.72395$. This is a rather high upper bound for the real efficiency and this loss in accuracy is the cost of using a simpler upper limit formula.

(iii) Repeating the procedure from Bejan (1987) under the framework of endoreversible thermodynamics one finds relationships generalizing Jeter, Spanner and Landsberg-Petela-Press efficiencies, respectively (Badescu, 1999). For example, the relation generalizing the last efficiency is:

$$\eta_{iii} \equiv 1 - \frac{4}{4-n}\left(\frac{T_0}{T_s}\right)^{n} + \frac{n}{4-n}\left(\frac{T_0}{T_s}\right)^4 \tag{3.13}$$

where n is a parameter related to the type of endoreversible thermal engine. In case of $n = 1$ (i.e. a Carnot engine) one finds of course from Eq (3.13) the efficiency η_{LPP}. Let us consider a CNCA engine ($n = 1/2$) as part of a power system converting the energy of direct solar energy on Mars. Then Eq (3.13) yields the upper limit $\eta_{iii} = 0.76286$. This rather high upper bound is comparable in size to η'_{ii}.

(iv). The case of gray thermal radiation is of special interest, as it is sometimes used as an approximation for diffuse solar radiation. Note that gray radiation is a particular case of diluted radiation. A general theory was first developed by Landsberg and co-workers (see for instance the review paper (Landsberg and Tonge, 1979)). However, important results were derived more recently and they will be summarized now. An accurate upper bound for the conversion efficiency of solar diluted radiation into work is given by (Badescu, 1990)

$$\eta_d \equiv 1 - \frac{4}{3}\left(f_d \frac{T_0}{T_s}\right) + \frac{1}{3}\left(f_d \frac{T_0}{T_s}\right)^4 \tag{3.14}$$

Here the dimensionless factor f_d can be obtained from Eqs (10) and (14) of (Badescu, 1990):

$$f_d \equiv \chi(\varepsilon_d)\left(\frac{\Omega_d}{\pi}\right)^{-1/4}\left(1-\frac{\Omega_d}{4\pi}\right)^{-1/4} \tag{3.15}$$

where Ω_d is the solid angle subtended by the source of diluted radiation source while $\chi(\varepsilon_d)$ is a function of the radiation dilution factor ε_d. This function was calculated in Landsberg and Tonge (1979). It is such that $\chi(1)=1$ and can be approximated for small ε_d by:

$$\chi(\varepsilon_d) = 0.9652 - 0.2777\ln\varepsilon_d + 0.0511\varepsilon_d \tag{3.16}$$

The above general formulation yields elegant results in the particular case of solar scattered radiation. The upper bound for the efficiency of converting into work the energy of multiply scattered solar radiation was derived in Badescu (1991) and is given by

$$\eta_{d,i} \equiv 1-\frac{4}{3}\left(f_{d,i}\frac{T_0}{T_s}\right)+\frac{1}{3}\left(f_{d,i}\frac{T_0}{T_s}\right)^4 \tag{3.17}$$

where the factor $f_{d,i}$ this time depends on the number i of scatterings (i=1,2,3,4). It is computed by

$$f_{d,i} \equiv \left(\frac{\Omega_{d,i}}{\pi}\right)^{-1/4}\left(1-\frac{\Omega_{d,i}}{4\pi}\right)^{-1/4}\prod_{j=1}^{i}\chi(\varepsilon_{d,i}) \qquad (i=1,2,3,4) \tag{3.18}$$

In case the scattered radiation is not subsequently concentrated, the following relations apply for the dilution factors and solid angles of i-scattered solar radiation

$$\varepsilon_{d,1} = \frac{\Omega_s}{\pi} \qquad \varepsilon_{d,i} = \frac{\Omega_s}{\pi}\left[\prod_{j=1}^{i-1}\chi^4(\varepsilon_j)\right] \qquad (i=2,3,4) \tag{3.19}$$

$$\Omega_{d,i} = 2\pi\left\{1-\left[1-\frac{\Omega_s}{\pi}\prod_{j=1}^{i}\chi^4(\varepsilon_{d,i})\right]^{\frac{1}{2}}\right\} \qquad (i=1,2,3,4) \tag{3.20}$$

where Ω_s is the solid angle subtended by the Sun. The half-angle $\delta_{d,i}$ of the cone subtended by the source of i-scattered radiation can be easily computed from:

$$\Omega_{d,i} = 2\pi(1-\cos\delta_{d,i}) \tag{3.21}$$

The effective temperature $T_{d,i}$ of i-scattered solar radiation is given by:

$$T_{d,i} = \frac{T_s}{\prod_{j=1}^{i} \chi(\varepsilon_{d,i})} \qquad (3.22)$$

Table 3.1 shows results obtained in case of i-scattered solar radiation on Mars. One can see that after a single scattering (i.e. i=1) solar radiation is still strongly anisotropic (compare $\delta_{d,1} = 3.061°$ with the half-angle $\delta_s = 0.1749°$ of the cone subtended by the Sun). After four scatterings solar radiation is isotropic and an observer at the ground would see a uniformly brilliant sky. The effective temperature of the scattered radiation decreases by increasing the number of scatterings, as expected. However, the upper bound efficiency $\eta_{d,i}$ predicted by Eq (3.10) yields the same value, whatever the number of scatterings. This is a consequence of keeping the energy flux constant during the scattering process.

Table 3.1. Upper bounds for the conversion efficiency of i-scattered solar radiation into work on Mars

Number i of scatterings	$\varepsilon_{d,i}$	$\Omega_{d,i}$	$\delta_{d,i}$ (deg)	$T_{d,i}$ (K)	$\eta_{d,i}$	$C_{\max,i}$	$\eta_{d,conc,i}$
1	0.000009 32	0.00897	3.06	1377.0	0.084	350.70	0.760
2	0.002851 00	0.41880	21.04	531.2	0.084	7.76	0.393
3	0.128900 00	2.99700	58.47	344.8	0.084	1.38	0.130
4	0.726500 00	6.28300	90.00	316.0	0.084	1.00	0.084

$\varepsilon_{d,i}$ - dilution factor. $\Omega_{d,i}$ and $\delta_{d,i}$ - solid angle and half-angle of the cone subtending the source of radiation, respectively. $T_{d,i}$ - effective temperature. $\eta_{d,i}$ - upper bound efficiency. $C_{\max,i}$ - maximum concentration ratio. $\eta_{d,conc,i}$ - upper bound efficiency for concentrated i-scattered radiation. Sun temperature $T_s = 5760\ K$; average temperature on Mars surface $T_0 = 248\ K$.

For small i-values scattered solar radiation is generally anisotropic. Consequently, it can be concentrated. The maximum concentration ratio $C_{\max,i}$ is given by:

$$C_{\max,i} = \left[\frac{\Omega_{d,i}}{\pi}\left(1 - \frac{\Omega_{d,i}}{4\pi}\right)\right]^{-1} \qquad (3.23)$$

The upper bound efficiency $\eta_{d,conc,i}$ for fully concentrated scattered radiation is given again by Eq (3.17) with $f_{d,i}$ replaced by $f'_{d,i}$ given by:

$$f'_{d,i} \equiv \prod_{j=1}^{i} \chi(\varepsilon_{d,i}) \quad (i = 1,2,3,4) \tag{3.24}$$

The upper bound efficiency $\eta_{d,conc,i}$ lies between 0.760 and 0.084 for single and four scatterings, respectively (Table 3.1). Generally, the effective temperature of scattered radiation is lower, and the upper bound efficiency are higher, on Mars than on Earth (compare present Table 3.1 with Table 1 of Badescu (1991)). This is mainly caused by the lower ambient temperature on Mars.

(v) The upper bounds at (i) to (iii) above apply to any source of blackbody undiluted or diluted radiation. Even more accurate (but also more complex) upper bounds of η were derived in case of solar energy applications (Badescu, 1998d, 2000). These upper bounds are again Carnot-like efficiencies in the sense that they do not involve the absorber temperature but are expressed just in terms of the two heat reservoir temperatures. However, they take into account a large number of details concerning the conversion process as for example, the dilution factors of solar and ambient radiation, the polarization degree of solar and absorber emitted radiation and the thermal and optical properties of the selective absorber (convective and conductive heat loss coefficients, concentration ratio, effective transmittance-absorptance product and emittances). These details increase of course the accuracy of the efficiency formula. As an example we shall use the accurate upper bound efficiency Eq (36) of Badescu (1998d). For the purposes of this chapter we shall simplify it to take account just on the selective properties of the absorber (i.e. its absorptance a and emittance e). Then, the upper bound efficiency is given by:

$$\eta_v = a\left[1 - \frac{4}{3}\left(f\frac{T_0}{T_s}\right) + \frac{1}{3}\left(f\frac{T_0}{T_s}\right)^4\right] \tag{3.25}$$

where

$$f = \left(\frac{e}{a}\right)^{1/4} \sin^{-1/2}\delta \tag{3.26}$$

For an usual *selective* flat-plate solar collector with $a = 0.9$ and $e = 0.1$, Eq (3.25) yields $\eta_v = 0.3724$ which is considerably higher than the upper bound efficiency $\eta_i = 0.08393$ we previously found in case of black-body (*non-selective*) solar collectors.

Finally note that the results presented in (i),(ii) and (v) correspond to solar direct radiation at normal incidence (i.e. Sun zenith angle $\theta_0 = 0$). In the general case the geometric factors in Eqs (3.9), (3.10) and (3.26) should be multiplied by $\cos^{-1/4}\theta_0$

(Landsberg and Badescu, 2000). For oblique radiation incidence ($\theta_0 > 0$) this procedure will yield lower estimates for the upper bound efficiency.

4. Endoreversible Carnot Solar Power Plants

In section 3 we presented simple accurate upper bound formulas for the efficiency of solar power plants operating on Mars. Much more involved models are necessary to further increase the accuracy of results. In this section a solar power plant model based on the Carnot endoreversible cycle is presented. This approach keeps the advantage of generality but also takes into account the irreversibilities associated to the heat transfer at the hot and cold components of the plant.

Solar Power Plant Model

Generally, a solar power plant consists of a solar collector, a thermal engine and a heat storage system. A solar collector mainly consists in mirror, absorber (receiver), transparent cover(s) and thermal insulation. The mirror is missing in case of flat - plate collectors while for some concentration systems the receiver is not protected by transparent cover(s). The solar power plant analysed in this section is designed to operate during the whole Martian year. Consequently, it consists of a *flat-plate selective* solar collector coupled to a thermal engine. No heat storage system is considered in this approach.

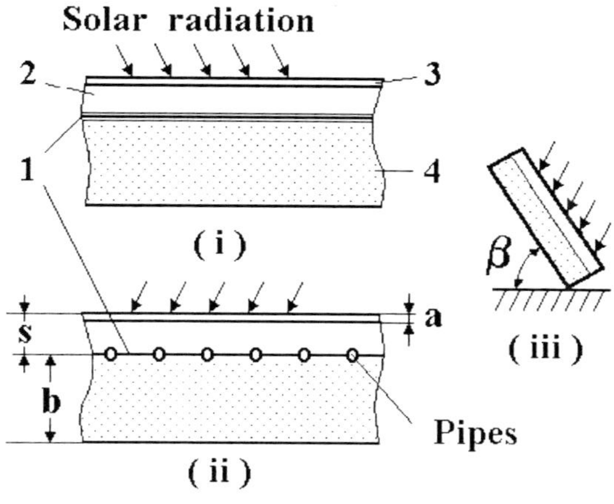

Figure 4.1. Solar collector. (i) longitudinal section; (ii) transversal section; (iii) lateral view. 1 - absorber plat; 2 - carbon dioxide layer; 3 - transparent cover; 4 - bottom thermal insulation.

Solar Collector Model

A selective flat - plate solar collector is shown in Figure 4.1. Its main components are: the absorber plate (1) with pipes for the working fluid and a transparent cover (3) of thickness a. The gap between (1) and (3) (thickness s) is filled with carbon dioxide at the Martian atmospheric pressure (carbon dioxide represents 95 % of the Martian atmosphere). The bottom thermal insulation (4) has a thickness b. The collector tilt angle is β. One denotes by T_C and T_a the absorber plate and ambient temperatures, respectively. Other temperatures related to collector operation are defined in Table 4.1.

Table 4.1. Various temperatures related to solar collector operation

T_C	absorber plate temperature
T_{si}	temperature of lower surface of transparent cover
T_{ss}	temperature of higher surface of transparent cover
T_a	atmosphere temperature
$T_{C,si}=(T_C+T_{si})/2$	average fluid temperature between absorber plate and lower surface of transparent cover (see Figure 4.2)
$T_{s,ave}=(T_{si}+T_{ss})/2$	average transparent cover temperature
$T_{ss,a}=(T_{ss}+T_a)/2$	average temperature of the atmosphere boundary layer near the collector transparent cover
$T_{iz}=(T_C+T_a)/2$	average bottom insulation temperature

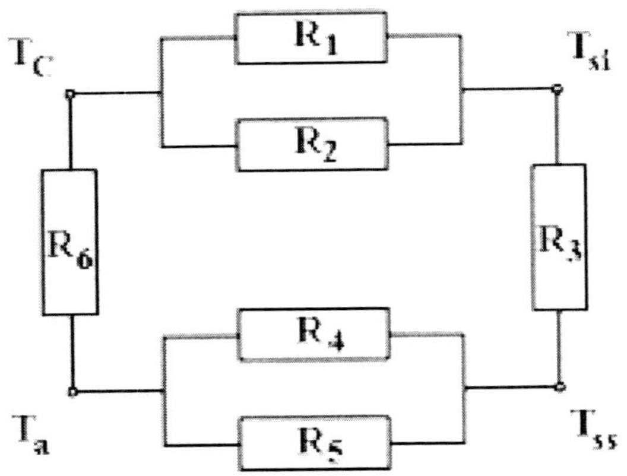

Figure 4.2. Solar collector thermal resistances. T_a - ambient temperature; T_C - absorber plate temperature; T_{si} - temperature of lower surface of transparent cover; T_{ss} - temperature of upper surface of transparent cover.

Direct beam (b), diffuse (d) and ground reflected (r) fluxes of solar energy are incident on the collector. Their magnitude depends upon collector's tilt and orientation. Generally, the incident and absorbed solar energy density fluxes (φ_{inc} and φ_{abs}, respectively) are given by

$$\varphi_{inc}=G_b+G_d+G_r \tag{4.1}$$

$$\varphi_{abs} = (\tau\alpha)_b G_b + (\tau\alpha)_d (G_d + G_r) \tag{4.2}$$

Here G 's denote solar irradiances at collector level while $(\tau\alpha)$'s are effective transmittance - absorptance products.

Table 4.2: Thermal resistances associated to solar collector operation (see Figures 4.1 and 4.2)

Thermal Resistance	Process
R_1	Convection and conduction between absorber plate (1) and transparent cover (3)
R_2	Radiation between absorber plate (1) and the lower surface of the transparent cover (3)
R_3	Conduction through transparent cover (3)
R_4	Radiation between the higher transparent cover surface (3) and Martian atmosphere
R_5	Convection between the higher transparent cover surface (3) and Martian atmosphere
R_6	Conduction through the bottom thermal insulation(4)

The collector heat losses will be studied by using the thermal resistance method (Eaton and Blum, 1975) (see Figure 4.2 and Table 4.2). Then, the flux of heat losses from the collector towards the ambient is given by:

$$q_{tot} = U_L (T_C - T_a) = (T_C - T_a) / R_{tot} \tag{4.3}$$

where U_L is the overall heat losses coefficient while R_{tot} is the total thermal resistance. U_L is given by (see Figure 4.2):

$$U_L = \frac{1}{R_{tot}} = \frac{1}{R_{1\text{-}5}} + \frac{1}{R_6} \tag{4.4}$$

where $R_{1\text{-}5}$ is computed from:

$$R_{1\text{-}5} = \frac{R_1 R_2}{R_1 + R_2} + R_3 + \frac{R_4 R_5}{R_4 + R_5} \tag{4.5}$$

Details concerning the evaluation of thermal resistances follow. The thermal resistance R_1 is computed from

$$R_1 = \frac{s}{<\lambda_{CO_2}>_{C,si} Nu_{C,si}} \tag{4.6}$$

where $<\lambda_{CO_2}>_{C,si}$ is thermal conductivity of carbon dioxide and $Nu_{C,si}$ is Nusselt number, both of them evaluated at the average fluid temperature $T_{C,si}$. To compute $Nu_{C,si}$, a number of criterial relationships developed for solar collectors were used (Table 4.3(a)) (Eaton and Blum, 1975). For cases not covered by Table 4.3(a) we used the general relationships from (Stefanescu et al, 1982), (Table 4.3(b)). Prandtl and Grashoff numbers in Table 4.3 were computed with:

$$Pr = \frac{<c_{p,CO_2}>_{C,si} <\eta_{CO_2}>_{C,si}}{<\lambda_{CO_2}>_{C,si}} \; ; \; Gr = \frac{g_M \, p_{CO_2}^2 \, s^3 \left(T_C - T_{si}\right)}{<\eta_{CO_2}>_{C,si} R_{CO_2} T_{C,si}} \tag{4.7}$$

where g_M is gravitational acceleration on Mars while p_{CO_2} and R_{CO_2} are the pressure and gas constant for the carbon dioxide between absorber plate and transparent cover. In Eq. (4.7) the specific heat at constant pressure ($<c_{p,CO_2}>_{C,si}$), the dynamic viscosity ($<\eta_{CO_2}>_{C,si}$) and the thermal conductivity ($<\lambda_{CO_2}>_{C,si}$) of carbon dioxide were interpolated at the average temperature $T_{C,si}$. In computations the thermal properties of carbon dioxide from (Pop et al, 1987) were used.

Table 4.3. (a) Relationships for convection heat transfer between absorber plate and transparent cover (Eaton and Blum, 1975)

(a)	$Nu_{C,si} = \left(1 - \frac{\beta}{2\pi}\right) Nu_H + \frac{\beta}{2\pi} Nu_V$	
	$Nu_H = 0.195 \cdot Gr^{0.25}$	for $2 \cdot 10^3 \le Gr < 4 \cdot 10^5$
	$Nu_H = 0.068 \cdot Gr^{0.33}$	for $Gr \ge 4 \cdot 10^5$
	$Nu_V = 0.180\, Gr^{0.25} (s/L)^{0.11}$	for $1 \cdot 10^4 \le Gr < 2 \cdot 10^5$
	$Nu_V = 0.065\, Gr^{0.33} (s/L)^{0.11}$	for $2 \cdot 10^5 \le Gr < 11 \cdot 10^6$
(b)	$Nu_{C,si} = 1$	for $Gr \cdot Pr \le 1 \cdot 10^3$
	$Nu_{C,si} = 0.105 \cdot (Gr \cdot Pr)^{0.30}$	for $1 \cdot 10^3 < Gr \cdot Pr \le 1 \cdot 10^6$
	$Nu_{C,si} = 0.400 \cdot (Gr \cdot Pr)^{0.20}$	for $1 \cdot 10^4 \le Gr < 2 \cdot 10^5$
	$Nu_{C,si} = 0.180 \cdot (Gr \cdot Pr)^{0.25}$	for $Gr \cdot Pr > 1 \cdot 10^{10}$

H and V denote a horizontal and vertical solar collector; β [rad] and L [m] are tilt angle and collector length, respectively. (b) General relationships for convection heat transfer between two plane parallel surfaces (Stefanescu et al, 1982, p. 160)

The thermal resistance R_2 is evaluated with:

$$R_2 = \frac{1/\varepsilon_1 + 1/\varepsilon_2 - 1}{2\sigma\left(T_C^2 + T_{si}^2\right)T_{Csi}} \tag{4.8}$$

where ε_1 and ε_2 are the emittance of absorber plate and lower surface of transparent cover, respectively (Duffie and Beckman, 1974) and σ is Stefan-Boltzmann constant.

The thermal resistance R_3 is given by:

$$R_3 = a/\lambda_s \tag{4.9}$$

The temperature dependence of transparent cover's thermal conductivity (λ_s) was described by using an interpolation parabolic function whose coefficients were obtained by using experimental data. The average transparent cover temperature $T_{s,ave}$ was used in computations.

The thermal resistance R_4 was evaluated with:

$$R_4 = \frac{1/\varepsilon_2}{2\sigma\left(T_{ss}^2 + T_a^2\right)T_{ss,a}} \tag{4.10}$$

where $T_{ss,a}$ is the average temperature of the atmosphere boundary layer near the collector transparent cover.

Because of the wind, the heat transfer between the transparent cover and Martian atmosphere is mainly by forced convection. Thus, the thermal resistance R_5 was computed from:

$$R_5 = \frac{L}{<\lambda_{CO_2}>_{ss,a} Nu_{ss,a}} \tag{4.11}$$

where L is collector's length on wind direction and $<\lambda_{CO_2}>_{ss,a}$ is carbon dioxide thermal conductivity at temperature $T_{ss,a}$. In Eq (4.11) $Nu_{ss,a}$ depends on the critical length X_T which separates the laminar and turbulent wind flows over the collector. The relationships of Table 4.4 were used. The Reynolds and Prandtl numbers were evaluated with:

$$Re = \frac{p_{CO_2} W L}{<\eta_{CO_2}>_{ss,a} R_{CO_2} T_{ss,a}}, \quad Pr = \frac{<c_{p,CO_2}>_{ss,a} <\eta_{CO_2}>_{ss,a}}{<\lambda_{CO_2}>_{ss,a}} \tag{4.12}$$

where W is wind speed and the thermal properties of the atmosphere ($<\eta_{CO_2}>_{ss,a}$, $<c_{p,CO_2}>_{ss,a}$ and $<\lambda_{CO_2}>_{ss,a}$) were interpolated at the average temperature $T_{ss,a}$.

The thermal resistance R_6 was evaluated with:

$$R_6 = b/\lambda_{iz} \tag{4.13}$$

where λ_{iz} is bottom insulation thermal conductivity, computed at the average temperature T_{iz}.

Table 4.4: Relationships for forced convection heat transfer between transparent cover and atmosphere (Stefanescu et al, 1982, p. 126)

$X_T = 5 \cdot 10^5 \dfrac{<\eta_{CO_2}>_{ss,a} R_{CO_2} T_{ss,a}}{p_{CO_2} W}$	
$Nu_{ss,a} = 0.335 \cdot Re^{0.50} \cdot Pr^{0.50}$	for $L \le X_T$ (laminar flow)
$Nu_{ss,a} = 0.036 \cdot Re^{0.80} \cdot Pr^{0.75}$	for $L > X_T$ (turbulent flow)

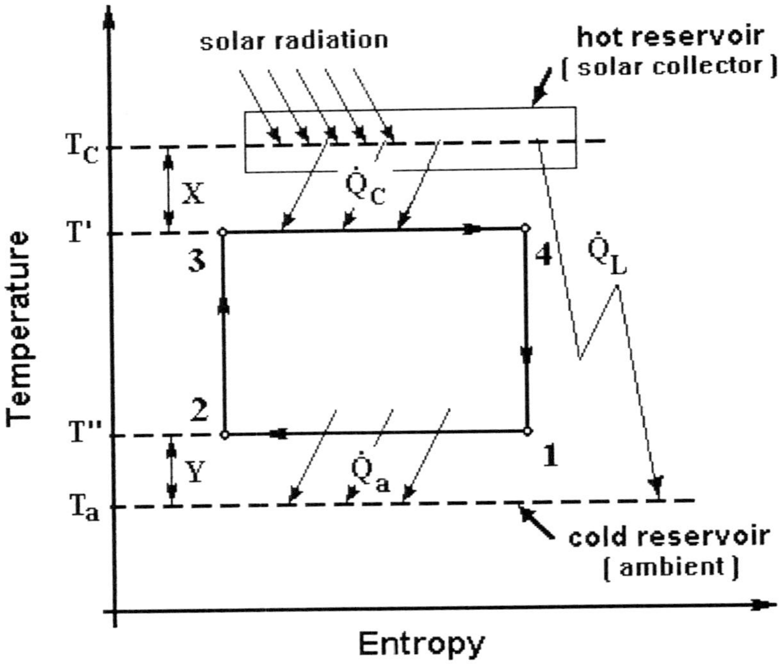

Figure 4.3. Temperature - entropy diagram for an endoreversible Carnot cycle (for notations see text).

Endoreversible Carnot engine Model

A variety of working fluids were proposed to be used in space solar engines. They include mercury, potassium and rubidium in the early studies (Menetrey, 1963) or organic fluids and mixtures of noble gases more recently (Angelino and Invernizzi (1993), Prisnjakov et al (1991)). The later seem to be more appropriate in case of Martian solar power plants.

The largest and smallest working fluid temperatures are denoted T' and T'', respectively (Figure 4.3). Generally, $T_C > T'$ and $T'' > T_a$. Hence, both

$$X \equiv T_C - T' \text{ and } Y \equiv T'' - T_a \tag{4.14}$$

are positive quantities.

The flux of solar energy ϕ_{abs} absorbed by the collector is

$$\phi_{abs} = \varphi_{abs} \ A_C \tag{4.15}$$

where A_C is collector surface area. Part of this flux is transferred to the working fluid ($\dot{Q}_C$) while the other part constitutes the flux of thermal losses to the ambient ($\dot{Q}_L$). Then, the solar collector steady-state energy balance is:

$$\phi_{abs} - \dot{Q}_C - \dot{Q}_L = 0 \tag{4.16}$$

The heat fluxes $\dot{Q}_C$ and $\dot{Q}_L$ can be written as:

$$\dot{Q}_C = h_C \, A_C \, X \text{ and } \dot{Q}_L = U_L \, A_C (T_C - T_a) \tag{4.17}$$

where h_C and U_L are appropriately defined overall heat transfer coefficients between the solar collector and the working fluid and ambient, respectively.

The Carnot engine partially converts the flux $\dot{Q}_C$ into power. Thus:

$$\dot{Q}_C = \dot{W} + \dot{Q}_a \tag{4.18}$$

where $\dot{W}$ is the output power and $\dot{Q}_a$ is the thermal energy flux finally reaching the surroundings. There is no further increase of entropy during work production. Consequently

$$\frac{\dot{Q}_C}{T'} + \frac{\dot{Q}_a}{T''} = 0 \tag{4.19}$$

The heat flux $\dot{Q}_a$ can be written as

$$\dot{Q}_a = h_a \ A_a \ Y \tag{4.20}$$

where A_a and h_a are appropriately defined engine radiator heat transfer area and overall heat transfer coefficient between working fluid and ambient, respectively.

Performance Indicators

Four performance indicators may be considered in relation to Martian solar plant operation (Badescu et al, 1999, 2000a, 2000b). First, there is the Carnot engine efficiency defined as:

$$\eta_{engine} \equiv 1 - \frac{T''}{T'} \tag{4.21}$$

Second, there is the output power $\dot{W}$, evaluated from:

$$\dot{W} = h_C \ A_C \ X - h_a \ A_a \ Y \tag{4.22}$$

Here we used the Eqs (4.17), (4.18) and (4.20). Third, there is the system efficiency, defined as:

$$\eta_{system} \equiv \frac{\dot{W}}{\varphi_{abs} A_C} \tag{4.23}$$

Finally, the solar (or sun-to-user) efficiency is defined as:

$$\eta_{solar} \equiv \frac{\dot{W}}{\phi_{abs} A_C} \tag{4.24}$$

Information about some of these indicators is reported in the following.

Optimisation

The output power $\dot{W}$ will be maximised now. The weight of space power plants should be kept at minimum. Following previous studies (see e.g. Badescu (1994)) we express the mass of the total heat transfer area (i.e. the mass M_C of the solar collector and the mass M_a of the engine radiator) as a k fraction from the total mass M_{tot} of the space system:

$$M_C + M_a = k \ M_{tot} \tag{4.25}$$

One denotes d_C and d_a the superficial mass density of collector and radiator, respectively. One can simply write:

$$M_C = d_C A_C , \quad M_a = d_a A_a \tag{4.26}$$

One can use $d_C = d_a$ as a reasonable assumption. By using eqs (4.25) and (4.26) one derives the following constraint to be fulfilled:

$$A_C + A_a = A \equiv k \frac{M_{tot}}{d_a} \tag{4.27}$$

Other constraints (say F_1 and F_2) are the Eqs (4.16) and (4.19). By using (4.14), (4.17) and (4.20) they become:

$$F_1 \equiv \frac{h_C \; A_C X}{T_C - X} - \frac{h_a (A - A_C)}{T_a + Y} = 0 \tag{4.28}$$

$$F_2 \equiv \varphi_{abs} - U_L (T_C - T_a) - h_C X = 0 \tag{4.29}$$

The power $\dot{W}$ is considered here as a function of T_C, X, Y and A_C. The Lagrange function L associated to $\dot{W}$ is:

$$L \equiv \dot{W} + \lambda_1 F_1 + \lambda_2 F_2 \tag{4.30}$$

where λ_1 and λ_2 are multipliers. The maximum output power is obtained by solving the six equations shown in Table 4.5.

Table 4.5. Equations to be solved for power maximisation (for details see text)

1.	$\frac{\partial L}{\partial T_C} = -\lambda_1 \frac{h_C \; A_C X}{(T_C - X)^2} - \lambda_2 U_L = 0$
2.	$\frac{\partial L}{\partial X} = h_C A_C + \lambda_1 \frac{h_C \; A_C T_C}{(T_C - X)^2} - \lambda_2 h_C = 0$
3.	$\frac{\partial L}{\partial Y} = -h_a (A - A_C) - \lambda_1 \frac{h_a (A - A_C) T_a}{(T_a + Y)^2} = 0$
4.	$\frac{\partial L}{\partial A_C} = h_C X + h_a Y + \lambda_1 \left(\frac{h_C X}{T_C - X} + \frac{h_a Y}{T_a - Y} \right) = 0$
5.	$\frac{\partial L}{\partial \lambda_1} = \frac{h_C \; A_C X}{T_C - X} - \frac{h_a (A - A_C) Y}{T_a + Y} = 0$
6.	$\frac{\partial L}{\partial \lambda_2} = \varphi_{abs} - U_L (T_C - T_a) - h_C X = 0$

Table 4.6 shows the solution that applies for $K > 1$. There, the dimensionless parameters K and f are defined as:

$$K = \frac{h_a}{h_C}\left(1 + \frac{h_C}{U_L}\right), f = \frac{\varphi_{abs}}{U_L T_a} \tag{4.31}$$

Table 4.6 and Eq (4.22) allow writing the maximum power as:

$$\dot{W}_{\max} = \frac{h_a A}{T_a} Y_{opt}^2 \tag{4.32}$$

The optimum solar efficiency $\eta_{solar,\,opt}$ can be derived by using Eqs (4.24), (4.32) and Table 4.6.

Table 4.6. Optimum parameters for maximum output power (for details see text)

1.	$T_{C,opt} = \left(1 + f - \frac{1 + f - (1+f)^{1/2}}{K^{1/2}\left(1 + K^{1/2}\right)} \frac{h_a}{U_L}\right) T_a$
2.	$X_{opt} = \frac{(1+f) - (1+f)^{1/2}}{K^{1/2}\left(1 + K^{1/2}\right)} \frac{h_a}{h_C} T_a$
3.	$Y_{opt} = \frac{(1+f)^{1/2} - 1}{1 + K^{1/2}} T_a$
4.	$A_{C,opt} = \frac{K^{1/2}}{1 + K^{1/2}} A$

Results and Discussions

One of the goals of this chapter is to perform a rough comparison between the performance of PV cell power systems and dynamic solar power plants. In case of PV cell systems we had in mind the Pathfinder's Sojourner. Sojourner is a small (11.5 kg), six - wheel robotic vehicle built at Jet Propulsion Laboratories. Sojourner was landed on Mars aboard the Pathfinder spacecraft on July 4, 1997. In the same sol she began to traverse the Martian terrain, perform science and technology experiments, and transmit images and data back to the Lander space craft. Sojourner's equipment (computers, motors, radio modem) was mainly powered by a lightweight 0.34 kg solar array of receiving surface area 0.22 m^2. The PV cell system was designed to provide Sojourner with around 16 W of electric power at noon on Mars. The landing place was in Ares Vallis at 19.17° N and 33.21° W on the U.S. Geological Survey (USGS) cartographic network (Golombek et al, 1997). Pathfinder landed in late northern summer (areocentric longitude $L_s = 143°$) and operated for 83 sols.

To allow comparison with Sojourner performance an appropriately designed dynamic power plant equipped with a flat - plate selective solar collector is analyzed here (see Table

4.7). A number of assumptions are accepted as follows. A simple relation is used for the effective transmittance - absorptance products:

$$(\tau\alpha)_b = (\tau\alpha)_d = \tau_{cs}\alpha_{ps} \tag{4.33}$$

where τ_{cs} and α_{ps} are the transparent cover transmittance and absorber absorptance, respectively, both of them for short (solar) wavelengths. The collector's transparent cover is made up of crystal (light flint glass). Its thermal conductivity lies between 0.691 $W(mK)^{-1}$ at -100 °C and 1.025 $W(mK)^{-1}$ at +100 °C (Lide, 1991). The ratio 0.34/11.5=0.0296 between the PV array weight and total Sojourner weight can be used as a first estimate for the coefficient k appearing in Eq. (4.27). Preliminary tests were performed to determine an optimum collector area $A_{C,opt}$ close to Sojourner PV cell surface area (0.22 m^2). Finally, the value A=0.23 m^2 was used in computations. It corresponds to a radiator superficial mass density d_a= 1.48 kg/m^2 in Eq (4.27). It is rather close to the value d_a=2.3 kg/m^2 accepted in (Angelo and Buden, 1991) in case of the radiator of a nuclear power satellite but only half of the value d_a=3 kg/m^2 accepted in (Mozjorine et al, 1991) for a solar space power station.

Table 4.7. Details about selective flat - plate solar collector design ($A_C = 0.22\ m^2$)

Quantity	Notation	Value
Number of transparent covers	N	1
Transparent cover thickness	a	0.003 m
Distance between transparent cover and absorber plate	s	0.045 m
Bottom thermal insulation thickness	b	0.1 m
Short wavelengths transparent cover transmittance	τ_{cs}	0.82
Short wavelength transparent cover absorptance	α_{ps}	0.90
Long wavelengths (IR) absorber plate emittance	ε_1	0.10
Long wavelengths (IR) transparent cover emittance	ε_2	0.88
Thermal conductivity of bottom insulation (polyurethane)	λ_b	0.024 $W/(mK)$

The distance between VL 1 and Pathfinder Lander is around 815 km. Thus, one expects quite similar meteorological and actinometric features in both places.

Four strategies of collecting solar energy were considered in preliminary tests (Badescu, 1998a). The following two strategies will be used in this section: (i) horizontal collector - strategy H; (ii) the collector tilt and orientation are continuously adjusted to keep the receiving surface perpendicular on Sun's rays - strategy P. Strategy H is easier to use while strategy P gives the higher power output most time of the year.

Solar Collector Operation

Solar collector thermal losses mainly depend on the heat transfer regime between absorber plate and transparent cover (separated through the layer 2 of thickness s in Figure 4.1). We considered the case of a horizontal solar collector. Three design solutions for the system solar collector - thermal engine were analysed (Table 4.8). They correspond to three different performance levels. For analysis we have chosen the middle of a clear sky autumn sol, when the thermal losses are rather high (sol 301 VL1, 14.04 local solar time (LST)).

Table 4.8. Three design solutions for the system solar collector - thermal engine

Case	h_a $[W/(m^2K)]$	h_C $[W/(m^2K)]$	Thermal engine performance level
(a)	1	1	Low efficiency
(b)	10	10	Medium efficiency
(c)	500	50	High efficiency

(h_a = heat transfer coefficient between the working fluid passing through the radiator of the heat engine and ambient (see Eq (4.20)); h_C = heat transfer coefficient between solar collector and working fluid (see Eq (4.17)).

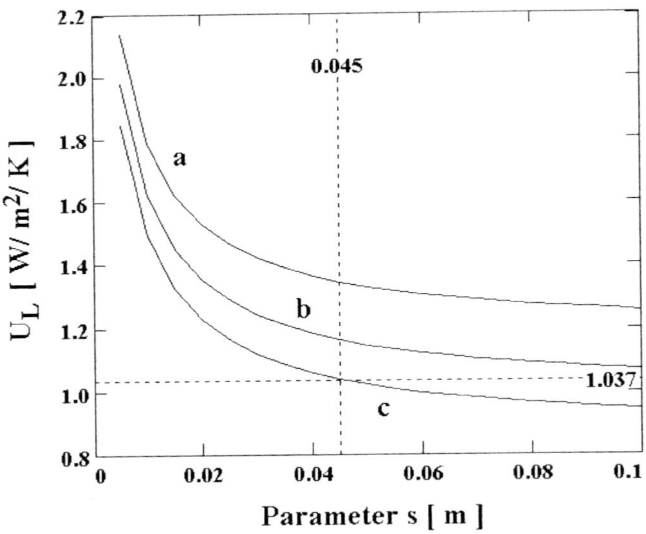

Figure 4.4. Overall heat loss coefficient U_L as a function of parameter s (distance between absorber plate and transparent cover - see Figure 4.1). Three design solutions (a), (b) and (c) were considered (see Table 4.8). (VL1 site, Autumn, sol 301, 14.04 LST).

The dependence of the overall heat loss coefficient U_L on the distance s between absorber plate and transparent cover is shown in Figure 4.4 for the three cases of Table 4.8. The heat losses decrease by increasing s. This is specific to the conduction heat transfer regime. The thickness s is limited, however, by economic reasons. In the following we use

s =0.045 m and the high efficiency engine (case (c) in Table 4.8). Note that the chosen value of s is about two times larger than the usual value for Earth placed solar collectors.

For Earth placed solar collectors the heat transfer between absorber plate and transparent cover is mainly by convection. This prompted various technical solutions meant to keep convection at local level (the honeycombs and Francia cells are examples of solutions for convection suppression (see Meinel and Meinel, 1976, p. 404)). The convective heat transfer is activated when the product $Gr \cdot Pr$ exceeds 1000 (see Table 4.3(b)). We analysed the heat transfer above the absorber plate by using the whole set of available meteorological data (Table 2.1). The results show that the product $Gr \cdot Pr$ varies between 9.65 and 211.2, well below the threshold value of 1000. We draw the important conclusion that on Mars the thermal losses between absorber plate and transparent cover are exclusively by conduction.

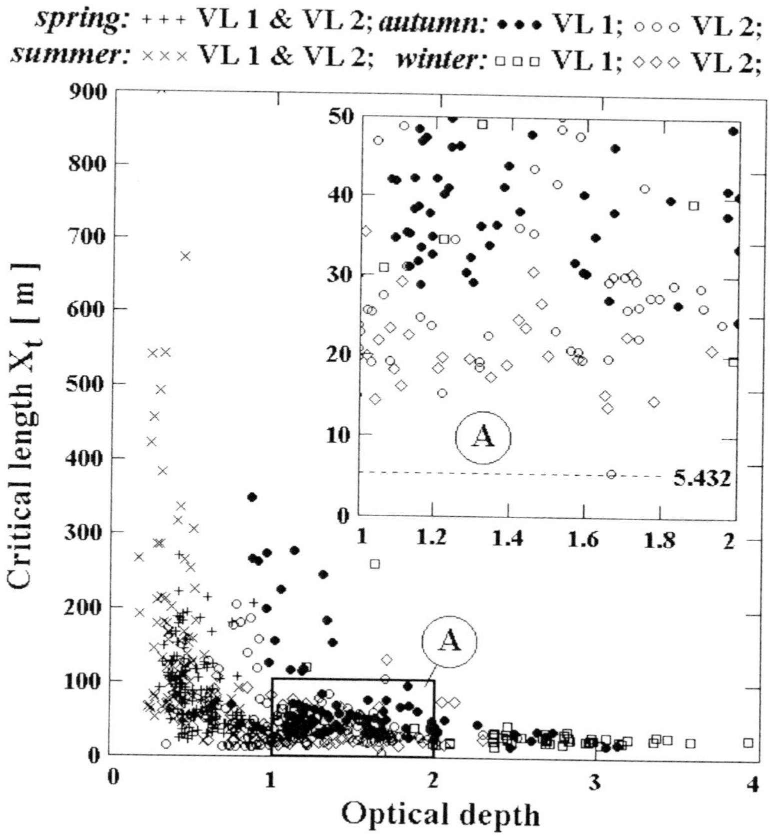

Figure 4.5. Critical length X_T (see Table 4.4) as a function of optical depth. A high efficiency engine was considered (see Table 4.8). All the available meteorological data were used.

The convection heat losses from transparent cover to atmosphere are mainly controlled by wind speed. The change from laminar to turbulent convection occurs in case the heat transfer surface exceeds a certain critical length X_T (see Table 4.4). The flow regime above the transparent cover was analysed by using the whole set of available meteorological data. Results are shown in Figure 4.5. The minimum value of the critical length X_T is 5.43 m.

Consequently, the convection heat transfer above the transparent cover is laminar. The critical length X_T is smaller during winter and larger during summer (Badescu et al, 2000a). No significant difference between VL1 and VL2 sites is observed (note that a larger number of meteorological values are available for VL2 - see Table 2.1). The critical length obviously decreases when the optical depth increases. During the "clear sky" days $(\tau < 1)$ the critical length values have an important dispersion. Scattering diminishes during the dust storm days $(\tau > 1)$.

For the design solution we selected (case (c) in Table 4.8), the values of the overall heat loss coefficient U_L range between about 0.5 and 1 $Wm^{-2}K^{-1}$ (Figure 4.6). Broadly, these values are comparable with those of Earth located vacuum solar collectors (see e.g. Meinel and Meinel, 1976, p. 387). The dispersion of the U_L values is slightly smaller during winter and higher during autumn (Badescu et al, 2000a). No obvious dependence of U_L on the optical depth is observed. However, the dispersion of the U_L values is larger for values $\tau < 1$, i.e. during the "clear" days. Also, Figure 4.6 shows that U_L doesn't depend significantly on the latitude of the solar collector.

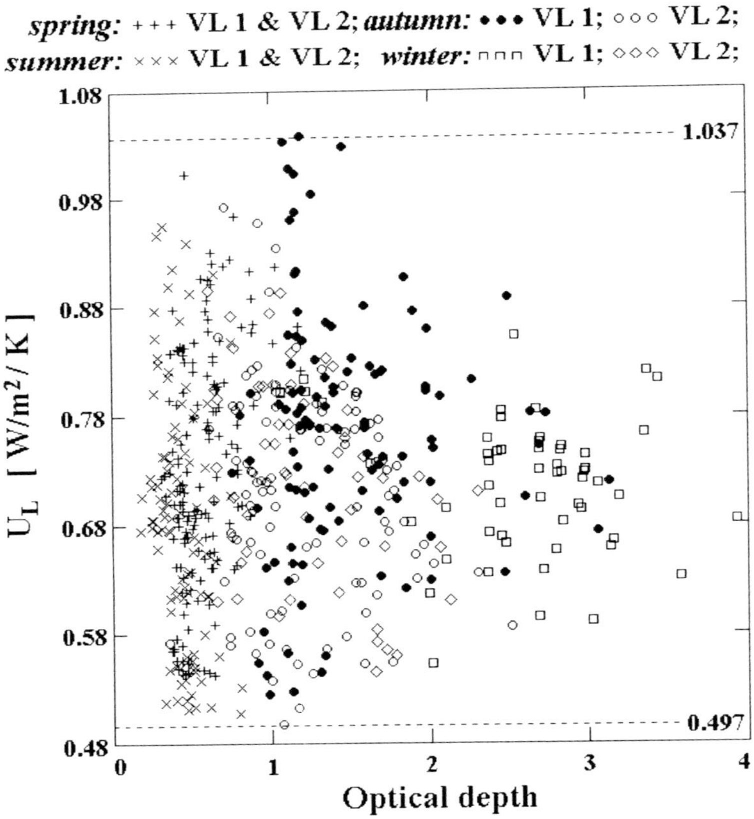

Figure 4.6. Overall heat loss coefficient U_L as a function of optical depth. A high efficiency engine was considered (see Table 4.8). All the available meteorological data were used.

Thermal Engine Operation

An important influence on performance have the parameters h_a and h_C. For analysis we have chosen the middle of a clear sky autumn sol (sol 301 VL1, 14.04 local solar time (LST)). Some results are shown in Figure 4.7 for the maximum power provided by the thermal engine. The influence of both h_a and h_C is more important on the maximum power provided by the heat engine that on the overall efficiency (Badescu et al, 2000a).

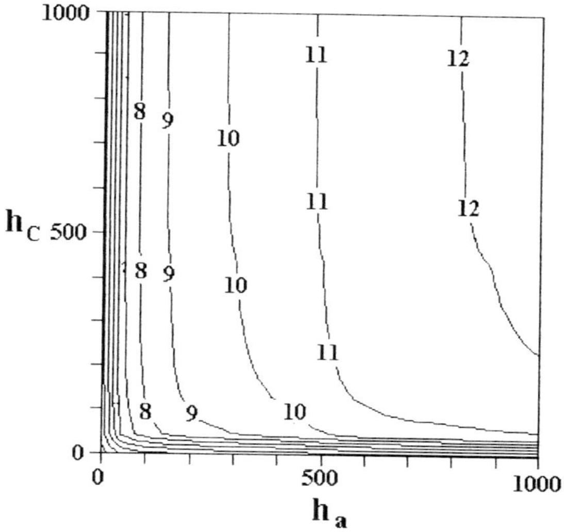

Figure 4.7. Maximum power $\dot{W}_{max}$ provided by the thermal engine as a function of the overall heat transfer coefficients h_a and h_C. Horizontal solar collector (strategy H). Unconcentrated solar radiation was considered (VL1 site, Autumn, sol 301, 14.04 LST)

Generally, the influence on performance is strong for low values of both heat transfer coefficients h_a and h_C (less than 100 $Wm^{-2}K^{-1}$. The performances are practically not dependent on the higher values of h_C. The influence of h_a on performance is more significant. The strategy of collecting solar radiation has a rather weak influence on the dependence of performance on h_a and. h_C. As expected, strategy P leads to higher performance (Badescu et al, 2000a).

The maximum output power could be as high as 12 W and 13 W in case of strategies H and P of collecting solar radiation, respectively (Badescu et al, 2000a). This is lower than the designed power of Sojourner PV arrays (16 W at noon). However, this comparison put in disadvantage the solar dynamic system as we used input meteorological data from late Autumn at VL1 site while Sojourner was designed to operate during Martian summer. This is a consequence of the fact that none of the 22 complete records available during summer at VL1 site (Table 2.1) corresponds to solar noon.

Performance Dependence on atMospheric Optical Depth

Previous results showed that the dynamic solar power plant performance depends significantly on the heat transfer coefficients h_C and h_a. Practical situations should normally lye between the cases (b) (i.e. medium efficiency engine) and (c) (i.e. high efficiency engine) in Table 4.8. These two cases will be used to provide lower and upper bounds for actual performance. To test the performance of the thermal engine - solar collector combination we simulated its operation at both VL1 and VL2 sites. We analyzed both the horizontal collector (strategy H) and the collector oriented perpendicularly on sun's rays (strategy P), respectively (Badescu et al, 2001a). At VL1 site we used meteorological data from summer, autumn and winter, year 1. All available data were used to simulate the operation at VL2 site. Some results are shown in Figure 4.8 (for solar efficiency) and Figure 4.9 (for output power).

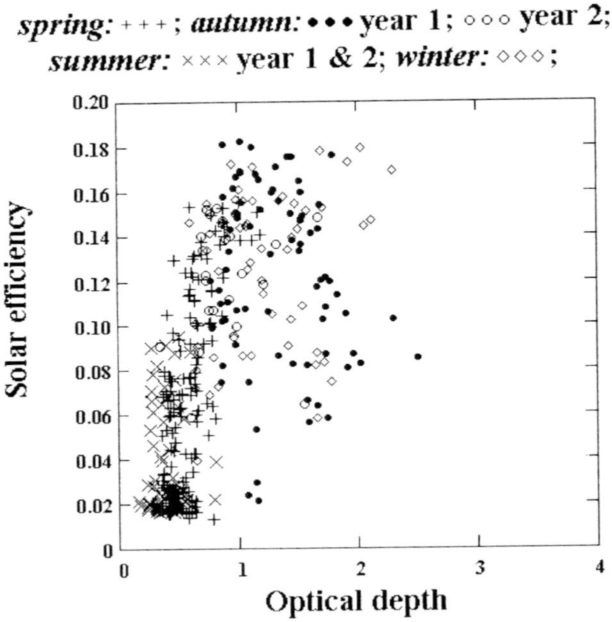

Figure 4.8. Dependence of solar efficiency $\eta_{solar,opt}$ on the atmospheric optical depth at VL2 site. High efficiency engine (see Table 4.8) and horizontal solar collector. All the available meteorological data for VL2 site were used.

The influence of latitude on performance is obvious. When the collector is horizontal, the solar efficiency is generally smaller at VL1 site as compared to VL2 site (Figure 4.8). When a P collector is considered, the efficiency increases (at small τ's) but it is still smaller at VL1 site that at VL2 site (Badescu et al, 2001a). The maximum power provided by a horizontal collector at VL1 site shows some interesting features (Badescu et al, 2001a). Generally, it is smaller than in case of VL2 site. However, there are some situations (for $\tau \approx 1$) when the meteorological effects compensate the latitudinal effects and the output power is quite similar for both VL1 and VL2. If a P collector is used, the power increases for small τ's but slighter at VL1 site than at VL2 site (Figure 4.9).

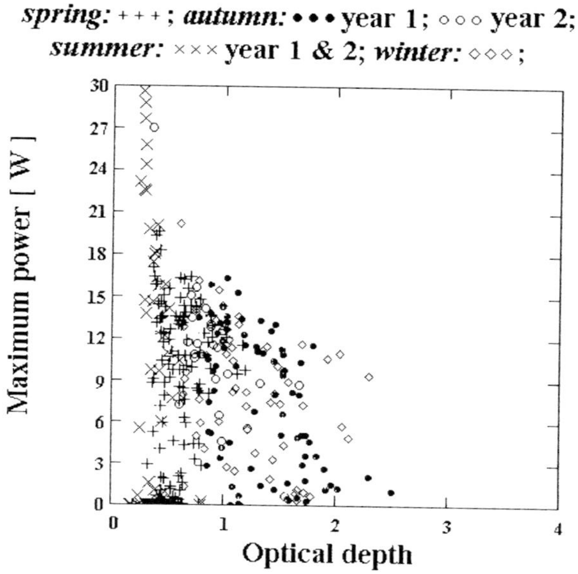

Figure 4.9. The same as Figure 4.8 in case of the maximum output power $\dot{W}_{max}$. High efficiency engine and P solar collector.

In case the medium efficiency engine is coupled to a horizontal collector, the efficiency does not exceed 0.13 at VL2 site (Badescu et al, 2001a). Generally, it is lower during summer, when the horizontal orientation is surely a good option (Badescu, 1998a). It is higher during winter, when the optical depth is higher, indicating dust storms occurrence. During dust storms the incident solar radiation is mainly diffuse and the horizontal orientation is the best strategy for radiation collection (Badescu, 1998a). In case a medium efficiency engine is coupled to a P collector, the maximum efficiency values lye around 0.13 (Badescu et al, 2001a). The efficiency increases at small optical depth values, which generally correspond to summer. The centre of the output data cloud is placed around 0.09-0.10. The P strategy does not diminish the performance at larger optical depth.

For a horizontal collector attached to a high efficiency engine, the solar efficiency is as high as 0.18 (Figure 4.8). The qualitative features pointed out previously maintain. The lower sun-to-user efficiency corresponds to lower optical depth. If a P collector is considered, the solar efficiency increases during summer and spring but does not significantly exceed 0.18 (Badescu et al, 2001a). The centre of the output data cloud is placed around 0.15, higher than in case of the horizontal collector. Again, the P strategy does not diminish the performance. This strategy keeps constant the solar efficiency at higher optical depths but improves the performance as compared to the H strategy at smaller optical depths.

The power provided by a system consisting in a horizontal collector and a medium efficiency engine does not exceed 7 W (Badescu et al, 2001a). Generally, it is lower for small optical depth. Indeed, the flux of direct solar energy is higher at smaller optical depth and the horizontal orientation is not recommended in this case. At higher optical depth direct solar radiation diminishes and diffuse radiation increases. As a result, the horizontal orientation is close to optimum. This explains the higher values of $\dot{W}_{max}$ around $\tau = 1$. For further increase of τ the diffuse radiation flux diminishes. Consequently, $\dot{W}_{max}$ decreases in this case. If a P collector is considered, the solar efficiency is higher during summer and spring, when the optical depth is small (Badescu et al, 2001a). Generally, the efficiency increases as compared to the H strategy. At high optical depth the solar efficiency is comparable for both H and P strategies.

Using a better engine with a horizontal collector leads to output power up to 13 W for τ around 1 (during sols of spring, autumn and winter) (Badescu et al, 2001a). For small values of τ the output power is smaller. The P strategy is recommended in combination with high efficiency engines (Figure 4.9). In this case the sun-to-user efficiency during sols with small optical depth increases strongly, exceeding 25 W. The centre of the output data cloud is placed around 13 W. The P strategy can be used at high optical depth, too.

Performance Dependence on the Level of Solar Irradiance

The dependence of solar efficiency on the level of incident global irradiance was studied in Badescu et al (2001a). In case of a horizontal collector operating at VL2 site, a slightly non-linear relationship exists between solar efficiency and incident global irradiance for both the medium and the high efficiency engines. When the P collector is considered, the dependence is more complicated.

When a combination medium efficiency engine - horizontal collector is considered, the efficiency is higher (whatever the incident irradiance is) during autumn and winter and lower during summer. This rather surprising fact is valid for the P collector, too. It is probably due to the influence of the ambient temperature, which is lower during autumn and winter. The situation is different in case of the high efficiency engine. Then, the solar efficiency is higher during summer, even if it corresponds to lower incident irradiance. This remark is even more obvious when a P collector is considered. A more uniform solar efficiency values distribution is obtained in this case.

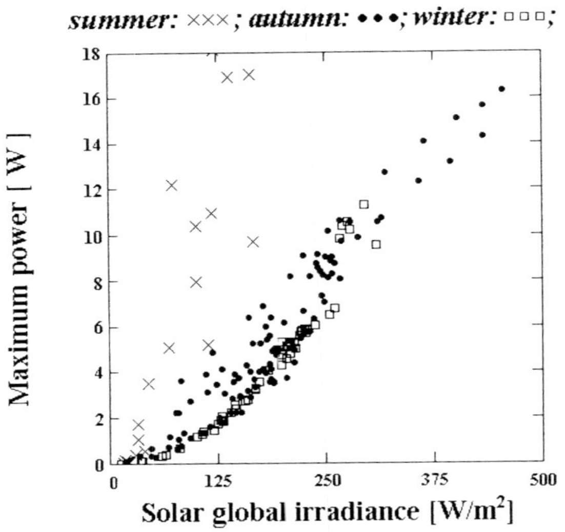

Figure 4.10 Dependence of maximum output power $\dot{W}_{max}$ on atmospheric optical depth at VL1 site. High efficiency engine (see Table 4.8).and P solar collector. Meteorological data for summer, autumn and winter year 1 were used.

In case of a horizontal collector, the efficiency increases by increasing the irradiance level, whatever the type of engine is. For the P collector, the efficiency depends on the irradiance level in a more interesting manner. Thus, the lower bound of the efficiency increases with the irradiance level, but the upper bound practically does not depend on the irradiance level. This is valid for both the medium and high efficiency engines.

During autumn and winter the dependence of η_{solar} on the level of incident global irradiance is nearly similar for both the horizontal and the P collector operating at VL1 site. Important differences exist during summer. The P collector is more effective, as expected. The influence of irradiance level on solar efficiency decreases drastically beyond a certain threshold value (about 300 W/m^2). This applies to both types of solar collectors.

Figure 4.10 shows some results for the output power at VL1 site. Meteorological data from year 1 were used. In case of a horizontal collector the dependence of the maximum output power on the level of global irradiance is not differentiated upon season (Badescu et al, 2001a). When a P collector is considered, the same incident irradiance value leads to an output power obviously higher during summer. For a horizontal collector the output power increases when the input irradiance increases (Badescu et al, 2001a) This remark maintains for the P collector except for summer results (Figure 4.10).

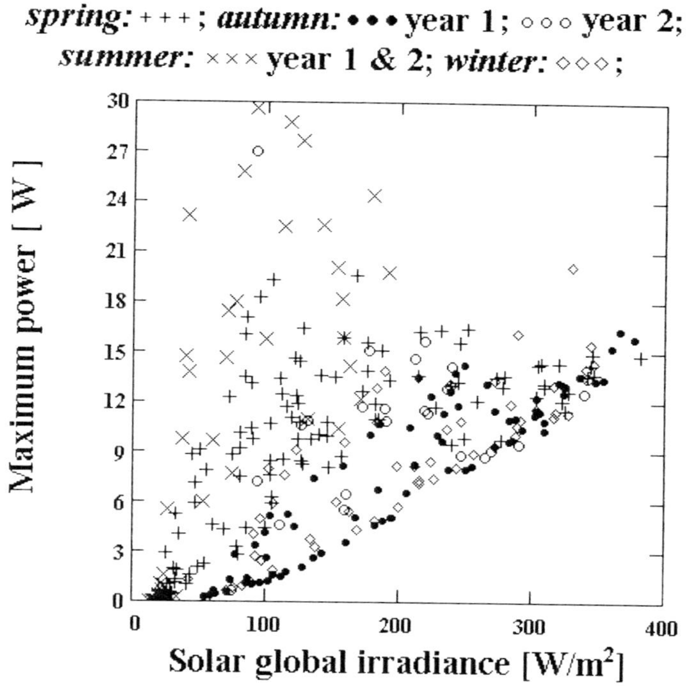

Figure 4.11 Dependence of the maximum output power $\dot{W}_{max}$ on the level of solar global irradiance incident on a horizontal surface at VL2 site. High efficiency engine (see Table 4.8) and P solar collector. All the available meteorological data for VL2 site were used.

The power provided by a system based on a horizontal collector increases at VL2 site with the level of solar irradiance, as expected (Badescu et al, 2001a). The increase is almost proportional in case of the medium efficiency engine and slightly non-linear for the high efficiency engine. Generally, there is no obvious dependence on season. In case of the P collector, the output power obviously depends on season, with a maximum during spring and summer and a minimum during autumn and winter. This is true for both the medium and high efficiency engines (Figure 4.11).

The seasonal influence is similar for both solar efficiency and output power in case of a medium efficiency engine - horizontal collector combination (Badescu et al, 2001a). This is still valid when the collector is connected to a high efficiency engine. However, in this case the power values have a larger dispersion, which increases slightly with increasing the irradiance.

In case of a medium efficiency engine and a P collector, the influence of the season is stronger on power than on solar efficiency (Badescu et al, 2001a). This is even more obvious if one connects a P collector to a high efficiency engine. The high solar efficiency values during spring and autumn at smaller irradiance values are not the main cause of the higher output power (which is obtained in summer (Figure 4.11)). For a medium efficiency engine, a P collector is recommended mainly during summer and spring (which are dust storm - free seasons) (Badescu et al, 2001a). One finds higher output power values as compared to the case when a horizontal collector is used. If the summer and spring values would be neglected, one can see that choosing a P instead of an H collector leads to a relatively small increase in power during the other two seasons (autumn and winter). This is valid for the high efficiency engine, too (Figure 4.11). However, the increase in power by using a P collector during seasons without dust storms is more spectacular. Compare the power values as high as 20 W obtained at lower input irradiance by using a P collector (Figure 4.11) with the power values of at most 12-13 W obtained at high irradiance values by using a horizontal collector (Badescu et al, 2001a).

Diurnal Variation of Performances

The input data file we prepared in Section 2 contains only a few sols with complete records covering appropriately the whole daylight time. The most part of the sols has one to three records only associated to time periods close to sunrise, noon or sunset. In particular, no summer sol with complete records covering the whole day was found either for VL1 or for the VL2 site. For each season, we selected those sols that are richer in available data. In case of VL1 these sols are 301 (autumn) and 328 (winter). Note that no available data refers to spring (Table 2.1). In case of VL2 the sols selected are 420 (spring) and 872 (autumn). The sol 406 (winter) is also available but no reference to it will be made here, as the winter conditions are relatively similar for both Viking Lander sites.

Various efficiencies defined in Section 4.1.3 were reported in Badescu et al (2000b) in case of sol 301 (Martian autumn) at VL1 site. During most of the time the optimum engine efficiency $\eta_{engine,\ opt}$ lies around 33 %. The efficiency is decreasing suddenly near sunset. Generally, the optimum engine efficiency is less dependent on the type of engine and on the strategy of collecting solar radiation. In exchange, the system efficiency $\eta_{system,\ opt}$ is strongly dependent on engine quality and less dependent on the strategy of solar collection. It lies around 0.2 for the high efficiency engine and around 0.15 for the medium efficiency engine. When the high efficiency engine is considered, the decrease in performance is more abrupt near sunset. The optimum solar efficiency $\eta_{solar,\ opt}$ is smaller than the optimum system efficiency (Figure 4.12) but exhibits the same qualitative features. Generally, it doesn't exceed 15 % for the high efficiency engine and 10 % for the medium efficiency engine. The strategy of collecting solar radiation has not a strong influence on $\eta_{solar,\ opt}$. In exchange, the strategy significantly influences the maximum output power $\dot{W}_{\max}$ (Figure 4.13). If the optimum solar efficiency is rather constant during the day, the maximum power's time variation is closely related to the temporal variation of solar irradiance. This is more obvious for the high efficiency engine. The maximum output power can exceed 15 W for the high efficiency engine and doesn't exceed 8 W for the medium efficiency engine. This proves that properly designed thermal power plants are comparable in performance with PV-based power systems (one reminds that Sojourner's design output power is 16 W at solar noon). Moreover, the previous analysis puts in disadvantage the solar dynamic system as we used here input meteorological data from late autumn at VL1 site while Sojourner was designed to operate during Martian summer.

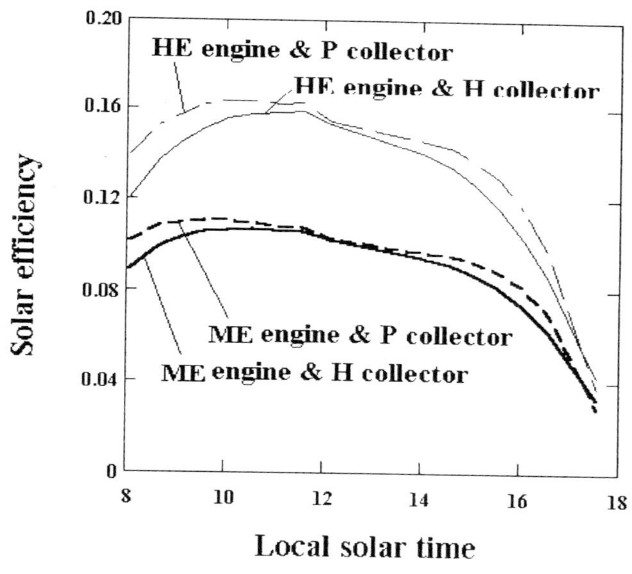

Figure 4.12 Optimum solar efficiency $\eta_{solar,\ opt}$ during sol 301 (autumn) at VL1 site. Medium efficiency engine (ME) and high efficiency engine (HE) (see Table 4.8). Strategy H - horizontal collector; Strategy P - collector permanently kept perpendicular on Sun's rays.

The hourly variation of solar collector's optimum temperature is shown in Figure 4.14. It is less dependent on the strategy of collecting solar radiation and a little more dependent on engine quality. The high efficiency engine leads to a smaller optimum collector temperature. Generally, the collector temperature exceeds with more than 100 K the ambient temperature. The hourly variation of $T_{C,\ opt}$ is stronger than that of the ambient temperature. This proves that $T_{C,\ opt}$ is mainly controlled by the level of incident solar radiation.

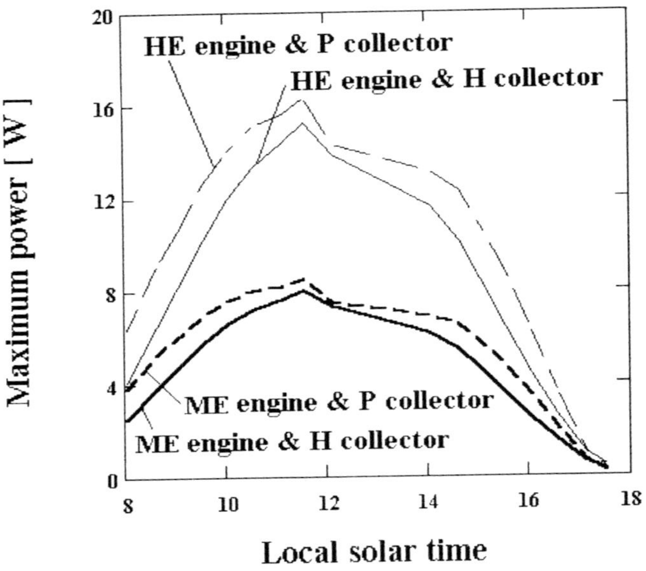

Figure 4.13. The maximum output power $\dot{W}_{max}$ during sol 301 (autumn) at VL1 site. For details see Figure 4.12.

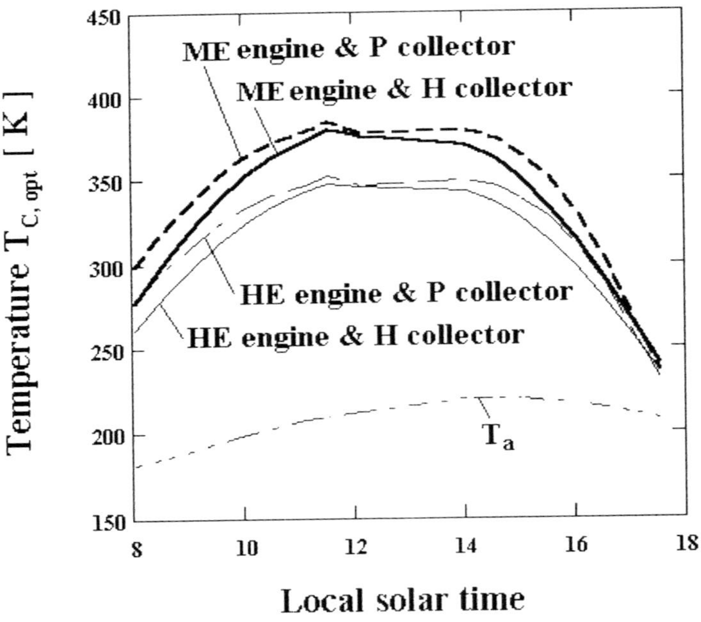

Figure 4.14. Optimum solar collector temperature $T_{C,\ opt}$ associated to solar thermal power plant operation during sol 301 (autumn) at VL1 site. For details concerning the power plant see Figure 4.12.

The optimum temperature of the working fluid in the cold part of the engine (T'_{opt}) is well correlated with the time variation of the ambient temperature (Badescu et al, 2000b). T'_{opt} is smaller in case of the high efficiency engine, when the influence of the strategy of collecting solar radiation is obvious. This influence is less important in case of the medium efficiency engine. In exchange, the optimum temperature of the working fluid in the worm part of the engine (T''_{opt}) is better correlated with the hourly variation of the solar radiation (Badescu et al, 2000b). Both the engine quality and the strategy of collecting solar radiation are important factors. Generally, T''_{opt} is close to $T_{C,opt}$.

Sol 328 belongs to winter at VL1 site. The main indicators of the solar thermal power plant are shown in Badescu et al (2000b). The optimum solar efficiency values are rather close to the autumn values. The quality of the engine is very important and the strategy of collecting solar radiation is less important. This is due to the fact that sol 328 belongs to a dust storm period (Badescu, 1998a). The maximum output power is, however, much smaller than during autumn. At noon it reaches 10 W in case of the high efficiency thermal engine but doesn't exceed 5 W in case of the medium efficiency engine. The maximum output power time variation is stronger than that of solar efficiency but this is less obvious than in autumn. The daily variation of the optimum collector temperature is rather weak. Generally, the high efficiency engine leads to a lower optimum collector temperature. Also, the optimum temperature is lower for a collector perpendicular on Sun's rays.

Sol 420 belongs to springtime at VL2 site. In this case, the optimum solar efficiency depends significantly both on engine quality and on the strategy of collecting solar radiation (Badescu et al, 2000b). In case of a P collector, the optimum solar efficiency is nearly constant during the day. It varies rather much for the horizontal collector (with a maximum at noon). The optimum solar efficiency values are quite close to those from autumn and summer at VL1 site. The maximum output power shows features similar to those of the optimum solar efficiency. Generally, the P collector assures a more constant maximum output power during the day. The engine quality is very important. The high efficiency engine connected to a P collector can ensure a maximum output power around 15 W during most part of the day. This is close to the design power of the PV system of Mars Pathfinder's Sojourner. When a P solar collector is considered, the optimum collector temperature shows a remarkable constancy during the day.

Sol 872 corresponds to autumn at VL2 site. The optimum solar efficiency shows an interesting variation during the day. This is more obvious for a P collector (Badescu et al, 2000b). In case of the horizontal collector, the optimum solar efficiency has a maximum around the noon. The optimum solar efficiency is higher in case of the high efficiency engine, as expected. It ranges from 8 to 16 %. The maximum output power shows similar features. Generally, the better performance in the morning corresponds to the stronger solar irradiance during that part of day (Badescu, 1998a). Note that a horizontal collector assures a constant power during the day. The high efficiency engine should be used in combination with a P collector. A maximum output power of 15 W can be provided only a short time period during the day. In case of a P collector, the optimum temperature is rather constant until 14.00 local solar time (LST). When a horizontal collector is considered, the optimum temperature has a maximum reached between 12.00 and 14.00 LST. Generally, $T_{C\ opt}$ is lower than in the cases above and this can be correlated to the smaller ambient temperature value.

Correlation between Maximum Output Power and Optimum Solar Efficiency

The dependence of the maximum output power on the optimum solar efficiency at VL1 site is shown in Badescu et al (2000b). Meteorological data from year 1 were used. Let us consider the case of the medium efficiency engine. The maximum output power increases when the optimum solar efficiency increases, as expected. There is little dependence on season, except for a few summer values in case of the P collector, when the solar efficiency is obviously higher. Broadly speaking, there is little difference between the maximum output power provided by systems based on H and P collectors, respectively. The highest optimum solar efficiency is 0.11 and the maximum output power doesn't exceed 10 W. When the high efficiency engine is considered, the influence of optimum solar efficiency on the maximum output power is important. The highest optimum solar efficiency is 0.16 and the output power can be as high as 15 W for an H collector and 17 W for a P solar collector.

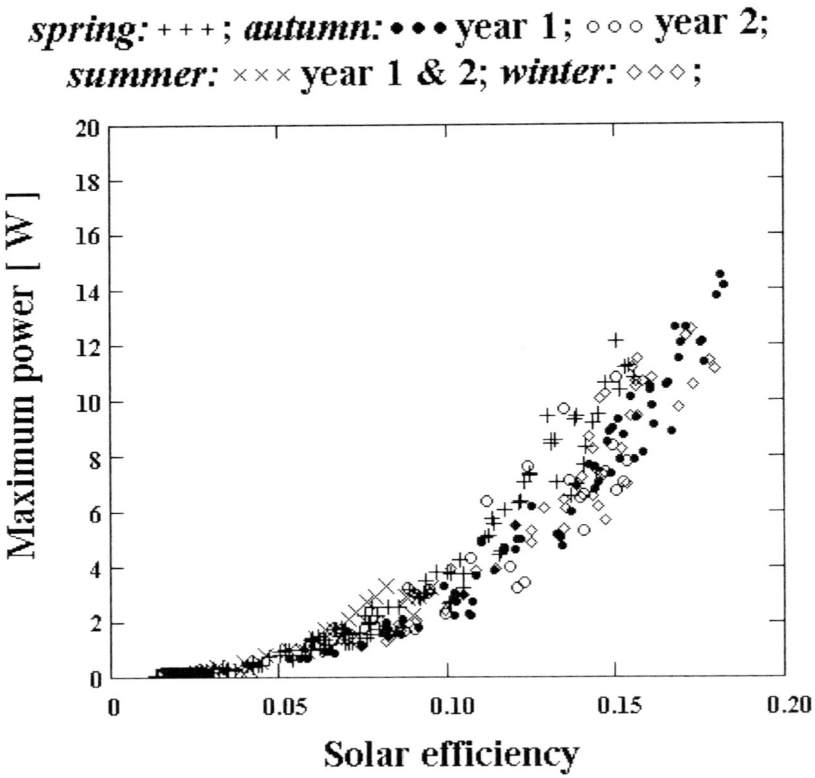

Figure 4.15 Dependence of the maximum output power $\dot{W}_{max}$ on the optimum solar efficiency $\eta_{solar,\ opt}$ at VL2 site. High efficiency engine (see Table 4.8) and horizontal solar collector. Meteorological data for summer, autumn and winter (year 1) and spring, summer and autumn (year 2) were used.

Figure 4.15 shows some results obtained at VL2 site (Badescu et al, 2000b). All the available data for the two years of VL2 operation were used in computations. Generally, the dependence of the maximum output power on the optimum solar efficiency is non-linear. In case of a horizontal collector, the dependence seems to indicate a quadratic dependence ($\dot{W}_{max} \propto \eta^2_{solar,\ opt}$). The coefficient of proportionality is smaller in case of the medium efficiency engine (Badescu et al, 2000b). When the P collector is considered, the dependence $\dot{W}_{max}(\eta_{solar,\ opt})$ is more complicate, but keeps the quadratic feature. The P collector has a more compact dispersion of the optimum solar efficiency values. It ranges from 8 to 12 % in case of the medium efficiency engine and between 13 and 19 % in case of the high efficiency engine.

In all cases a certain optimum solar efficiency threshold (around 5 %) must be exceeded in order the system provide useful power. The dependence of the maximum output power on optimum solar efficiency is stronger in case of the high efficiency thermal engine than in case of the medium efficiency engine, on one hand, and in case of the P collector than for the horizontal collector, on the other one hand. It seems no obvious difference exists between the performances of the power plant in the two years of VL2 operation.

5. Solar Stirling Engine

A more realistic solar thermal engine is considered in this section. It consists of a solar collector - Stirling engine combination. The Stirling engine is indeed very attractive due to its operation at low temperature difference between the two heat reservoirs. This could enable power production during the Martian dust storm period, when both the incoming solar energy flux and collector temperature are small. Preliminary results were briefly reported in Badescu et al (2001b) where the diurnal solar plant operation was studied. One concluded that the efficiency of a usual (Earth based designed) Stirling solar engine diminishes with about 8% when operates under Martian weather conditions. However, the recommended volumetric ratio lies between 1 and 2.6 and the best thermal agent is helium, in agreement with conclusions based on Earth made experiments. The model previously proposed in Badescu et al (2001b) is now presented. A number of improvements are also included (Badescu, 2004).

Solar Engine Model

The solar engine consists of a selective flat-plate solar collector coupled to a Stirling engine with partial heat regeneration. No heat storage system is considered in this approach.

Solar Collector Model

First, we shall consider the flat-plate solar collector (Figure 5.1). Its main components are the collector plate and a transparent cover of thickness a. The gap of thickness s between these two components is filled with carbon dioxide at Martian atmospheric pressure. Solar direct and diffuse radiation penetrates the transparent cover and is absorbed by the collector plate. The thermal energy generated in the solar collector plate is transferred to the thermal agent at the hot head of the Stirling engine and later-on part of it is converted into work.

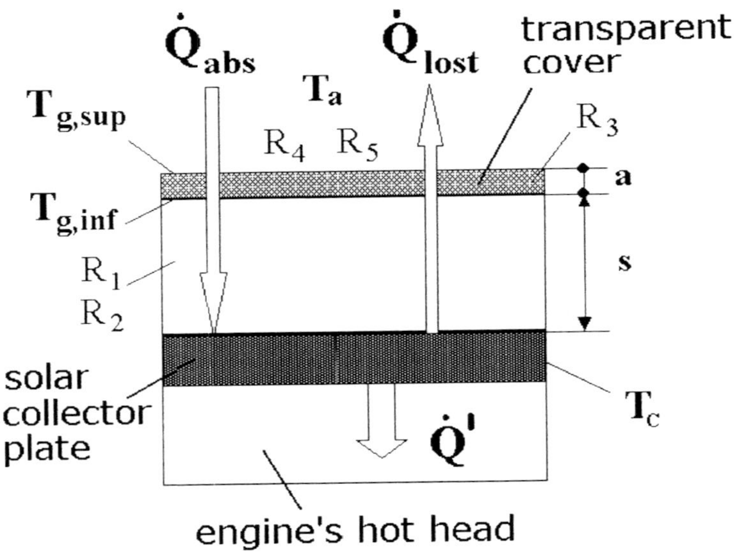

Figure 5.1. The solar collector of the Stirling engine.

Steady state energy balance for the solar collector yields

$$\dot{Q}_{abs} - \dot{Q}' - \dot{Q}_{lost} = 0 \tag{5.1}$$

Here $\dot{Q}_{abs}$ is the thermal energy flux generated into solar collector plate, $\dot{Q}_{lost}$ is the heat flux lost towards the ambient and $\dot{Q}'$ is the heat flux transferred to the thermal fluid inside the Stirling engine. The thermal energy flux generated into the solar collector plate is given by

$$\dot{Q}_{abs} = \varphi_{abs} A_c \tag{5.2}$$

where A_c is the area of collector plate surface. The heat flux lost towards the ambient is given by the usual Newton relationship:

$$\dot{Q}_{lost} = U_L A_C (T_C - T_a) \tag{5.3}$$

where U_L is the overall heat loss coefficient, while T_C and T_a are collector and ambient temperature, respectively. Note that T_C is an averaged value over collector surface and normally is not constant in time.

Seven temperatures are associated to solar collector operation (Badescu, 2004). With a single except (i.e. the average temperature of bottom insulation) they have the meaning explained in Section 4.1.1 (Tables 4.1 and 4.2). The collector heat losses could be studied by using the thermal resistance method (see section 4.1.1). Then, the flux of heat losses $\dot{q}_{lost}$ per unit collector surface area is defined by:

$$\dot{q}_{lost} \equiv \frac{\dot{Q}_{lost}}{A_c} = U_L\left(T_C - T_a\right) = \frac{T_C - T_a}{R_{tot}} \tag{5.4}$$

where R_{tot} is the total thermal resistance given by

$$R_{tot} = \frac{R_1 R_2}{R_1 + R_2} + R_3 + \frac{R_4 R_5}{R_4 + R_5} \tag{5.5}$$

Note that R_{tot} here is equivalent to R_{1-5} in the model of solar collector of section 4.1.1. Therefore, Figure 4.2 and Table 4.2 show the meaning of the thermal resistances R_1 to R_5. Details concerning the computation of these thermal resistances are similar to those presented in Section 4.1.1 and will be not repeated here.

Stirling Engine Model

During the 1980s Prof. I. Kolin at the University of Zagreb and Prof. J. Senft at the University of Wisconsin started research on low temperature differential Stirling engines and developed the first Stirling engine running with a temperature difference below 20 degrees Celsius (Senft, 1996). Meanwhile a rather large number of research or commercial Stirling engines using unconcentrated solar radiation were built, with powers from a few to tens of Watts (Badescu, 2004).

A simple Stirling engine model is used at this stage of the analysis. It was first developed in Howell and Bannerot (1977) and sligthly improved in Badescu (1992). The model is based on two simplifying assumptions. First, one neglects the heat losses during the thermal transfer from the solar collector to the working fluid inside the engine. Second, one neglects the friction of the moving parts. These make of course the results reported below to be upper bounds for the performance of a real solar Stirling system. The Stirling cycle efficiency $\eta_{Stirling}$ is defined as:

$$\eta_{Stirling} = \frac{P}{\dot{Q}'} = \frac{\eta_{max}}{1 + D\eta_{max}} \tag{5.6}$$

where P is the output mechanical power while $\eta_{max} = 1 - T_a / T_C$ is the maximum efficiency of the cycle in case of perfect heat regeneration. In case the thermal agent is an ideal gas the parameter D is given by (Howell and Bannerot (1977), Badescu (1992)):

$$D = \frac{x}{(k-1)\ln \varepsilon_V} \tag{5.7}$$

Here $x(\in [0,1])$ is the heat regeneration factor, k is the adiabatic exponent of the thermal agent and ε_V is the volumic ratio (i.e. the ratio of the extreme values of the thermal agent total volume). $x=0$ means ideal heat regeneration. In this case the efficiency of the Stirling engine Eq (5.6) equals the Carnot efficiency.

Use of Eqs (5.1),(5.2),(5.3) and (5.6) yields the following dependence of the power provided P on the collector plate temperature T_C:

$$P(T_C)=\frac{1-T_a/T_C}{1+D(1-T_a/T_C)}[\varphi_{abs}-U_L(T_C-T_c)]A_C \qquad (5.8)$$

Note that U_L is not a constant but depends on T_C (among other variables) (see Figure 5.4). Consequently, a simple numerical procedure to maximize directly the power P given by Eq (5.8) was used here, as follows. One monotonously increases the plate temperature T_C starting from the ambient temperature. For any value of T_C one evaluates first the overall heat loss coefficient U_L (Eq (5.4)) and then the output power P (Eq (5.8)). The maximum power P_{max} and the appropriate optimum plate temperature $T_{C,opt}$ are finally selected from the series of results.

The way of using the output power of the Stirling engine is not relevant for this study as the key aspect here is the comparison with the output power of a similarly sized PV-based system. Consequently, electric generators and/or gear boxes are not considered, neither efficiencies for these (possible) components are included in the model.

Use of the power P allows to define the solar energy conversion efficiency (or solar-to-user efficiency) as $\eta_{sol} \equiv P/(\varphi_{inc}A_C)$. The maximum solar efficiency is of course given by

$$\eta_{sol,\max}=P_{\max}/(\varphi_{inc}A_C) \qquad (5.9)$$

Note that in the case analysed here both η_{sol} and P have their maximum for the same value $T_{C,opt}$.

Results and Discussions

A horizontal solar collector will be considered. This is the easier-to-use strategy, that gives for many periods of time a performance less than 10% smaller than the strategies involving more sophisticated, orientable, solar collectors (Badescu et al, 2000a, 2000b, 2001a). Details about solar collector design are given in Table 4.7 (where the information about the bottom insulation should be omitted of course). The values of the effective transmittance- absorptance products were computed with $(\tau\alpha)_{dir}=(\tau\alpha)_{dif}=\tau\alpha$.

Three Stirling engine thermal agents are usually considered: carbon dioxide (adiabatic exponent $k = 1.33$), air ($k = 1.4$) and helium ($k = 1.66$). Helium gives the best performance and is our choice for this section. A simple analysis proves that the Stirling engine efficiency increases by increasing the compression ratio ε_V. Here the value $\varepsilon_V = 2$ is adopted as a good compromise between better performance and smaller engine size. Increasing the heat regeneration parameter x from the ideal value $x = 0$ diminishes the engine output power. Here the value $x = 0.2$ is accepted, in agreement with current terrestrial practice.

Results of computations performed by using the meteorological data during the autumn and winter of the first year at VL1 site are shown in Figure 5.2. It is obvious that both the maximum power $P_{\max}$ and the maximum solar efficiency $\eta_{sol,\max}$ increase around the solar noon. However, $\eta_{sol,\max}$ seems to be less dependent on solar time than $P_{\max}$. The maximum power of the solar Stirling engine around the noon is comparable in magnitude with the designed power of Sojourner's PV cells system (i.e. 16 W). Also, the solar efficiency is as large as 18 %, which is close to the design value of Sojourner's PV cells efficiency (about 18 %). Note, however, that the horizontally placed collector is a good solar energy collection strategy especially during autumn and winter. Indeed, results not shown here prove that during spring and summer the solar Stirling engine performance is slightly worse. During the warm season other strategies of collecting solar energy are much more effective (Badescu, 1998a).

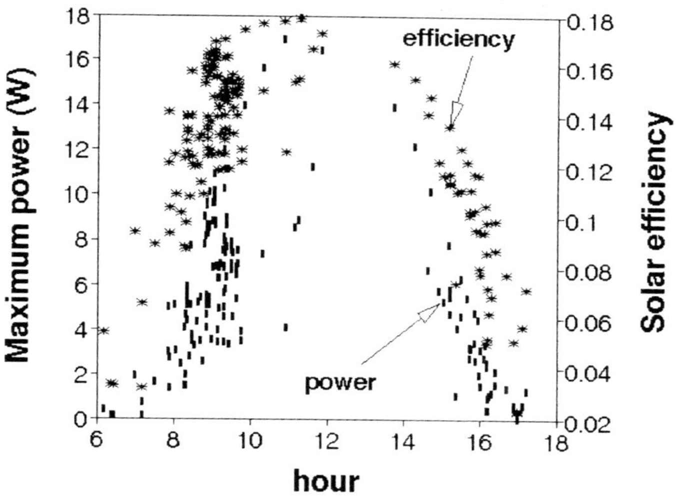

Figure 5.2. Dependence of maximum power $P_{\max}$ and maximum solar efficiency $\eta_{sol,\max}$ on local solar time (in Earth hours). Results obtained by using the meteorological data during the autumn and winter in the first year at VL1 site.

The large dispersion of the results in Figure 5.2 is mainly due to the influence of optical depth. The performances of solar Stirling engine obviously decrease by increasing the atmospheric optical depth (Badescu, 2004). The maximum efficiency $\eta_{sol,\max}$ is less dependent on the optical depth than the maximum power $P_{\max}$. Note than under-unitary optical depth values are usually associated with "clear sky" conditions while during the dust storm the optical depth has larger values (Badescu, 2001).

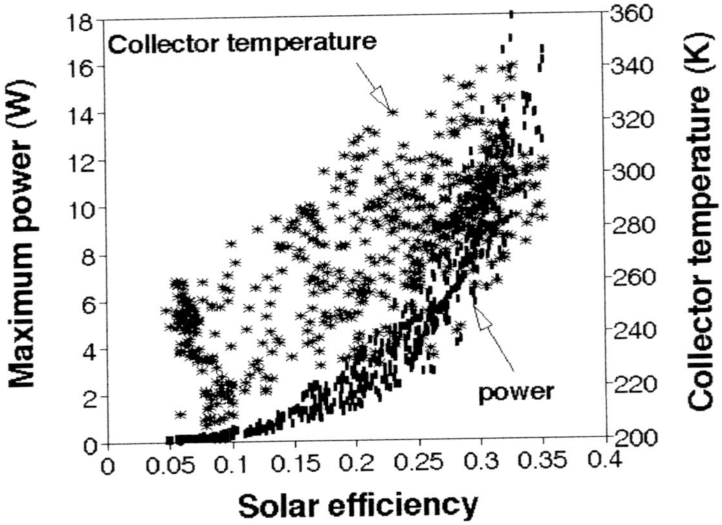

Figure 5.3. Dependence of the maximum power $P_{\max}$ and optimum collector temperature $T_{C,opt}$ on maximum solar efficiency $\eta_{sol,\max}$. All available meteorological data were used (see Table 2.1).

The dependence of the maximum output power $P_{\max}$ on optimum solar efficiency $\eta_{sol,\max}$ is shown in Figure 5.3. All the meteorological data of Table 2.1 were used. The maximum output power increases when the optimum solar efficiency increases. Generally, the dependence of $P_{\max}$ on $\eta_{sol,\max}$ seems to indicate a quadratic relationship. This is in agreement with previous results obtained in case of solar Carnot engines. A certain optimum solar efficiency threshold (around 5 %) must be exceeded in order the system provide useful power. Again, this confirms previous results reported in section 4. The optimum solar collector temperature $T_{C,opt}$ increases by increasing the solar efficiency, as expected. The larger dispersion of the $T_{C,opt}$ values for the same value of the solar efficiency is mainly due to the difference in the solar time.

Figure 5.4 shows that the overall heat loss coefficient U_L increases by increasing the collector temperature. The same set of meteorological data as in case of Figure 5.3 was used. The values of U_L range between 0.3 and 0.7 $Wm^{-2}K^{-1}$. Broadly, these values are smaller than those of Earth located flat-plate vacuum solar collectors (see e.g. Benz and Beikircher (1999) where the experimentally derived overall heat loss coefficient is about 1.25 $Wm^{-2}K^{-1}$) during a rather heavy utilization as steam production). They are slightly smaller than the results obtained during simulation of the Carnot solar engine in section 4 (U_L values between 0.5 and 1 $Wm^{-2}K^{-1}$). This is to be expected as an (ideal) Carnot engine normally operates at higher collector temperature than the Stirling engine considered here. The dependence of U_L on collector temperature justifies a posteriori the numerical optimization procedure we used in this section.

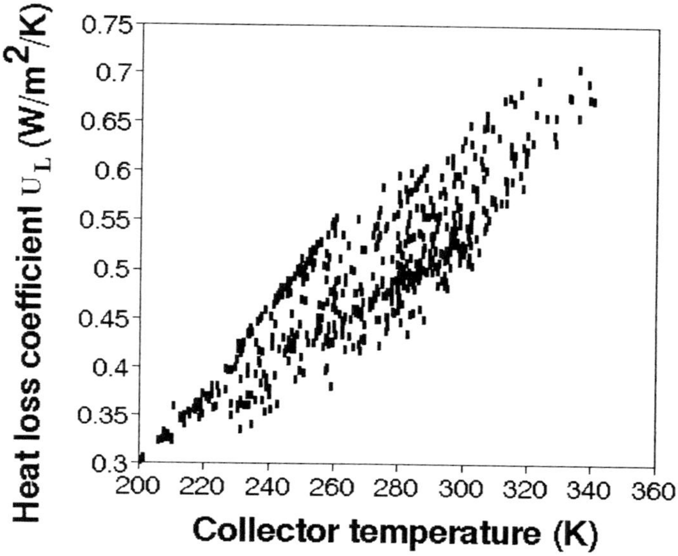

Figure 5.4. Dependence of overall heat loss coefficient U_L on collector temperature. All available meteorological data were used (see Table 2.1).

6. Conclusion

The chapter focuses on solar power generation on Mars surface. A "dynamic" solar power plant (which consists in a solar collector - thermal engine combination) is proposed as an alternative for the more usual photovoltaic cells.

Upper bounds for the efficiency of solar thermal power plants operating in the Martian environment are evaluated in section 3. Three different theories are usually quoted in literature to compute the maximum work that can be extracted from a given amount of thermal radiation energy. Using an original thermodynamic argument here we showed that these theories do not contradict each other but they predict upper bounds of different accuracy degree for the amount of work provided by a real conversion system. As far as the traditional thermodynamic approach is considered, the theory proposed by Jeter gives the exergy of thermal radiation. The above theories predict too high efficiencies for radiation energy conversion into work to be of practical interest. Much more accurate simple upper bounds were already proposed in literature. For reader convenience they are summarized in section 3.2 and applied to power plants operating on Mars under both direct and diffuse solar illumination. Both black-body and selective absorbers were considered.

A more elaborated model uses an endoreversible Carnot cycle to describe solar engine operation in section 4. The solar power plant is designed to operate during the whole Martian year. Consequently, it contains a *selective flat - plate* solar collector. A detailed model of collector heat losses towards the ambient is developed. The optimization procedure developed in section 4.1 is based of finite-time thermodynamics methods. The following two strategies of collecting solar energy were analysed in this chapter: (i) horizontal collector - strategy H; (ii) the collector tilt and orientation are continuously adjusted to keep the receiving surface perpendicular on Sun's rays - strategy P. The first strategy is easier to built and use while the later strategy provides higher collected solar energy most time of the year. A medium and a high efficiency thermal engine were analysed. Practical situations should normally lye between these two cases. All the computations were performed for a solar collector comparable in size with Pathfinder's Sojourner (solar energy collection area of 0.22 m^2). During the application we used as input data values of atmospheric pressure and temperature, wind speed and atmospheric optical depth measured at Viking Landers sites. The main results are:

(1) The heat losses between the solar collector absorber plate and transparent cover are exclusively by conduction. This differs from the Earth-based solar collector where the main heat loss mechanism is convection.
(2) The convection heat transfer above the transparent cover is laminar. This differs from the Earth-based solar collectors where turbulence is the main convection mechanism.
(3) For the design solution we selected, the values of the overall heat loss coefficient U_L range between about 0.5 and 1 $Wm^{-2}K^{-1}$. Roughly, these values are comparable with those of Earth located vacuum solar collectors. The dispersion of the U_L values is slightly smaller during winter and higher during autumn. No obvious dependence of U_L on the optical depth is observed. However, the dispersion of the U_L values is larger for values during the "clear" days. U_L does not depend significantly on the latitude of the solar collector.
(4) Two parameters are used to quantify the heat transfer inside the thermal engine. They are h_a (the heat transfer coefficient between the working fluid passing through the radiator of the heat engine and ambient; see Eq. (4.20)) and h_C (the heat transfer coefficient between solar collector and working fluid (see Eq (4.17)). The influence of h_a on performance is more significant.
(5) The strategy of collecting solar radiation has a rather weak influence on the dependence of performances on h_a and h_C. As expected, strategy P leads to higher performances.
(6) The dynamic solar power plants equipped with selective flat - plate collectors could provide power comparable to that of similar-size PV cell systems.
(7) The influence of latitude on performance is obvious. Generally, the solar efficiency is smaller at VL1 site as compared to VL2 site for both strategies of collecting solar radiation. In most cases the maximum power provided by a horizontal collector at VL1 site is smaller than at VL2 site. However, in some situations the meteorological effects compensate the latitudinal effects and the output power is quite similar at both VL1 and VL2 sites.

(8) In case a medium efficiency engine is coupled to a horizontal collector, the solar efficiency does not exceed 0.13 at VL2 site. Generally, it is lower during summer and higher during winter's dust storms. In case the medium efficiency engine is coupled to a P collector, the solar efficiency increases during summer.
(9) The solar efficiency is as high as 0.18 in case of a horizontal collector attached to a high efficiency engine. If the high efficiency thermal engine is connected to a P collector the solar efficiency increases significantly during summer and spring but does not exceed 0.18.
(10) Using a high efficiency engine coupled to a horizontal collector leads to output power up to 13 W during spring, autumn and winter.
(11) The P strategy is recommended mainly during summer and spring in combination with high efficiency engines. In this case the output power could be as high as 25 W.
(12) During an autumn day at VL1 site (sol 301) the optimum engine efficiency is around 33 %. The system efficiency lies around 20 % for the high efficiency engine and around 15 % for the medium efficiency engine. The optimum solar efficiency doesn't exceed 15 % and 10 %, respectively, in the two cases. The maximum output power can exceed 15 W and doesn't exceed 8 W for the high and medium efficiency engine, respectively. .
(13) During a winter dust-storm day at VL1 site (sol 328) the maximum output power is much smaller than during autumn. At noon it reaches 10 W in case of the high efficiency thermal engine but doesn't exceed 5 W in case of the medium efficiency engine.
(14) During a spring day at VL2 site (sol 420) the optimum solar efficiency values are quite close to those from autumn and summer. The high efficiency engine connected to a solar collector kept perpendicular on Sun's rays can ensure a maximum output power around 15 W during most part of the day. This is close to the design power of the PV system of Mars Pathfinder's Sojourner.
(15) During an autumn day at VL2 site (sol 872) the optimum solar efficiency ranges from 8 to 16 % in case of a good quality engine. The high efficiency thermal engine should be used in combination with a solar collector kept perpendicular on Sun's rays. A maximum output power of 15 W might be provided a short time period during the day.
(16) When a horizontal solar collector is considered, the dependence of the maximum output power on optimum solar efficiency seems to be quadratic at both VL1 and VL2 sites. When a collector perpendicular on Sun's rays is considered, this dependence is more complicate, but keeps the quadratic feature.
(17) A certain optimum solar efficiency threshold (around 5 %) must be exceeded in order the system provide useful power.
(18) The dependence of the maximum output power on the optimum solar efficiency is stronger in case of a high efficiency thermal engine than in case of a medium efficiency engine, on one hand, and in case of a collector kept perpendicular on Sun's rays than in case of a horizontal collector, on the other one hand.
(19) No obvious difference exists between power plant performances in the two years of VL2 operation.

A solar Stirling engine based on a horizontal selective flat-plate converter is analyzed in section 5. A numerical optimization procedure was used to maximize the power provided by the engine. The main results are as follows:

(1) The values of the overall heat loss coefficient U_L range between about 0.3 and 0.7 $Wm^{-2}K^{-1}$.
(2) The solar energy conversion efficiency at noon could be as high as 0.18 at VL2 site during autumn and winter. Generally, it is lower during spring and summer.
(3) The output power at noon produced by the Stirling engine at VL2 site was as high as 16 W during autumn and winter. It obviously decreases during the dust storm periods.
(4) Properly designed solar Stirling engines equipped with horizontal flat-plate collectors could provide output power comparable in magnitude to that of PV cell systems. However, this conclusion should be treated with caution as it is based on figures referring to an existing PV system, on one hand, and to a simplified solar Stirling engine model, on the other one hand.
(5) To increase the output power and the efficiency, the design parameters of Stirling engines operating on Mars should be different from those of terrestrial engines. For example, one expects the compression ratio to exceed the usual value $\varepsilon_V = 2$ while a smaller regeneration factor than $x = 0.2$ to be implemented.

ACKNOWLEDGMENTS

I am indebt to my father, Mihail (Misu) Badescu, who first showed me the rules. I thank Prof. Gheorghe Popescu and Dr. Monica Costea (Polytechnic University of Bucharest) and Prof. Michel Feidt (University of Nancy) for previous collaboration.

REFERENCES

Aldrich, A.D. In Mars: Past, present and future; Pritchard E.B.; Ed.; *Progress in Astronautics and Aeronautics;* AIAA: Washington, DC, 1992; Vol 145, pp 3-11.

Angelino G.; Invernizzi C. *J Solar Energy Engng*. 1993, 115, 130-137.

Angelo J.A.; Buden D. Jr, The nuclear power satellite (NPS) - key to a sustainable global energy economy and solar system civilization; *Proc. SPS 91*; Paris, 1991; pp 117 - 124.

Badescu V.; *J. Sol. Energy Engng*. 1988a, 110, 349.

Badescu V.; *Entropie* 1988b, 145, 41-45.

Badescu V.; *J Phys D* 1990, 23, 289-292 .

Badescu V.; *J Phys D* 1991, 24, 1882-1885.

Badescu V.; *Int J Energy* 1992, 17(6), 601-607.

Badescu V.; *Space Technol* 1994, 14(5), 331-337.

Badescu V.; *Acta Astronautica* 1998a, 43 (7-8), 409-421.

Badescu V.; *Acta Astronautica* 1998b, 43(9-10), 443-453.

Badescu V.; *Phys Lett A* 1998c, 244, 31-34.
Badescu V.; *J Phys D* 1998d, 31, 820-825.
Badescu V.; *J. Non-Equilib. Thermodyn.* 1999, 24, 196-202.
Badescu V.; *Int. J Solar Energy* 2000, 20, 149-160.
Badescu V.; *Renewable Energy* 2001, 24, 45-57.
Badescu V.; *J. Sol. Energy Engng.* 2004, 126, 812-818.
Badescu V.; Popescu G.; Feidt M.; Model of optimized solar heat engine operating on Mars, *Proc of ECOS98*; Bejan A.; Feidt M., Moran M. J.; Tsatsaronis G.; Eds.; Nancy, France, 1998; pp 813-819.
Badescu V.; Popescu G.; Feidt M.; *Energy Conv Mngmnt.* 1999, 40, 1713-1721.
Badescu V.; Popescu G.; Feidt M.; *Renewable Energy* 2000a, 21, 1-22.
Badescu V.; Popescu G.; Feidt M.; *J British Interplanetary Soc.* 2000b, 53 (3/4), 131-144.
Badescu V.; Popescu G.; Feidt M.; *Acta Astronautica* 2001a, 49 (12), 667-679.
Badescu V.; Popescu G.; Feidt M.; Costea M.; *Termotehnica* 2001b, 5(1), 24-28.
Bejan A.; *Advanced engineering thermodynamics*; Wiley: New York, 1988.
Bejan A.; *J Sol Energy Engng.* 1987, 109, 46-51.
Benz N.; Beikircher T.; *Solar Energy* 1999, 65, 111-118.
Candau Y.; *Solar Energy* 2003, 75, 241-247.
Collozza, A.J.; *Preliminary design of a long - endurance Mars aircraft*, NASA CR185243, Sverdrup Technology Inc., Aerospace Technology Park, Brookpark, Ohio 44135, prepared for Lewis Research Center under Contract NAS 3-25266, April 1990.
Duffie J. A.; Beckmann W. A.; *Solar Energy Thermal Processes*; Wiley: New York, 1974.
Eaton C.B.; Blum H.A.; *Solar Energy* 1975, 17, 151-158.
Golombek M. P.; Cook R. A.; Economou T.; Folkner W. M.; Haldermann A. F. C.; Kallemeyn P. H.; Knudsen J. M.; Manning R. M.; Moore H. J.; Parker T. J.; Rieder R.; Schofield J. T.; Smith P. H.; Vaughan R. M.; *Science* 1997, 278 (5344), 1743-1748.
Hibbs B.D.; Mars rover feasibility study, *Final Report Aero Vironement*, Inc, Report AV -FR 89/7011, October 1989.
Hourdin F.; Forget F.; Talagrand O.; *J. Geophys. Res.*1995, 100(E3), 5501-5523. .
Howell J. R.; Bannerot, R.B.; *Solar Energy* 1977, 19, 149-153.
Jeter S. J.; *Solar Energy* 1981, 26, 231-236.
Landsberg P. T., Tonge G.; *J Appl Phys* 1979, 51, R1-R20.
Landsberg P.T.; Mallinson, J. R.; Thermodynamic constraints, effective temperatures and solar cells, In *Coll. Int. sur l'Electricite Solaire*, CNES: Toulouse, 1976, pp 27-35.
Landsberg P. T.; Badescu V.; *Europhys Lett.* 2000, 50(6), 816-822.
Lee S.W.; *Viking Lander Meteorology and Atmospheric Opacity Data Set Archive,* Volume VL-1001, Laboratory for Atmospheric and Space Physics, Campus Box 392, University of Colorado, CO 80309-0392, (10 July 1995).
Lide D. R.; Ed.; *Handbook of chemistry and physics*, 71st Edition; C.R.C. Press, 1991, pp 15-39.
Martin L. J.; Zurek R. W.; J. *Geophys. Res.*1993, 98(E2), 3221-3246.
McKissock, B.I.; Kohout L.L.; Schmitz P.C.; A solar power system for an early Mars expedition, *NASA Technical Memorandum 103219*, Lewis Research Center, Cleveland, Ohio, American Institute of Chemical Engineers, Summer National Meeting, August 19-23, 1990.

McLallian K.L. et al., The solar dynamic radiator with a historical perspective, *Proceedings of the 23rd International Energy Conversion Engineering Conference*, Denver, CO, vol 3, ASME, July 31-Aug 5 1988, pp. 335-340.

Meinel A. B.; Meinel M. P.; *Applied Solar Energy*, Addison-Wesley Publishing Company: Reading, 1976.

Menetrey W. R.; In *Introduction to the utilization of solar energy;* Zarem A. M.; Erway D. D.; Eds.; Mc Graw Hill: New York, 1963, p. 326.

Mozjorine Y.A.; Senkevich V.P.; Koval A.D.; Narimanov E.A.; Small - scale space power stations: feasibility and usage prospects; *Proc. SPS 91*, Paris, 1991, pp. 381 - 392.

Petela R.; J Heat Transfer 1964, 86, 187-192.

Petela R.; *Solar Energy* 2003, 74, 469-488.

Pop M. G.; Leca A.; Prisecaru I.; Neaga C.; Zidaru G.; Musatescu V.; Isbasoiu E. C.; *Indrumar -Tabele, monograme si formule termotehnice*; Editura Tehnica: Bucuresti, 1987, vol 1.

Pollack J. B.; Haberle R. M.; Murphy J. R.; Shaeffer J.; Lee H.; *J. Geophys. Res.*1990, 95, 1447-1473.

Press W. H.; *Nature* 1976, 264, 734-735 .

Prisnjakov V.; SPS interest and studies in USSR, In *Proc. SPS 91*, Power from space, Paris 27-30 August 1991, p 36.

Prisnjakov V. F.; Statsenko I. N.; Kondratjev A. I.; Markov V. L.; Petrov B. E.; Gabrinets V. A.; Developing space power Brayton systems with solar heat input. Research of working process of high temperature latent heat storage system, In *Proc. SPS 91*, Power from space, Paris 27-30 August 1991, p. 465-470.

Prisnjakov V. F.; Statsenko I. N.; Kondratjev A. I.; Markov V. L.; Petrov B. E.; Gabrinets V. A.; *Space Power* 1994, 13(3&4), pp. 135-144.

Secunde R.; Labus T. L.; Lovely R. G.; Solar dynamic power module design, In Proc 24th *International Energy Conversion Conf.*, IEEE: Piscataway NJ, 1989, Vol. 1, pp 299-307.

Senft J. R.; *An introduction to low temperature differential Stirling engines*; Moriya Press: River Falls, WI, 1996.

Spanner D. C.; *Introduction to thermodynamics*, Academic Press: London, 1964, p 218.

Stefanescu D.; Marinescu M.; Danescu A.; *Transferul de caldura în tehnica*; Editura Tehnica: Bucuresti, 1982, Vol. 1.

Weingartner S.; Blumenberg J.; Ruppe H. O.; *Space Power* 1994, 13 (1&2), 103-120.

Zurek R. W.; Barnes J. R.; Haberle R. M.; Pollack J. B.; Tillman J. E.; Leovy C. B.; In *Mars*; Kieffer H. H. et al; Eds.; University of Arizona Press, 1992, pp 835-933.

In: Solar Energy: New Research
Editor: Tom P. Hough, pp.199-238
ISBN 1-59454-630-4

Chapter 6

THERMOSYPHONS AND THERMODIODES FOR SOLAR ENERGY UTILIZATION

Kuan Chen
Department of Mechanical Engineering, University of Utah, SLC, UT 84112, U.S.A.
Wongee Chun
Department of Nuclear and Energy Engineering, Cheju National University, Cheju 690-756, Korea.

ABSTRACT

Thermosyphons are active heat transfer devices that utilize natural convection in a closed or open loop for effective heat transfer. The thermosyphon idea has been commonly used in domestic solar water heaters as well as other solar energy systems and heat transfer processes. Based on the working principle of thermosyphon, several thermodiode designs which allow heat to be transferred in the desired direction but block the heat flow in the opposite direction were developed and tested for solar heating applications. More recently thermodiodes with easily alterable direction of heat flow, so-called bi-directional thermodiodes, were invented and their performances tested and optimized for both summer-cooling and winter-heating of buildings and shelters. Presented in this chapter are a review of single-phase thermosyphons and thermodiodes applied to solar energy utilization and energy-efficient buildings, and the authors' and co-workers' research results in this area.

The authors' work includes heat transfer analyses and optimizations of natural convection loops of different configurations and operating conditions, different designs of bi-directional thermodiodes, and lab test results as well as year-round test data collected from experiments on energy–efficient construction modules under actual weather conditions in the United States and Korea. The performance and efficiencies of different thermodiode designs are compared and discussed.

THERMOSYPHONS

The thermosyphon is a heat transfer device of simple design and construction and consuming no electrical or mechanical work. It typically consists of a fluid-filled open or closed loop, and a heat source and sink inside the loop or on the loop surface (see Figure 1). Natural convection can be induced inside the loop for effective heat transfer due to the buoyancy effect resulting from the temperature difference in the working fluid. The buoyancy effect is the consequence of the presence of a body-force acceleration, gravitational or centrifugal for example, and density gradients in the fluid. The stratified fluid is unstable for certain angles between the body-force acceleration and the stratification direction, and a circulation of the working fluid will be induced throughout the loop when the buoyancy force exceeds a critical value.

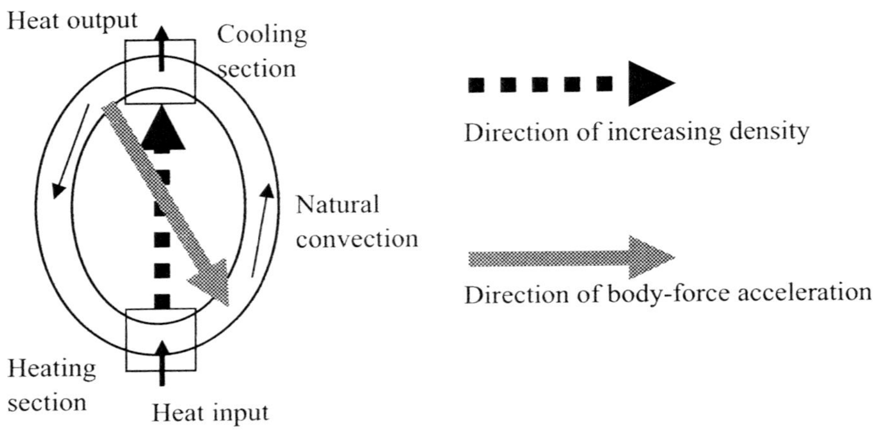

Figure 1. Natural convection induced in a thermosyphon.

The onset of flow circulation and thermal instability of natural convection loops are much more complicated than those of a fluid layer subjected to uniform heating or cooling [Rosenhead 1963; Turner 1973]. It can be proven in theory and experiments had verified that the heat source of a natural convection loop is not necessarily to be above the heat sink – a necessary condition for the induction of natural convection in a fluid layer between two surfaces of uniform thermal boundary conditions. As a matter of fact, the heat source is directly above the heat sink in the anti-gravity thermosyphons developed by the National Engineering Laboratory during the 1970s [Dunn and Reay 1994]. However, the induced convective heat transfer rate is generally the highest when the direction of the body-force acceleration and the direction of decreasing fluid density are parallel to each other. This explains why the storage tank of a natural circulation solar water heater is often installed directly above the collector panel as shown in Figure 2.

The design and performance of thermosyphon solar water heaters had been studied quite extensively in the 1970s and 1980s. Both simplified analytical models [e.g., Morrison and Tran 1984; Huang and Hsieh 1985; Norton et al. 1987] and rigorous numerical [Ong 1974, 1976] and hybrid (numerical/analytical) [Hobson and Norton 1988] schemes were developed to design and predict the performance of this type of solar water heaters. Experiments with thermosyphon solar water heaters have also been carried out under real climates [e.g., Shitzer et al. 1978; Kang, Kang, and Chun 2003] as well as simulated operating conditions in a lab [Hobson and Norton 1988].

Figure 2. Photo of a domestic solar water heater (courtesy of Korean Institute of Energy Research).

Applications of the thermosyphon idea are not limited to solar energy systems. Thermosyphons have also been used for cooling internal combustion engines [Kulhavy 1964], nuclear reactors [Hellman et al. 1956] and turbine blades [Ivanov and Lapin 1966]. More recent applications include two-phase, micro-structured thermosyphons (MSTS) for cooling desktop and laptop computers [A-STAR 2005], capillary-assisted thermosyphon (CAT) loops for shipboard electronics cooling [Larsen 2003], and so on. Since a pump or a fan to produce flow circulation is not needed, thermosyphons are active heat transfer devices. They are inexpensive and reliable, invulnerable to power outage or pump failure. On the other hand, a sufficiently large temperature difference and/or gravitational or centrifugal acceleration is needed to overcome the friction loss if too long or too small flow channels are used.

Without wick materials in the loop or tiny grooves fabricated on the loop inner surface, the construction of the thermosyphon is much simpler than that of the heat pipe. With proper design and/or through the use of phase–change materials, fairly high heat transfer rates can be achieved by a thermosyphon. It could be applied as effectively as a heat pipe to harnessing solar energy and no appreciable difference in performance arises from the selection of either heat transfer device [Chun et al. 1999]. Both single-phase thermosyphons and thermosyphons utilizing a phase-change fluid [Kang, Kang, and Chun 2003] have been developed for solar energy utilization. The latter generally has a higher heat transfer rate, but requires good sealing and strong construction to withstand the large pressure variations accompanied by the phase-change processes. Attention in this chapter will be focused on single-phase thermosyphons. The simple design and construction of the single-phase thermosyphon makes it easier to operate and analyze. Furthermore, the dominant thermal resistances in the low-temperature solar systems we investigated were found to be those associated with convection and radiation heat transfer on the thermosyphon surfaces. Improving the heat transfer rate inside the thermosyphon has little effect on the overall energy efficiency and system performance.

Despite its simple working principle, the single – phase thermosyphon can still have a high heat transfer rate if properly designed and operated. In their experimental investigation of liquid–filled closed loops, Madjeski and Mikielewicz [1971] demonstrated that the heat transfer rate of these "liquid –fins" (Figure 3) can be 300 times higher than that of copper fins of identical dimensions and configuration. The single-phase thermosyphon is attractive to simple and economical energy systems. Environmentally safe fluids such as water can be used in single-phase thermosyphons, and leakage and refill of the working fluid is easy to handle.

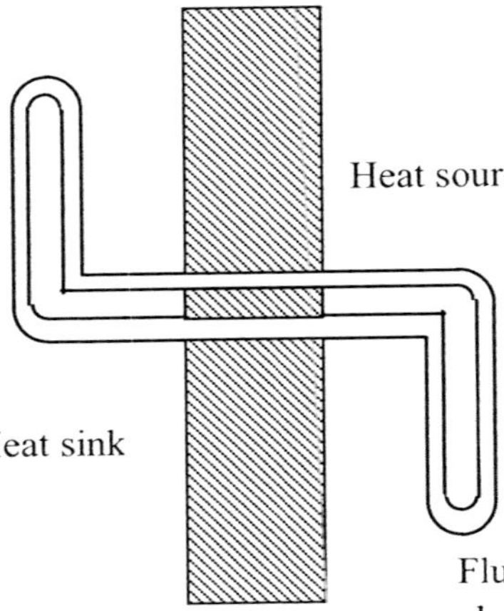

Figure 3. Liquid fin tested by Madjeski and Mikielewicz [1971].

The heating process of a single-phase thermosyphon can be divided into two intervals. During the start-up interval small, three-dimensional convection cellulars appear in the heating section, but the buoyancy force generated by the fluid density difference between the heating and cooling sections is not large enough to induce a large convection cell throughout the loop. The complicated three-dimensional flows in the heating section during start-up are more suitable for numerical simulations or empirical studies. Alternatively, the lumped-system approach in conjunction with a heat conduction analysis between the heat source and sink often yields an approximation solution with satisfactory accuracy for most low-temperature solar systems during start-up.

Other phenomena encountered in natural convection loops which are of great academic interest are the instability and secondary flows of the working fluid when subjected to a high energy input rate. The stability criteria, oscillatory flow, and flow reversal phenomena in various natural convection loops have been studied quite extensively, both analytically and experimentally [Keller 1966; Welander 1967; Creveling et al. 1975; Damerell and Schoenhals 1978; Greif, Zvirin, and Mertol 1979 a,b, and 1981; Kaizerman et al. 1981; Bau and Torrance 1981; Chen 1985 b]. Numerical simulations had also been carried out to investigate the evolutions of unstable flows in various natural convection processes. These problems have not received much attention in the past for solar energy applications, probably because the input energy fluxes for most solar energy systems with planar collectors are seldom high enough to exceed the stability criteria. Unstable and oscillatory flows observed in the experiments of natural convection loops with high energy inputs should be taken into consideration in the design and operation of thermosyphons to be applied to high-temperature solar systems.

After the short start-up period which is characterized by three-dimensional convection cellulars in the heating section, a convection cell throughout the loop is induced. Although today's powerful CFD (Computational Fluid Dynamics) codes and computers are capable of simulating natural convection processes under almost any operating conditions, a one-dimensional analytical solution often yields satisfactory accuracy for steady-flow natural convection in a closed loop. Analytical solutions also provide more physical insights about the flow and energy transfer processes involved. In addition, normalization and dimensional analysis become much easier if the flow can be approximated as one-dimensional. For these reasons, one-dimensional analytical solutions have been widely applied to thermosyphon designs and analyses, and to obtain the pertinent dimensionless parameters.

A closed-form solution was obtained in Chen's analytical study of convective heat transfer in a single-phase, closed-loop thermosyphon [1985 a]. Despite the simplifications and approximations made in this one-dimensional, steady-flow analysis, results and conclusions of Chen's study shed much light on the design and operation of closed-loop thermosyphons, and can be applied to other thermosyphon designs.

The closed–loop thermosyphon studied by Chen consists of two insulated vertical branches with two horizontal connections of uniform surface temperature, as shown schematically in Figure 4. The loop is filled with a fluid which remains in one phase during operation and the cross-sectional area of the flow channel is constant. The lower case s in the drawing is the one- dimensional coordinate along the loop and T_H and T_L are the loop surface temperatures of the heated and cooled sections, respectively. The space between the heat source and sink is well insulated.

For one-dimensional steady flow in a closed loop of constant flow cross-sectional area, the equation of motion, after invoking the Boussinesq approximation, can be expressed as :

$$\pi r^2 \rho_0 V \frac{d}{ds} \oint V ds = -\pi r^2 \oint dp$$

$$-g\pi r^2 \oint \rho \cos\alpha ds - 2\pi r \oint \tau ds \qquad (1)$$

where V is the mean fluid velocity in the loop, τ the wall shear stress, ρ and p the fluid density and pressure, α the angle between the flow direction and the vertical, g the gravitational acceleration, and r the tube inner radius, respectively. All fluid properties except for ρ are assumed constant and evaluated at a reference state.

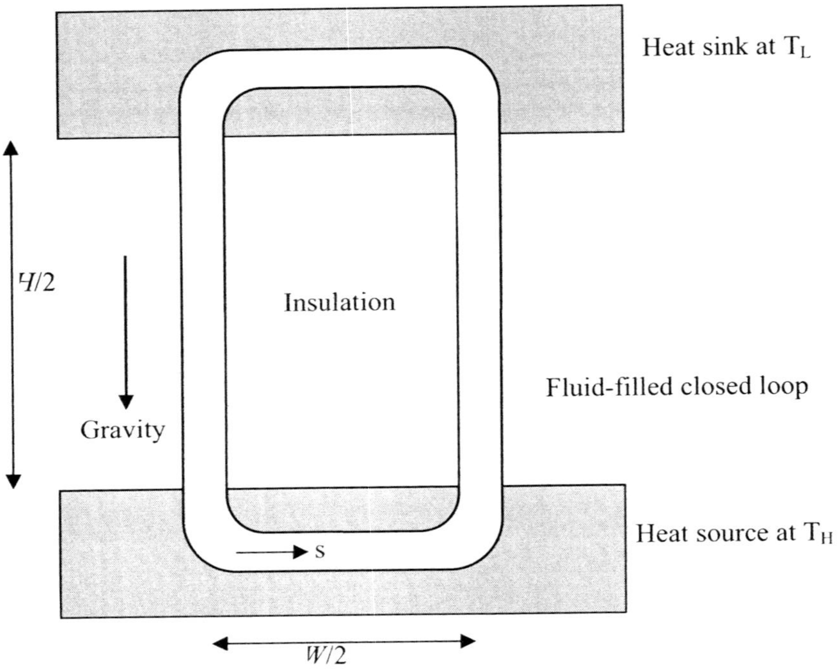

Figure 4. Rectangular – loop thermosyphon with insulated vertical branches.($\Delta T = T_H - T_L$)

If heat conduction in the axial direction can be neglected in comparison with convective heat transfer, the energy equation for the fluid flow becomes:

$$\pi r^2 \rho_0 c V \frac{dT}{ds} = 2\pi r h (T_w - T) \tag{2}$$

where T is the fluid mean temperature, c the fluid specific heat, and T_w the channel wall temperature. With the relationships between τ and V and between the heat transfer coefficient (h) and the mean velocity determined either empirically or theoretically, the mean velocity and mean fluid temperature can be calculated from the coupled momentum and energy equations for any heating and cooling conditions.

For the rectangular loop shown in Figure 4, the normalized governing equations involve only one dimensionless parameter:

$$P^0 = \frac{a(1+A)}{2^{4-b} Gr_r} \left(\frac{mW}{r \Pr} \right)^{\frac{2-b}{1-n}} \tag{3}$$

where A is the aspect ratio of the loop ($A = W/H$), and Pr and Gr_r the Prandtl and Grashof number (based on the flow channel radius), respectively. The normalized mean velocity and heat transfer rate can be expressed as :

$$V^0 = \left(\frac{\rho_0 crv}{mH\kappa A}\right)\left(\frac{2Vr}{\nu}\right)^{1-n} = (2P^0)^{\frac{1-n}{b-2}} \tag{4}$$

$$Q^0 = \frac{2Q}{\nu\rho_0 c\pi r\Delta T}\left(\frac{\rho_0 crv}{mW\kappa}\right)^{\frac{1}{1-n}} = (2P^0)^{\frac{1}{b-2}} \tag{5}$$

where ν and κ are the kinematic viscosity and thermal conductivity of the working fluid.

In the above equations a, b, m, and n are the coefficients and exponents in the correlations for wall shear stress and heat transfer coefficient:

$$\tau = \frac{a\rho_0 \nu^b V^{2-b}}{8r^b} \tag{6}$$

$$h = \frac{m\kappa}{2r}\left(\frac{2Vr}{\nu}\right)^n \tag{7}$$

The normalized mean fluid temperature is a function of V^o and s^o :

$$\theta = C_1 e^{\frac{-2s^0}{V^0}} + \frac{1}{2} \text{ for } s^0 = 0 \text{ to } \frac{1}{2}$$

$$= -C_1 e^{-2[s^0-(1+A)/(2A)]/V^0} - \frac{1}{2} \text{ for } s^0 = \frac{1+A}{2A} \text{ to } \frac{1+2A}{2A} \tag{8}$$

where

$$C_1 = -\frac{e^{-\frac{1}{V^0}} - 1}{1 - e^{-\frac{2}{V^0}}} \tag{9}$$

and s^o is the non-dimensional coordinate:

$$s^o = s/W \tag{10}$$

The computed non-dimensional temperature θ for laminar ($b = 1$, $n = 0$) and turbulent ($b = 0$, $n = 0.8$) flows in the rectangular loop are depicted in Figure 5. The mean fluid temperature increases when passing through the heating section($s^o = 0$ to $1/2$), and decreases when passing through the cooling section ($s^o = (1 + A)/(2A)$ to $(2A + 1)/(2A)$).

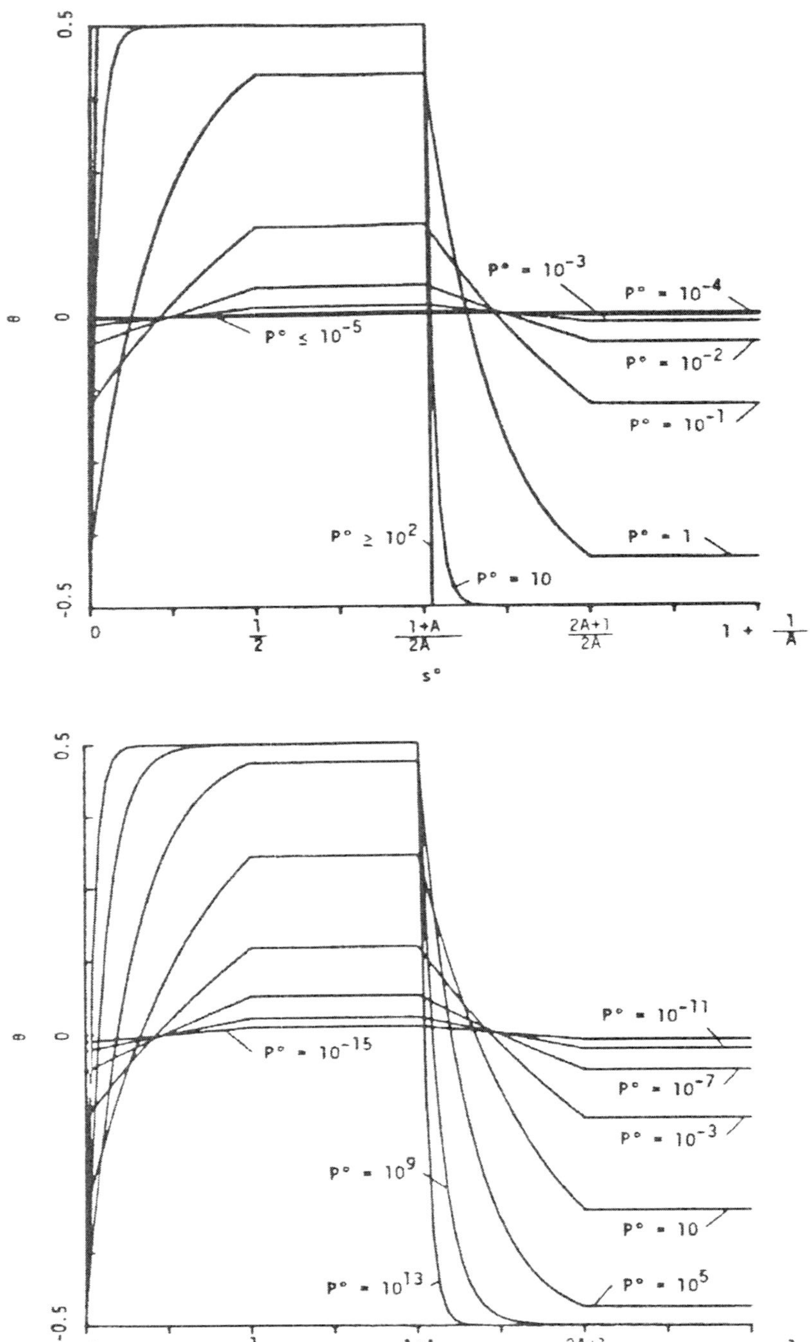

Figure 5. Fluid temperature variations along the rectangular-loop thermosyphon for different values of P^o. (top) Laminar flow, (bottom) Turbulent flow [Chen 1985a].

It remains unchanged in the two insulated branches connecting the heating and cooling sections if axial conduction and heat loss through the insulating materials are negligible. Temperature changes during heating and cooling depend on fluid properties, mean velocity, and the dimensions of the loop. Variations of the fluid temperature along the loop increase as P^o increases, but are always less than the difference between the heat source and sink temperatures. A high P^o value, which corresponds to long heat transfer sections, a small channel cross-sectional area, a low mean velocity, and so forth, will make the fluid temperature approaches the heat source or sink temperature very rapidly during heating or cooling.

It was pointed out in Chen's analytical study [1985 a] that the optimum thermosyphon design and operation is the one with

$$P^0 = 2^{\frac{n-b+1}{1-n}} \qquad (11)$$

for a given loop height. This corresponds to P^o between 1 and 10 in Figure 5(top) and between 10^5 to 10^9 in Figure 5(bottom). Too large a P^o value causes the fluid temperature to reach the heat source or sink temperature before leaving the heat transfer section. In the remainder of the heat transfer section there is no heat transfer but friction loss between the fluid and the channel wall. Conversely for too small a P^o value the fluid temperature variations during heat transfer are too small, resulting in too low a buoyancy force in the loop for strong natural convection. The optimal thermosyphon design and operation should make θ approach the heat source or sink temperature at the exit of the heating or cooling section. In this case the highest flow rate is induced and the heat transfer rate is the maximum for a given temperature difference ΔT and H. This principle about optimum loop configuration is applicable to other thermosyphons of different designs and heat transfer conditions; that is, the heating and cooling sections of any thermosyphon should be just long enough for the working fluid to approach its maximum or minimum possible temperature. Too short heat transfer sections produce little buoyancy force while too long heat transfer sections result in additional friction losses.

The above flow and heat transfer analyses can be applied to other closed–loop thermosyphons under steady-flow conditions. Different mean velocities in different flow channels can be determined from the conservation of mass requirement if the flow cross sectional area is not uniform or manifolds are used in the flow system. The vertical component of the gravitational acceleration should be used for the buoyancy force term in the momentum equation if the flow direction is not parallel to the vertical direction. If surface heat flux instead of surface temperature is specified, the energy equation should be revised as:

$$\rho_o c V \pi r^2\, dT/ds = q_w(s)\mathrm{P}(s) \qquad (12)$$

where q_w and P are the circumferentially-averaged surface heat flux and channel perimeter at location s.

In an open-loop thermosyphon the fluid leaving a flow channel is mixed with the fluid in a mixing chamber or a storage tank during heating and/or cooling. Due to the effective and rapid mixing of the hot and cold streams in the chamber or tank, a lumped system approach is often employed for the energy balance calculation. In this approximation the fluid temperature in the storage tank is assumed to be uniform at any instant of time, and is determined from an overall energy balance applied to the tank. Included in the overall energy balance calculation are the rate of increase in thermal energy storage in the tank, the energy transported by the inflow and exit streams, and heat losses from the tank surfaces to the surroundings. Since the heat loss rate is dependent on the storage tank temperature, an iterative scheme is often needed to determine the steady-state temperature of the fluid in the tank. Recourses must be made to CFD methods and codes if detailed information about the flow patterns and mixing between the hot and cold streams in the storage tank are needed [e.g., Chen et al. 1998]

THERMODIODES

In many solar energy applications as well as other heat transfer processes, it is necessary to direct the flow of heat in a desired direction but block the heat flow in the opposite direction. Heat transfer devices of this feature are termed thermodiodes – analogous to the electrical diodes which are commonly employed in electrical circuits for rectification. Heat pipe thermodiodes have been applied to fine temperature control of spacecraft components, electronics in missiles, and cryogenic applications [Williams 1978; Dunn and Reay 1994]. Thermodiodes with fast shut-down time can be used as a thermal switch. Thermodiodes with variable- conductance, which can be achieved by insulating part of the heat source or sink surface area, are routinely used in thermal control and precision temperature calibration on satellites and electronic devices and components. A good collection of research results and applications of heat pipe thermodiodes can be found in Dunn and Reay's book [1994].

The natural convection loops discussed in the preceding section are examples of single-phase thermodiodes. Except for liquid metals, thermal conductivities of most fluids are very low, thus the Peclect number (the product of the Reynolds and Prandtl numbers - it has the physical interpretation of the ratio of convective heat transfer to heat conduction in the flow direction) of a natural convection loop is typically very high under the designed conditions, and axial conduction is negligible. Consequently heat is transferred effectively by convection if the body-force vector in Figure 1 is pointing from the heat sink to the heat source. The thermodiode is under forward bias in this case. The thermodiode is said to be reverse biased if the body-force vector is pointing in the direction of increasing fluid density, and only heat conduction takes place in the loop between the heat source and sink.

The heat transfer rate of the rectangular-loop thermodiode (Figure 4) under forward bias has been discussed in the preceding section. The heat leakage rate of this diode under reverse bias can be estimated from:

$$Q_{rev} = 4\,\pi r^2\,\kappa \Delta T/H \tag{13}$$

if only heat conduction through the fluid in the two vertical branches takes place. Other thermodiode configurations designed and tested by the authors and co-workers include the bayonet diode [Chen et al. 1998] and the loop-and-tank diode [Chun and Chen 2002] depicted in Figures 6 and 7. The bayonet thermodiode is a variation of the tank-and-tongue liquid diode developed by Jones [1986] for space heating. More insulation was added to Jones original design for less heat conduction loss between the tank and the tongue, and a longer cavity was used in the bayonet diode for more surface area facing the outdoors. In addition, a guide vane is inserted in the cavity and the flow channel connecting the tank and the cavity to reduce heat exchange between the ascending and descending streams. The measured efficiency of the bayonet diode under forward bias was about the same as that of Jones tank-and-tongue diode [Chen et al. 1995; Chu 1996]. The advantages of the bayonet diode are a larger cavity surface area for more solar energy absorption, and a longer channel between the cavity and the storage tank. The latter alteration allows more thermal insulation to be added between the heat source and sink, and less heat conduction loss through the connecting channel when the diode is reverse biased.

The dynamic behavior of the bayonet diode under real weather conditions was investigated by Chen et al. [1998]. Measured solar flux and atmospheric and indoor temperatures in Salt Lake City, Utah, over a 16-hour period were used as the input data for their analytical model and numerical simulations. Shown in Figure 6 are the diode dimensions and coordinate system in their analysis. Heat conduction through the insulation was neglected. The analytical model for winter heating application of the bayonet thermodiode is briefly discussed below.

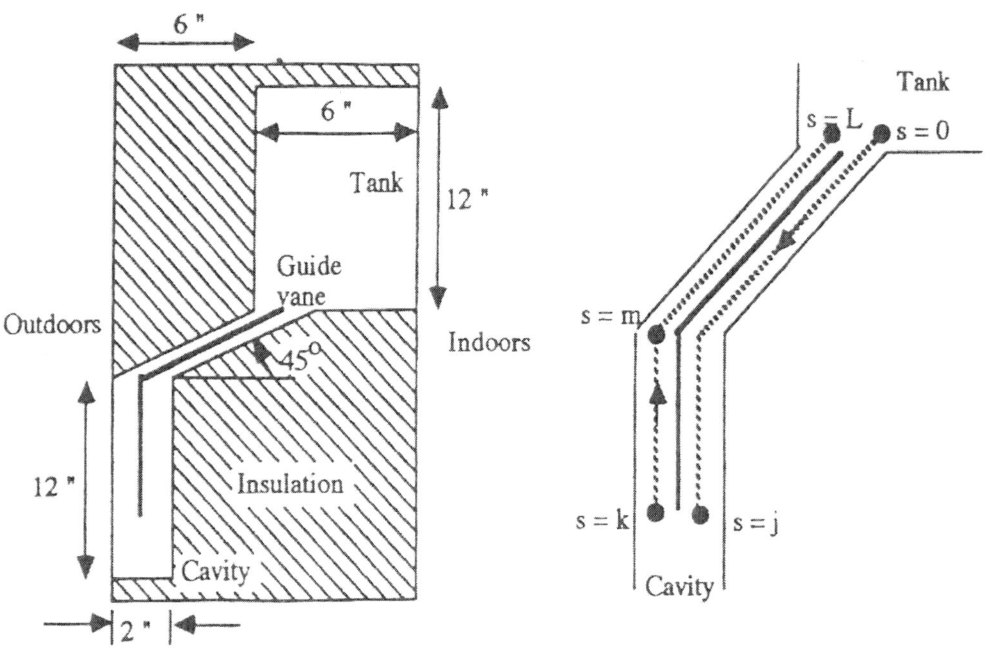

Figure 6. Configuration and coordinate system of the bayonet thermodiode investigated by the authors. Diode dimensions are in inches [Chen et al. 1998].

The analysis began from the instant when the sun started shining on the diode. The heating phase is divided into two intervals. The first one is a short start-up period during which the water in the cavity was colder than that in the tank, and was heated up by solar energy absorbed by the cavity surface. The fluid in the cavity and the inclined channel was broken into many small convection cells. There is a large but weak (slow-moving) convection cell in the storage tank induced by heat transfer through the connecting channel and natural convection on the right vertical wall of the tank, as predicted by Chen et al.'s numerical simulation using FLUENT and observed in Jones and Behnia et al.'s experiments [1986; 1987]. The assumption of one-dimensional flows in the connecting channel is not valid and heat transfer in the channel is conduction dominated despite the appearance of the small convection cells.

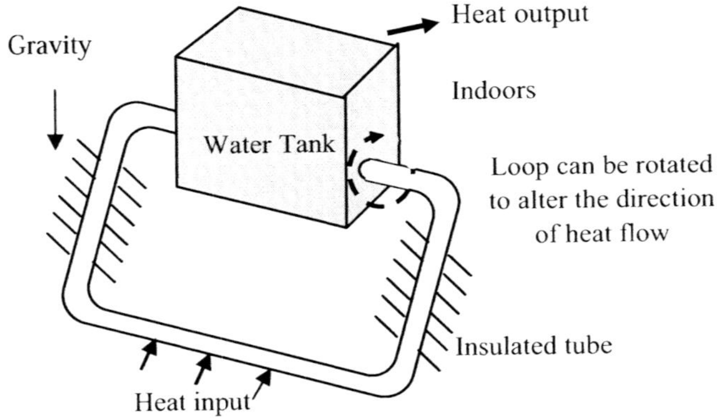

Figure 7. The loop-and-tank bi-directional thermodiode.

Since heat conduction through the connecting channel is much less than the solar energy absorbed by the cavity surface minus the heat loss from the cavity surface during the start-up period, the fluid temperature variation in the cavity can be calculated from:

$$m_v c \frac{dT_v}{dt} = A_v [q_v - U_0 (T_v - T_{\infty,0})] \tag{14}$$

The small heat transfer coefficient associated with natural convection along the tank surface justifies use of the lumped –system approximation for the fluid in the storage tank. That is, the convection resistance for heat transfer from the tank surface to the indoor air is much greater than the convection resistance of the fluid in the storage. The fluid temperature inside the tank was therefore assumed to be uniform and heat conduction through the connecting channel negligible during the start-up phase:

$$m_t c \frac{dT_t}{dt} = -A_t U_i (T_t - T_{\infty,i}) \tag{15}$$

In the above two equations m is the fluid mass, q_v the solar flux absorbed by the cavity surface, A the surface area, T_∞ the surroundings temperature, and U the overall heat transfer coefficient. Subscripts v, t, i, and o refer to the cavity and tank of the bayonet diode, and the indoors and the outdoors respectively.

Upon completion of the start-up period, the temperature of the fluid in the heated cavity equals the fluid temperature in the tank. More heat input to the cavity produces an unstable stratification in the thermodiode, and a circulation of the fluid between the cavity and the storage tank is induced. The flow in the connecting channel in the second phase of heating is modeled as one-dimensional, and the momentum and energy equations for a differential control volume in the flow passageway (see Figure 8) can be reduced to:

$$\rho_0\left(\frac{\partial V}{\partial t}+V\frac{\partial V}{\partial s}\right)=-\frac{\partial P}{\partial s}-\rho_0 g\beta(T-T_0)\cos\alpha-2\frac{\tau}{d} \tag{16}$$

and

$$\rho_0 c\left(\frac{\partial T}{\partial t}+V\frac{\partial T}{\partial s}\right)=\kappa\frac{\partial^2 T}{\partial s^2}+\frac{q}{d} \tag{17}$$

If the flow channel is well insulated,

$q = 0$ for $s = 0$ to j and m to L,

$= q_v - U_o\,(T - T_{inf,o})$ for $s =$ k to m.

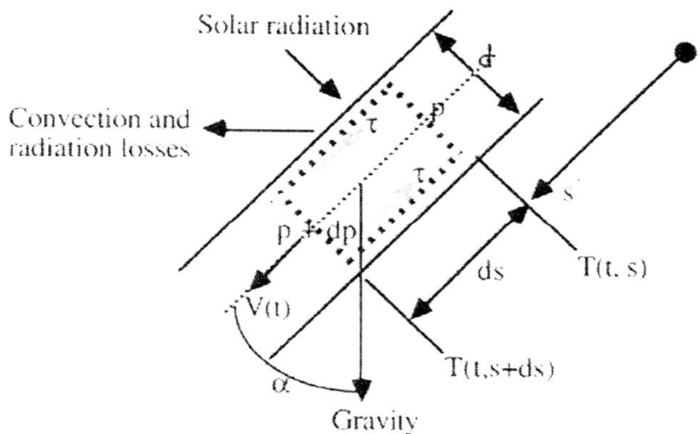

Figure 8. The differential control volume employed in the analytical model of the bayonet diode [Chen et al. 1998].

The axial conduction term is included in the energy equation in case the Peclet number drops close to or below unity. The analytical solution for a fully developed laminar flow between two parallel plates was used for the shear stress term in the momentum equation:

$$\tau = 6\rho_0\frac{V^2}{\mathrm{Re}_d} \tag{18}$$

where Re_d is the Reynolds number based on the channel height d.

The pressure gradient term in the momentum equation can be dropped by integrating the equation from $s = 0$ to L in Figure 6. The pressures at $s = 0$ and L in the tank are close to each other and thus can be cancelled if the fluid temperature in the tank is assumed to be uniform at any instant.

With a one-dimensional convection flow induced through the connecting channel during the heat –up phase, the energy balance analysis for the tank should include the convective heat transfer in the connecting channel :

$$m_t c \frac{dT_t}{dt} = \rho_0 c V w d (T_{s=L} - T_{s=0}) - A_t U_i (T_t - T_{\infty,i}) \tag{19}$$

where w is the channel width.

A lumped–system approximation can be employed for the storage tank during the cool-down phase too. With heat losses through the insulation and the connecting channel neglected, the energy equation for the tank during the cool-down phase is identical to Eq. (15).

For a short vane not fully extended to the bottom of the cavity, the equation for fluid in the lower portion of the cavity is :

$$m_{c,b} c \frac{dT_{c,b}}{dt} = \rho_0 c V w d (T_{s=j} - T_{s=k}) + A_{c,b} [q - U_0 (T_{c,b} - T_{\infty,o})] \tag{20}$$

where subscript c,b refers to the lower portion of the cavity.

The coupled, time-dependent momentum and energy equations were first solved explicitly, but the numerical computation failed to converge for solar flux greater than 20 W/m^2. Later an implicit model in which the mean fluid velocity was determined from the momentum equation by iteration was developed for stable calculations of high heat input rates.

After comparing the analytical solution with experimental data, the overall heat transfer coefficients associated with natural convection between the diode surfaces and the indoor and outdoor air were found to be:

$$U_i = 10 \text{ W/m}^2\text{-}^o\text{C} \tag{21}$$

and

$$U_o = 6.5 \text{ W/m}^2\text{-}^o\text{C}. \tag{22}$$

These coefficients are typical for natural convection heat transfer in air [e.g., Incropera and Dewitt 2002]. The higher overall heat transfer coefficient for indoor natural convection is the consequence of the use of fins on the indoor-facing surface.

A commercial CFD code FLUENT was also used for the two-dimensional, time-dependent heat transfer analysis of the bayonet diode. The grid used in FLUENT calculations was 14 x 21 for the tank, 6 x 8 for the connecting channel, and 6 x 11 for the cavity. A comparison of the analytical and numerical solutions and experimentally measured diode temperatures is given in Figure 9. The good agreement verifies the applicability of simple analytical models to heat transfer analyses of thermodiodes employed in solar energy systems. Similar analytical solutions were later applied to the transient analysis of the authors' loop-and-tank diode [2002]. Both the analytical solution and the numerical simulation indicate a longer guide vane in the cavity of the bayonet diode is preferred, as it can shorten the start-up time and produce higher buoyancy forces.

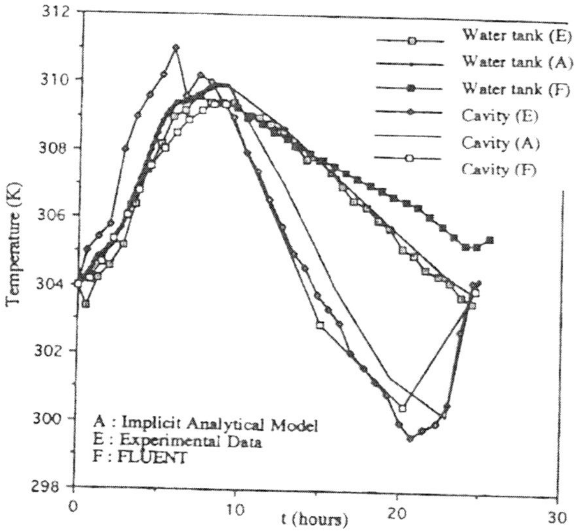

Figure 9. Measured and predicted temperature variations of the bayonet diode under real weather conditions in Utah [Chen et al. 1998].

Bi-Directional Thermodiodes

If the direction of heat flow of a thermodiode can be easily altered, the diode is called a bi-directional diode. Depicted in Figures 10-12 are a few bi-directional thermodiode designs devised by the authors from their research of solar systems and energy –efficient building envelopes. The plane-type bi-directional thermodiode (Figure 10) is composed of two vertical plates and several fluid-filled rectangular loops with their horizontal segments soldered to the plates. The joints between the horizontal segments and the branches connecting the horizontal segments can rotate freely. With one vertical plate fixed and the other plate free to move vertically, the direction of heat flow of this thermodiode can be easily reversed.

The equivalent thermal circuits of the plane-type thermodiode under forward and reverse bias are plotted in Figure 13 [Chen 1988]. When the heated horizontal segment of a rectangular loop is below the cooled segment, the diode is forward biased, and there are four thermal resistances in parallel for heat transfer between the two vertical plates (see Figure 13(a)): the conduction and convection resistances of the working fluid in the loops, and the conduction resistances of the walls of the inclined branches and the insulating materials between the vertical plates.

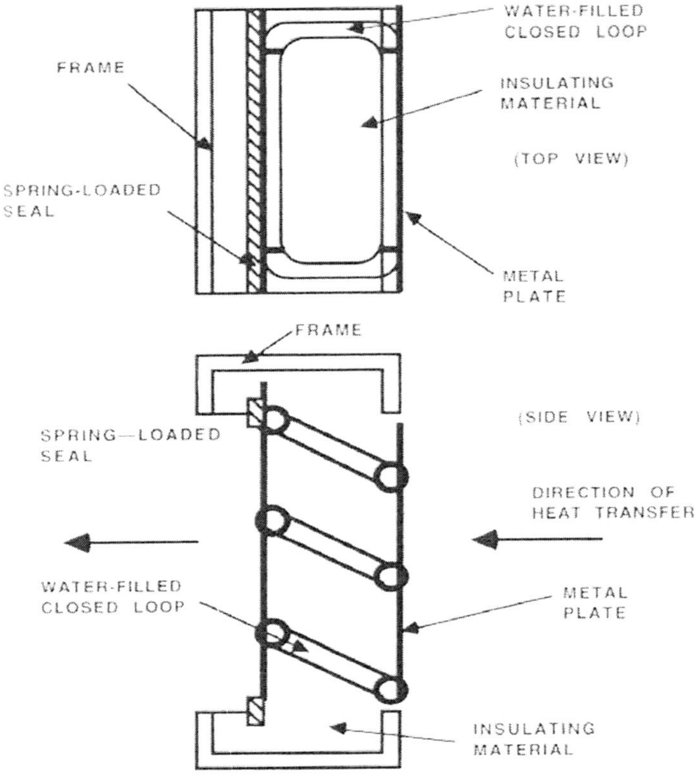

Figure 10. Plane-type bi-directional thermodiode [Chen 1988].

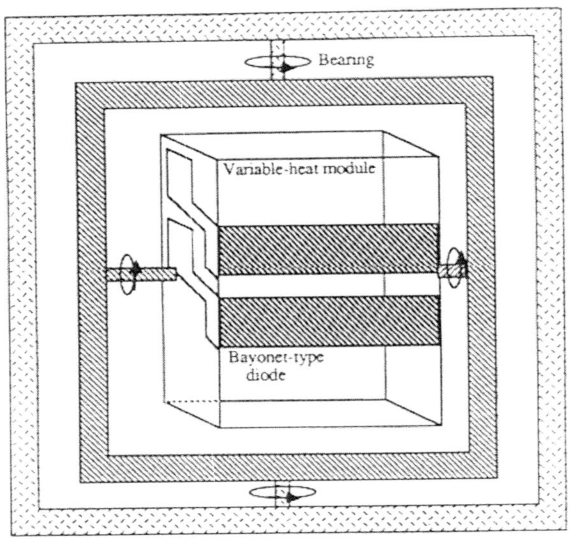

Figure 11. Three bayonet diodes mounted on a Gimbals mechanism for easy change of heat flow direction [Chen et al. 1995].

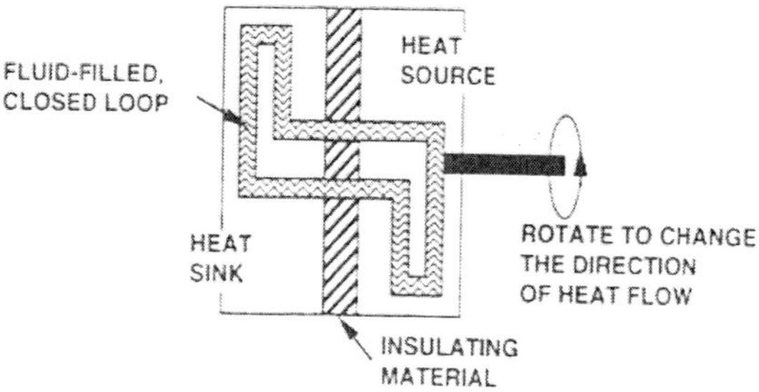

Figure 12. Another bi-directional thermodiode design based on the liquid-fin idea [Chen et al. 1995].

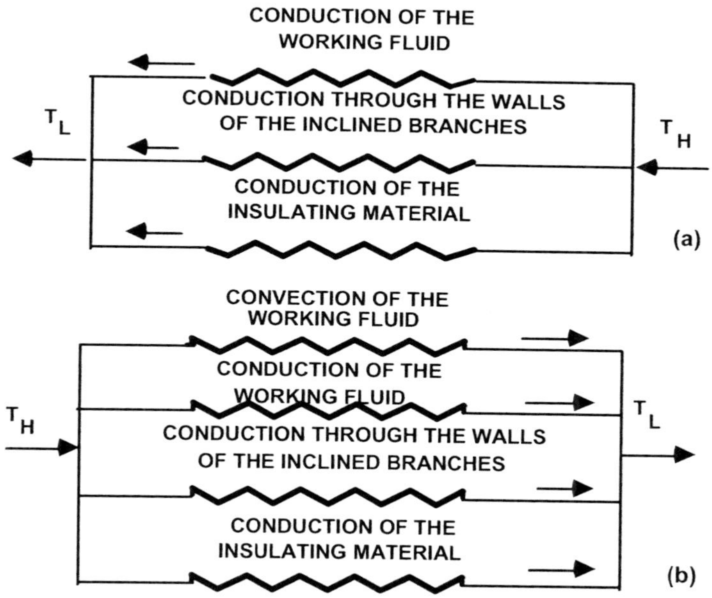

Figure 13. Equivalent thermal circuits and involved heat transfer modes of the plane-type thermodiode under (a) forward bias and (b) reverse bias [Chen 1988].

Since the convective resistance is much smaller than the other three resistances after natural convection has been induced throughout the loops, heat can be effectively transferred from the left to the right vertical plate. If the thermodiodes are reverse biased, that is, if the heated horizontal segment of a loop is above the cooled segment, there are no flow circulations throughout the loops. In this case heat transfer from the hot to the cold plate is by conduction through the insulating material, the fluid in the inclined branches, and the walls of the branches as represented by the three parallel conduction resistances in Figure 13(b).

Chen's analytical solution for natural convection in a rectangular loop [Chen 1985 a] was combined with a one-dimensional conduction analysis for the insulating material and the inclined branches to determine the heat transfer rates of this thermodiode under forward and reverse bias. The length of the horizontal segments was determined from the optimum width of the rectangular–loop thermosyphon. The resultant loop width for laminar flow is :

$$\left(\frac{W}{2}\right)_{opt} = \frac{(H^2 + 0.273HrRa_r\cos\alpha)^{\frac{1}{2}} - H}{2} \tag{23}$$

where the Rayleigh number, Ra_r, is the product of the Grashof and Prandtl numbers.

Shown in Figure 14 are the computed heat fluxes under forward (q+) and reverse (q-) bias, and the ratio of q+ to q- for a 20 °C temperature difference across the diode with water as the working fluid. The loops are made of tubes of 1 cm inner diameter with a vertical spacing of 5 cm between the loops. For small inclination angles, q+ is high due to the large gravitational acceleration. However the insulation between the two vertical plates is thin in this case, resulting in high q- and low current ratio q+/q-. Both q+ and q- change only slightly for inclination angles greater than 20 degrees if the loop width is allowed to vary with the inclination angle according to Eq. (23).

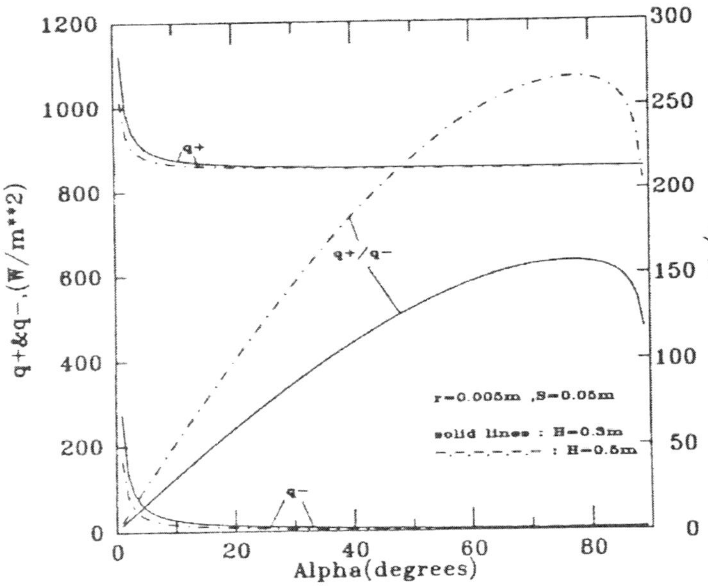

Figure 14. Heat flux ratio and heat fluxes of the plane-type bi-directional thermodiode under forward and reverse bias [Chen 1988].

The heat transfer rate becomes very small at large α since the optimum loop width decreases very rapidly as α approaches 90 degrees. The inclination angle that optimizes the heat transfer rate and the current ratio for this plot is somewhere between 40 to 70 degrees. Too large an inclination angle has too short heat transfer sections for high heat transfer rate while too small an angle has too high a heat leakage rate under reverse bias. Comparing the results of H = 0.3 and 0.5 m indicates that increasing the length of the inclined branches considerably increases the current ratio due to the increase in insulation between the two plates, but has little effect on q+ and q-. Whether the heat transfer rate will increase with H or not depends on the effect of the change in H on the optimum loop width determined from Eq. (23).

If the right vertical plate of the thermodiode in Figure 10 is in contact with the indoor air, the plate should be pushed up when the desired direction of heat flow is from the outdoors (left plate) to the indoors (right plate). For summer cooling application, the right vertical plate is pulled down and effective heat transfer only takes place from the indoors to the outdoors when the outdoor temperatures (the atmospheric temperature and the night sky temperature) are below the indoor temperature.

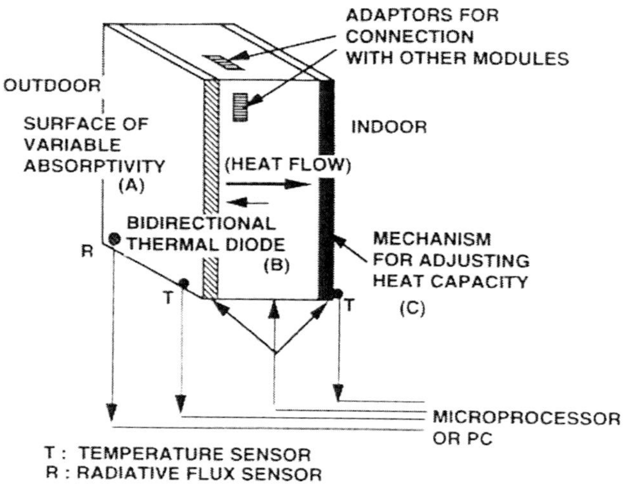

Figure 15. Smart construction module of variable direction of heat flow, surface absorptivity, and thermal capacitance [Chen et al. 1995; Chun and Chen 2002].

The bi-directional thermodiode was the major component of a "smart" construction module proposed and investigated by Chen, Chun, and co-workers in recent years for energy–efficient buildings and emergency shelters. A schematic diagram of the module is shown in Figure 15. The direction of heat flow, surface absorptivity, and heat capacity of the module all can be adjusted for maximum energy efficiency and/or maximum degree of comfort under various weather conditions. Operations of the module for winter heating and summer cooling are tabulated in Table 1. Solar energy is utilized in the winter for space heating. As a result the outdoor-facing surface is adjusted to be highly absorptive during the daytime and the direction of heat flow is from the outdoors to the indoors. Since a surface which absorbs radiation efficiently also emits radiation efficiently (this is known as Kirchhoff's law in thermal radiation), the outdoor surface should change to highly reflective to reduce radiation loss at night in the winter.

One method to alter the surface absorptivity is to paint the surface in black and white strips and to use rectangular plates pivoted vertically to the intersections of the black and white strips as depicted in Figure 16. One side of the rectangular plates is painted in black and the other side in white. The entire surface will turn into black or white when the plates are flipped to the right or to the left. Another design of variable surface absorptivity is to use a flexible sheet of highly reflective material. The sheet is wound on two cylindrical rollers and can be scrolled up and down to cover a black surface. Installation of a blind of reflective surface in front of a black surface serves the same purpose.

Table 1. Operation of the smart module.

	Winter Heating		Summer Cooling	
	Day	Night	Day	Night
(A) Variable direction of heat flow	From the outdoors to the indoors	Little heat leakage to the outdoors	Little heat leakage to the indoors	From the indoors to the outdoors
(B) Variable surface absorptivity	Highly absorptive	Highly reflective	Highly reflective	Highly absorptive
(C) Variable heat capacity	(Depends on indoor-outdoor temperature difference and the desired operation condition)			

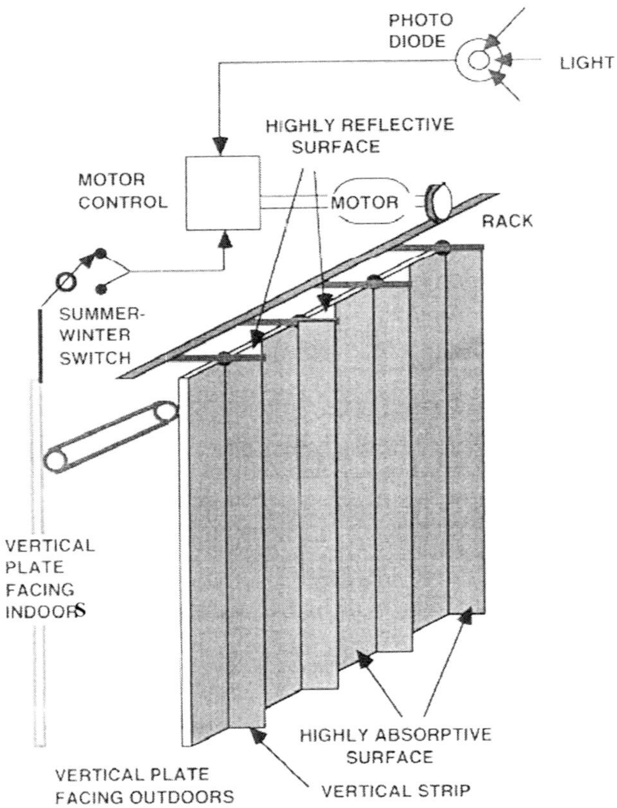

Figure 16. A smart module design using tilted loops to alter the direction of heat flow and vertical strips to change its surface absorptivity [Chen et al. 1995].

For winter heating, the thermal diode between the indoor and outdoor surfaces allows heat flow from the outdoors to the indoors when the outdoor surface is heated by solar flux, but blocks the flow of heat in the opposite direction at night. The setting of the direction of heat flow and surface absorptivity are just the opposite for summer cooling application. The outdoor surface is highly reflective during the day for less solar heating, but changes to highly absorptive at night to transfer heat to the sky by radiation. The desired direction of heat flow of the diode for summer cooling is from the indoors to the outdoors.

Operation of the mechanism that changes the heat capacity of the module depends on the weather conditions and the occupancy of the building. For winter heating, the heat capacity of the module should be set low when the building is occupied and a fast warm - up of the indoors is desired. A low heat capacity allows the module temperature to rise rapidly during heating. When the indoor temperature reaches a comfortable level, the heat capacity is adjusted high to store excess energy input for later use. When the outdoor surface is heated by solar energy but the building is not occupied, the heat capacity should be set high for immediate storage of solar energy. For summer cooling, a large heat capacity is needed to hold a large amount of energy when the building is occupied and the outdoor air temperature is high. A small heat capacity, on the other hand, allows the indoor temperature to quickly drop when the outdoor temperature is low. A few methods for altering the heat capacity of the smart module were proposed by the authors [Chen et al.1995] and are illustrated in Figures 17 and 18. The pressure of the phase-change material in the cylinder in Figure 17 can be altered by the movement of the piston, which changes the phase –change temperature. The pivoted partition in Figure 18 is well insulated and can change the amount of fluid in the tank that is in direct contact with the streams entering and leaving the tank during mixing, thus changing the heat capacity of the thermodiode. If heat conduction through the insulated partition is negligible, the diode heat capacity is reduced by 50 % when the partition is tilted by 45 degrees from its original position.

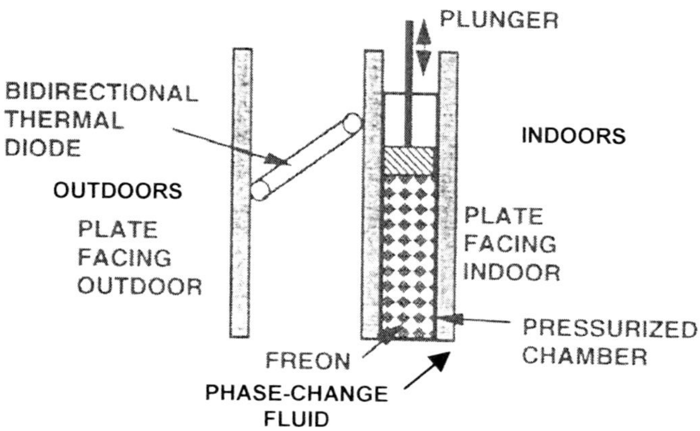

Figure 17. conceptual design of variable specific heat for the smart module [Chen et al. 1995].

The best combination of the controls of direction of heat flow, heat capacity, and surface absorptivity is determined by a PC (Personal Computer) or a microprocessor for maximum energy efficiency and/or most comfortable indoor temperature. The adaptors on the module allow several modules to be connected together and controlled by a PC or a microprocessor. Only one set of temperature and radiation sensors are needed and mounted on a master module (usually the center one). The smart module takes advantage of the rapid developments and significant cost drops in microprocessor and sensor technologies in the past two decades to automatically optimize the performances of building and shelter envelopes for different weather conditions.

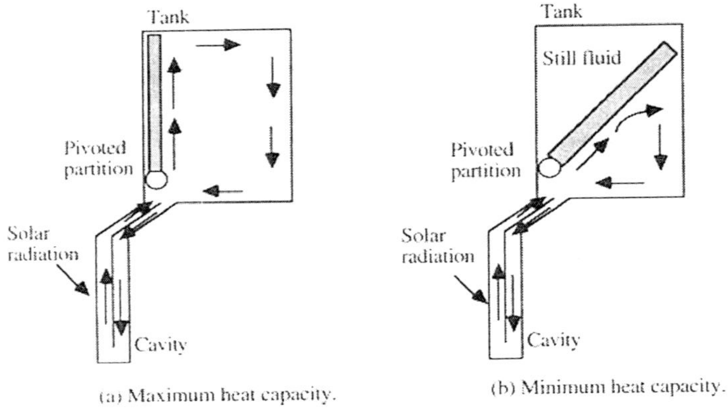

Figure 18. Water tank with a pivoted partition for adjustable thermal capacitance [Chen et al. 1995].

Unlike solar house design which is part of the building and cannot be moved freely from one location to another, the module can be mass - manufactured and stored in one place and shipped to another place to be quickly assembled as building or shelter envelopes. Several modules can be connected together and controlled by a microprocessor, with temperature and radiation sensors mounted on one master module to operate all the modules.

A detailed discussion of the design and application of the smart module and the outdoor test data collected in Utah from 1994 to 1995 can be found in Chen's report [Chen et al. 1995] and Chu's dissertation [1996]. The thermodiode tested in this project was the bayonet – type thermodiode. Lab test data demonstrated this thermodiode can reach an efficiency about 50 % which was among the highest of all single-phase thermosyphons. Switching from winter heating to summer cooling operation was done by manually turning the bayonet thermodiodes upside down in the spring. An alternative method to alter the direction of heat flow of the bayonet thermodiode is the crossed-channel design depicted in Figure 19. In this design several bayonet diodes are stocked up and the valves at the centers of the crossed flow channels control the direction of natural convection between the cavity and the storage tank of the diode. A prototype consisting of one pair of bayonet thermal diodes and a cylindrical valve was constructed to demonstrate the working principle of this diode design for alterable direction of heat flow (Figure 20). Since one of the diode cavities will make no contribution to convective heat transfer in this design (the top cavity in Figure 19 for winter heating, and the bottom one for summer cooling), this design is more suitable for a large number of diode pairs installed on a tall wall. Other designs of bi-directional thermodiodes were also proposed by the authors [Chen et al. 1995], and are shown in Figures 10 to 12. The one in Figure 12 was based on Jones liquid- fin design [1986]. The directions of heat flow of this diode and the plane-type diode tested by Chun and co-workers [2002] are much easier to change.

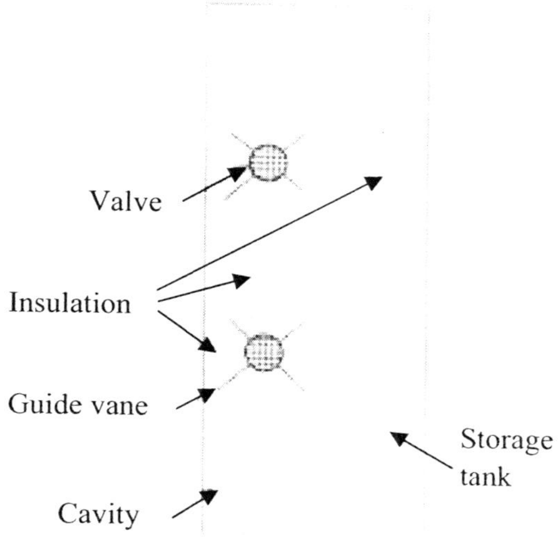

Figure 19. Bayonet diodes connected through a valve for easy change of the heat flow direction [Chen et al. 1995].

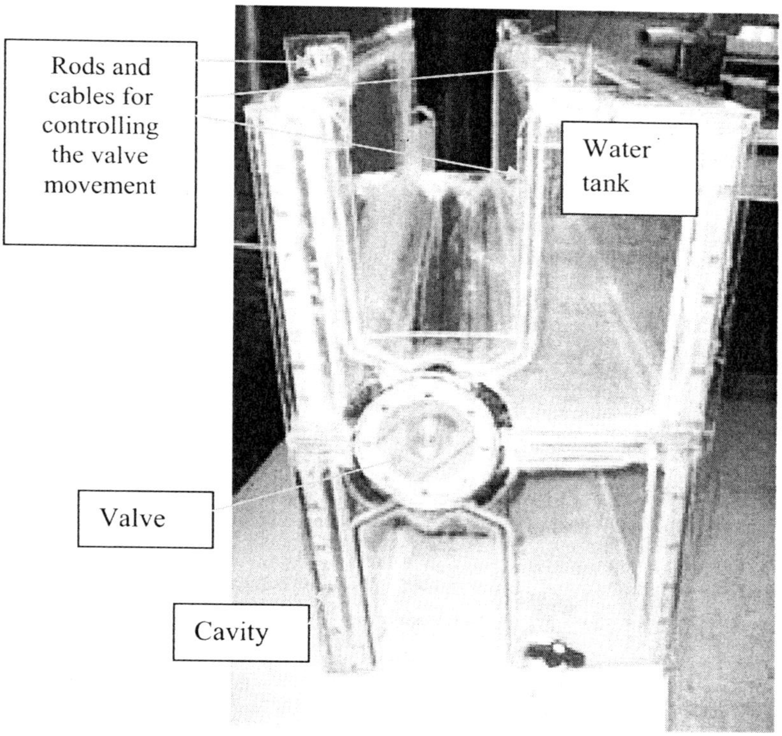

Figure 20. Photo of a bayonet diode pair connected through a cylindrical valve.

In the weather test performed in Utah, two identical sturdy sheds called Tuff Sheds, each was approximately 64" long, 64" wide, and 77" tall with a 48.5" by 72.5" door, were purchased and insulated according to the local building codes. The door of one of the Tuff

Sheds was replaced by a smart module with three to four bayonet thermodiodes on it. The two sheds were located side by side outside the south wall of North Salt Lake pilot plant which was a University of Utah research facility for refining tar sands into petroleum.

Presented in Figure 21 are the two Tuff Shed temperatures and the ambient temperature measured from October 22 to 29 in 1994. The shed temperatures were measured at the centers of the shed interiors by T-type thermocouples. The thermocouple used for ambient temperature measurements was shielded by a reflective coil to eliminate radiation from the surrounding environment.

The improvement in space heating through the use of the thermodiodes is quite evident in this plot. The temperature of the Tuff Shed with the smart module installed was nearly 10 °C warmer than that of the Tuff Shed without the smart module in the morning. During the daytime when the sun shone on the two sheds, all three temperatures (the two Tuff Shed temperatures and the ambient temperature) increased at about the same rate. There was an approximately two-hour phase lag between the ambient temperature and the Tuff Shed temperatures during heating. The phase lag between the two Tuff Shed temperatures was much less. This is different from the fluctuations of the two shed temperatures for which the shed with the smart module had a much higher heat capacity, thus exhibited much less fluctuations in its daily temperature variation. The maximum and minimum temperatures of the shed without the smart module and the ambient temperature were very close to each. This implies the major heat source and sink of the shed without the smart module was the ambient air. Solar radiation on the shed surface had little effect on its interior temperature.

The temperatures of the diode cavity and water tank measured during the same period were presented in Figure 22. The small differences between the maximum temperatures of the cavity and the water tank demonstrated the small thermal resistance of the bayonet thermodiodes under forward bias. The large differences between the minima of these two temperatures indicated little heat leakage when the diode was reverse biased.

The smart module were turned upside down in the summer of 1994 and tested for summer cooling application. Measured Tuff Shed, ambient air, and thermodiode surface temperatures from August 15 to 21 are presented in Figure 23. Just like the winter heating data, the temperature of the shed without thermodiodes was very close to the ambient temperature, but with a phase lag about two hours due to the thermal inertia of the shed mass. During the daytime the maximum temperature of the shed with the smart module was about 5 °C lower than that of the ambient air and the shed without the smart module. However, the energy absorbed by the water tanks during the daytime was released at night and the minimum temperature of the shed with the smart module was 2 to 3 °C higher than the minimum temperature of the ambient air and the minimum temperature of the other shed. The decrease in daytime cooling load due to the use of the smart module was partially cancelled by the increase in cooling load at night.

Inspection of the diode temperature plot, Figure 24, indicates the thermodiodes on the smart module functioned as desired in this summer cooling application. There was a large difference between the water tank temperature and the cavity temperature when the smart module was under reverse bias at daytime. Thus heat flow into the indoors was blocked. After the cavity temperature dropped below the water tank temperature, the diodes became forward bias and the water tank temperature started to drop at the same rate (but with a phase-lag due to its large thermal inertia) as the cavity temperature. The little improvement in the overall

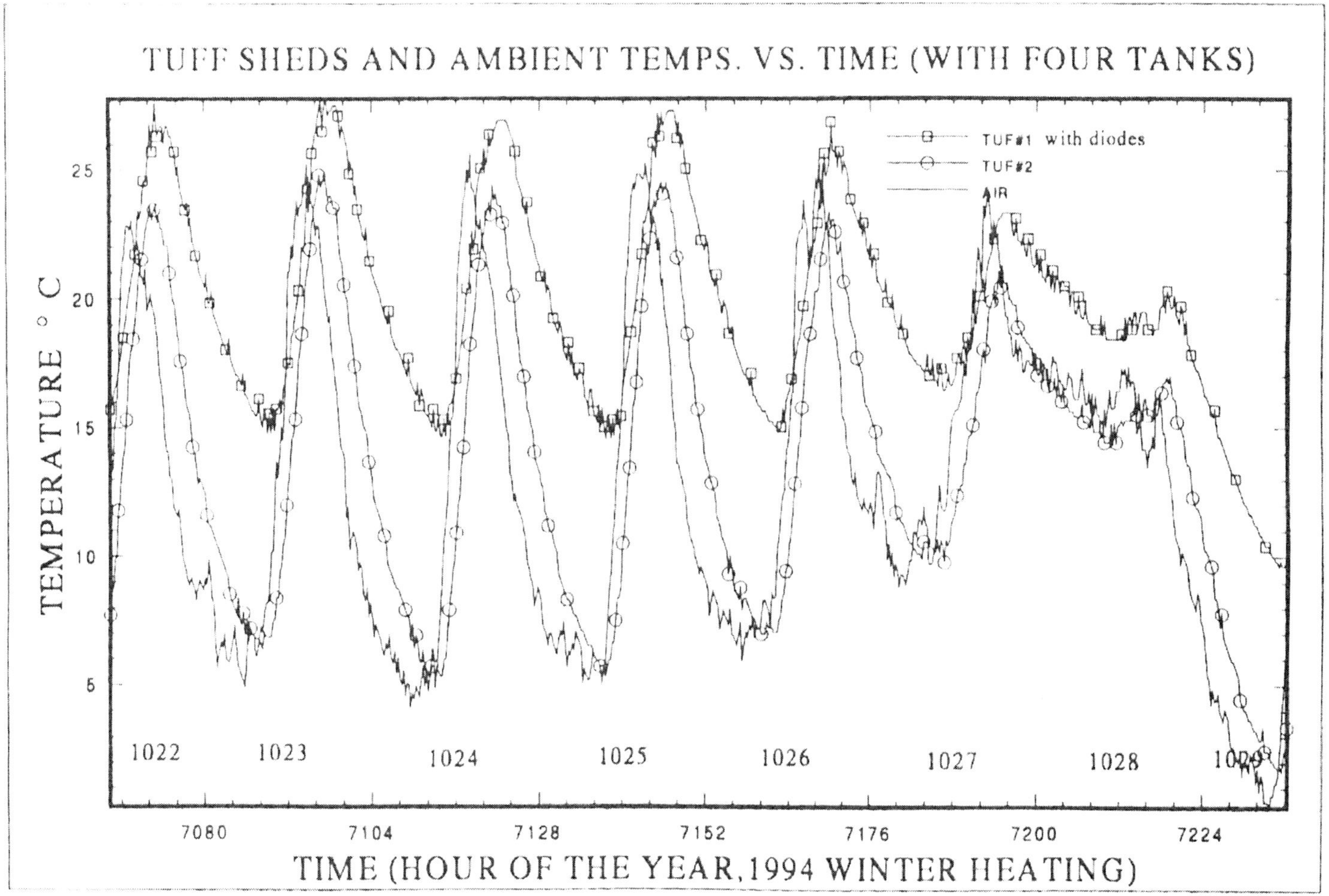

Figure 21. Shed interior temperatures and the ambient temperature measured in the winter test in Utah [Chen et al. 1995; Chu 1996].

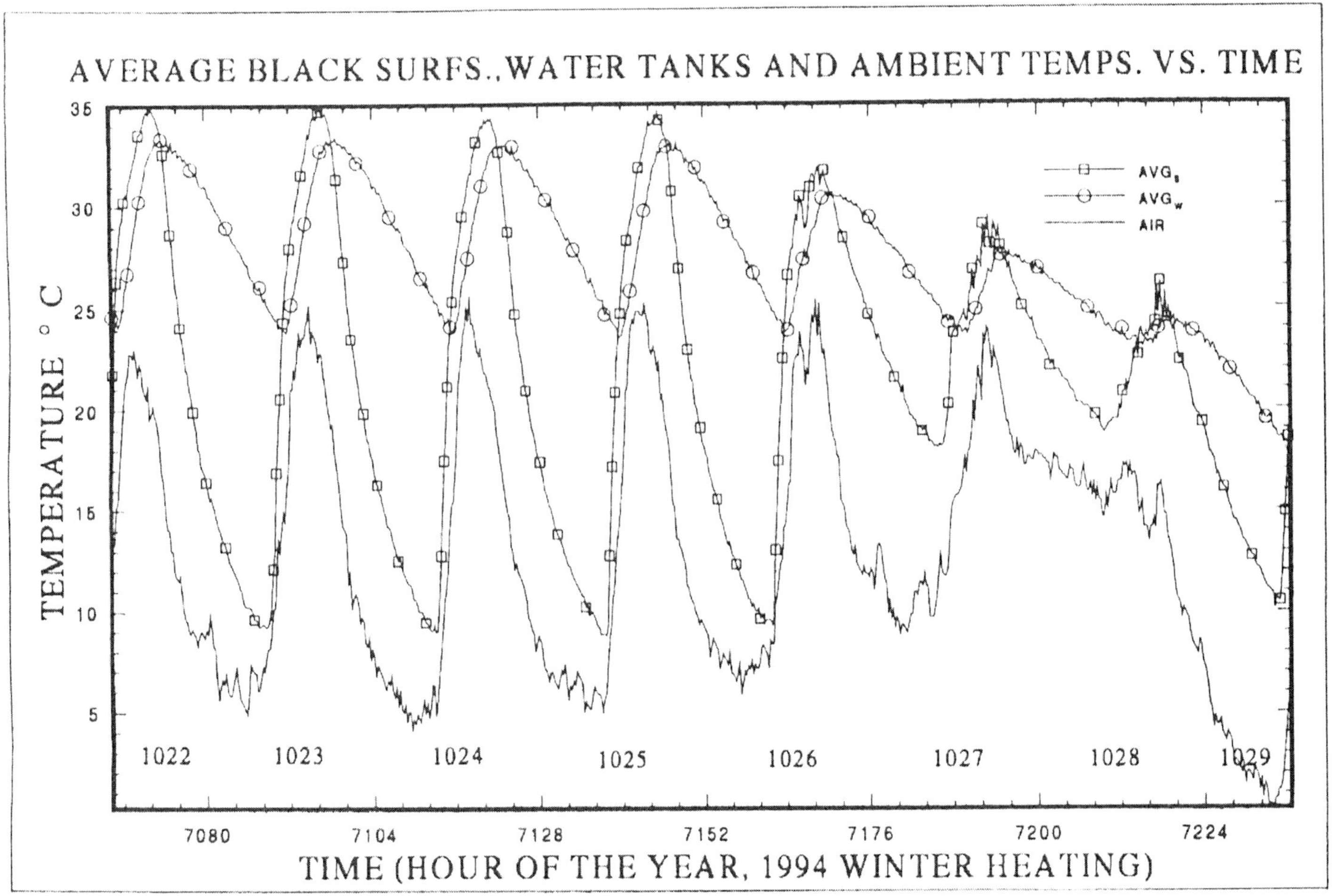

Figure 22. Ambient temperature and temperatures of the water tank and the cavity [Chen et al. 1995; Chu 1996].

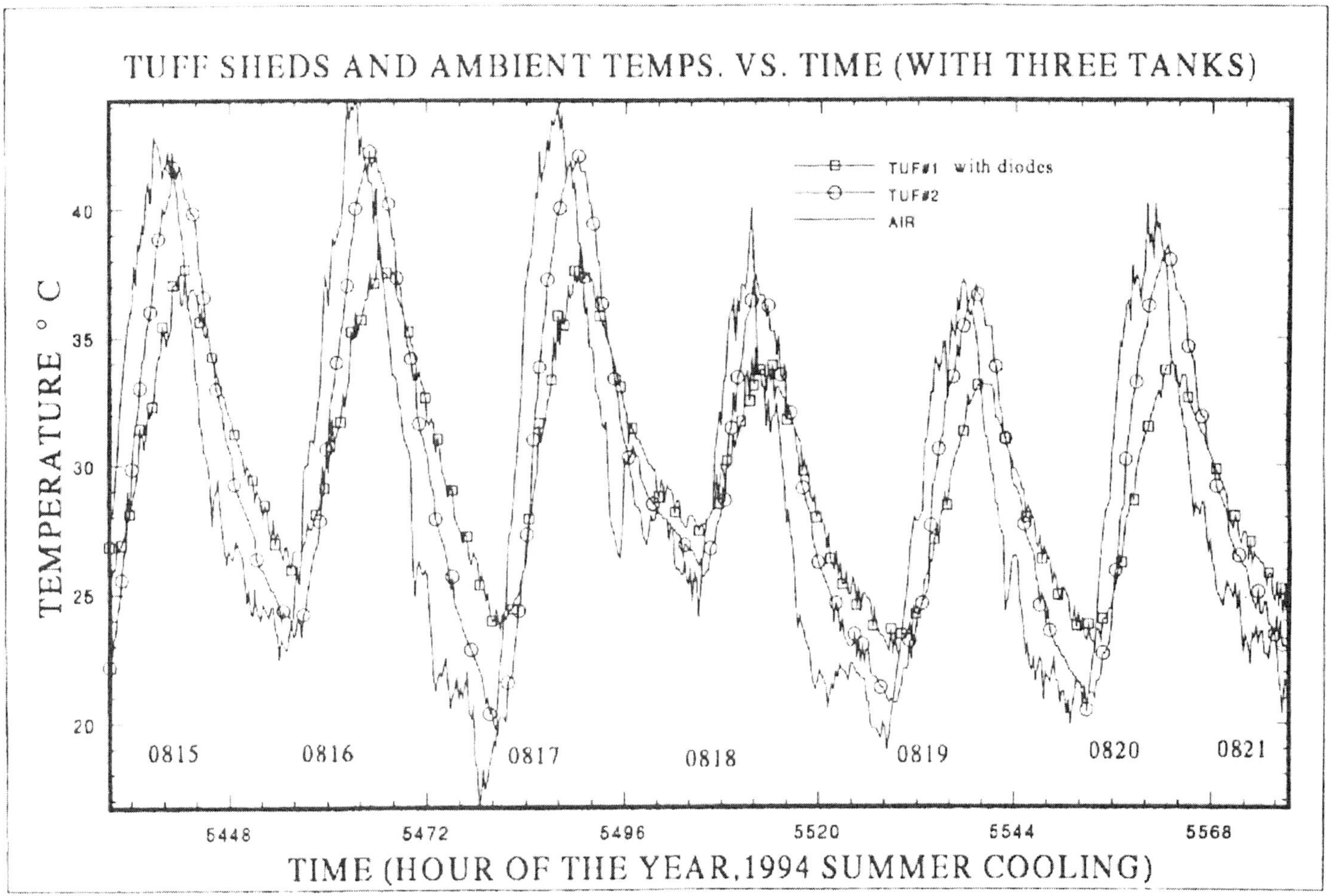

Figure 23. Shed interior temperatures and the ambient temperature measured in the summer cooling test in Utah [Chen et al. 1995; Chu 1996].

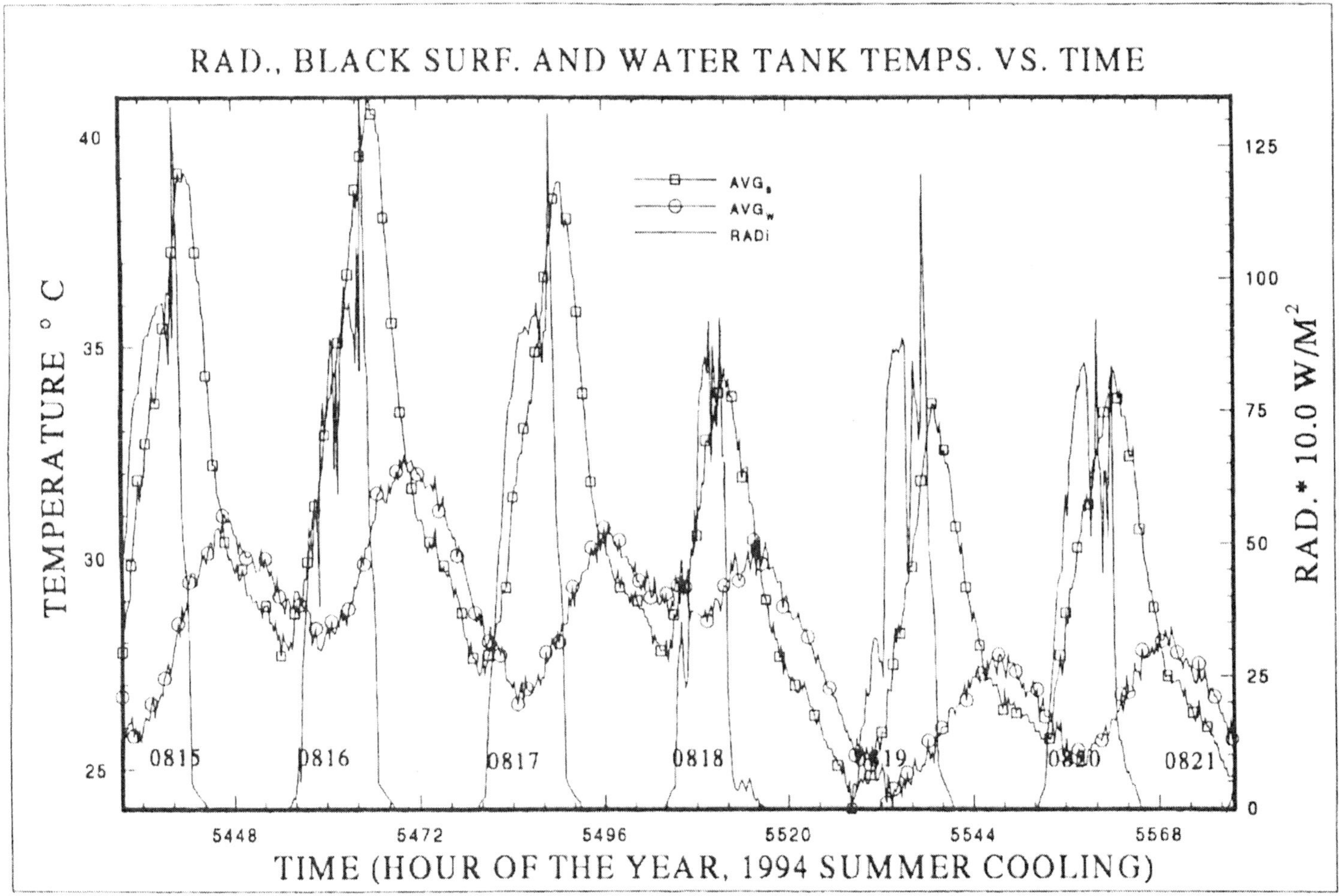

Figure 24. Solar flux and temperatures of the water tank and the cavity [Chen et al. 1995; Chu 1996].

daily cooling load through the use of the smart module is due to the low convection and radiation heat transfer rates from the outdoor surface at night. Installation of fins on the outdoor surface and tilting the cavity of the bayonet diode to obtain a less obstructed view of the night sky (see Figure 25) would improve the overall heat transfer rate and the performance of the smart module for summer cooling applications. Another remedy to alleviate the high nighttime cooling load when the smart module is used is to cover the indoor surface of the module with an insulating pad at night. The pad is removed in the morning after the water tank temperature drops below the ambient air temperature.

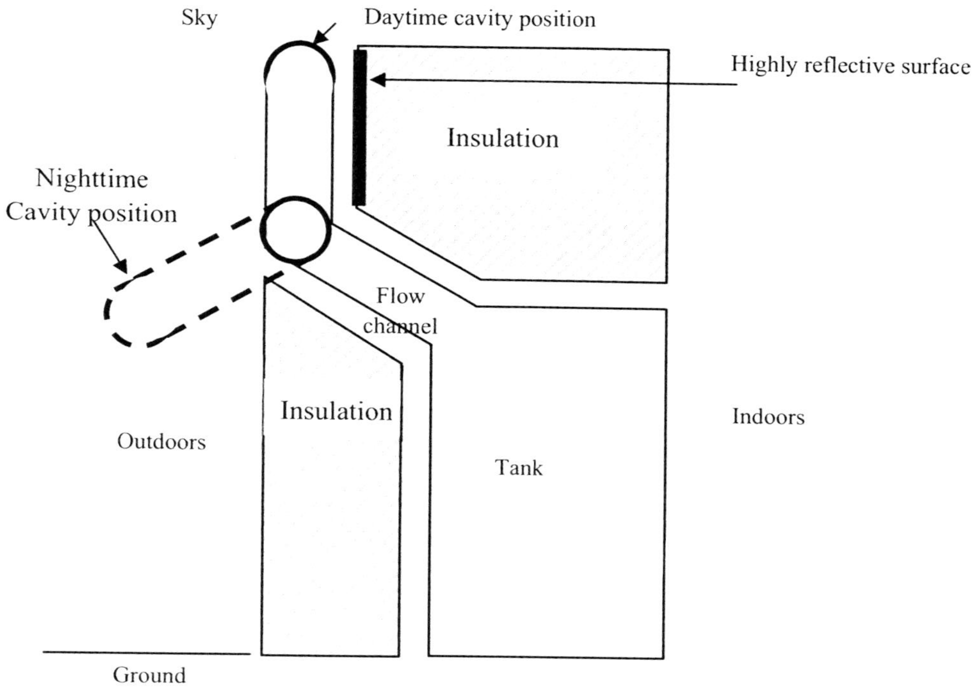

Figure 25. A revision of the bayonet diode design for improved summer cooling efficiency.

Without considerable increases in night sky radiation loss and convective heat transfer from the smart module surfaces, the suitable application of the smart module for summer cooling would be as a supplementary cooling system for office buildings which are not occupied in the evenings and nights. Another application is for combined daytime cooling and nighttime heating which are needed at early spring and late autumn.

After the preliminary study performed in Utah, research on the energy-efficient construction module was continued in Korea. Since 1995 Chun and co-workers have been developing and testing different component designs for the smart module. The bi-directional thermodiodes they used were the rectangular loop type (see Figure 10), and a blind was installed in front of the outdoor surface of the module to reduce solar energy absorption in the summer. Unlike the bayonet thermodiode used in Chen's outdoor tests, the flow channel diameter and loop dimensions of the rectangular-loop diode must be carefully designed for effective convective heat transfer throughout the loop. Chun and co-workers had successfully demonstrated the performance of the rectangular- loop diode is comparable to that of the bayonet diode when the right tube diameter and loop dimensions are used.

One major advantage of the rectangular- loop diode over the bayonet diode design is that the direction of heat flow can be easily reversed by moving one of the two vertical plates up or down. In the Solar Energy Exhibition for the 1997 Solar World Congress in Taejon, Korea, Chun and his students displayed a prototype smart module similar to the one in Figure 26 with a motor-driven indoor plate and a fixed outdoor plate for automatic direction change of the heat flow. Later their smart modules had undergone both lab and outdoor tests in Korea. Some of Chun and co-workers' research results are mentioned below. More detailed information can be found in Chun, Chen, and co-workers' recent publications.

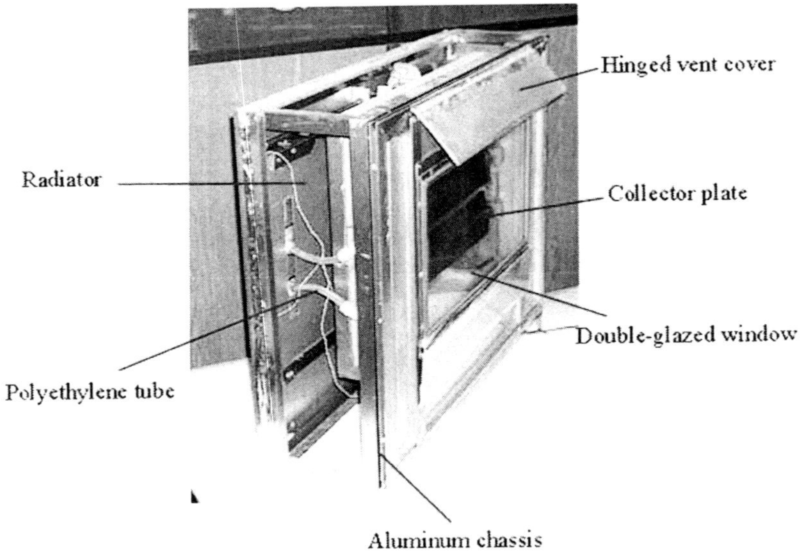

Figure 26. The bi-directional thermodiode system tested in Cheju, Korea [Chun, Chen, and Kim 2002].

Thin-walled copper tubes with an inner diameter of 0.0125 m was used as the horizontal segments of the rectangular loops in Chun and co-workers' plane-type bi-directional thermodiodes. The inclined branches connecting the two copper tubes were polyethylene tubes of approximately the same inner diameter. Each copper tube that faces the outdoors was soldered to a thin metallic sheet 0.12 m wide which is forged into a semicircle in the middle for good thermal contact with the copper tube. A water tank having the shape of a slab was incorporated in some of the modules Chun, Chen, and co-workers tested to increase the energy storage capacity of the radiator that faces the indoors. Analytical solutions similar to those in their study of the bayonet thermodiode were obtained to determine the time-dependent behaviors of the plane-type bi-directional thermodiodes with or without a small water tank.

Shown in Figure 26 is one of the plane-type bi-directional thermodiodes Chun et al. tested. The side walls and insulation of the diode were removed to reveal the details of the diode interior. Indoor test results of a diode equipped with a small water tank are shown in Figure 27 for two- hour heating followed by a three- hour cooling phase. Temperatures in these plots were measured by thermocouples whose locations are marked in Figure 28. The diode system in this test was heated by a radiant heater that produced a heat flux about 800 W/m^2 on the absorber plate during heating. The humps in the absorber plate and radiator inlet temperatures marked the onset of a throughflow in the loop. Before that, water in the horizontal segment of the loop was heated up very rapidly, but the buoyancy force was not sufficient to induce a throughflow, and the fluid temperatures in the radiator remained at their initial temperatures. After the onset of a global convection cell throughout the loop, the absorber plate and radiator inlet and outlet temperatures increased at the same rates until a steady state was reached. During the cool-down phase the radiant heater was turned off and the radiator temperatures dropped the slowest due to its large thermal capacitance.

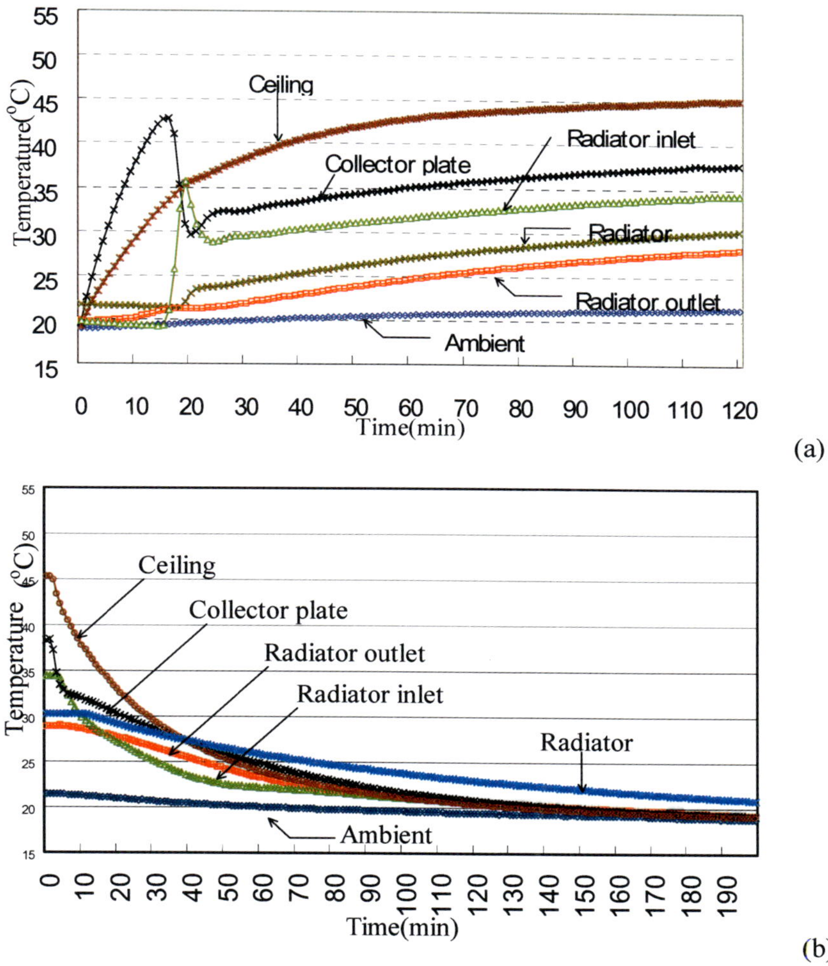

Figure 27. Indoor test results for a radiation flux of 800 W/m^2. (a) Heating phase. (b) Cooling phase [Chun and Chen 2002].

The same heat flux was applied to the absorber plate when the diode was reverse biased (the radiator plate in Figure 28 was pulled down). The nearly invariant radiator temperature in Figure 29 clearly demonstrated the diode design can effectively prevent heat flow in the undesired direction.

More detailed information about the indoor tests of the plane-type bi-directional thermodiode can be found in [Chun, Chen, and Kim 2002]. In this paper test results of different heat fluxes were compared and discussed. As might be expected, the time for the onset of a throughflow decreased from 20 to 10 minutes as the heat flux on the absorber plate increased from 400 to 800 W/m^2. The steady-state temperature of the radiator inlet increased from 34 °C for 400 W/m^2 heat flux, to 48 °C for 800 W/m^2 heat flux.

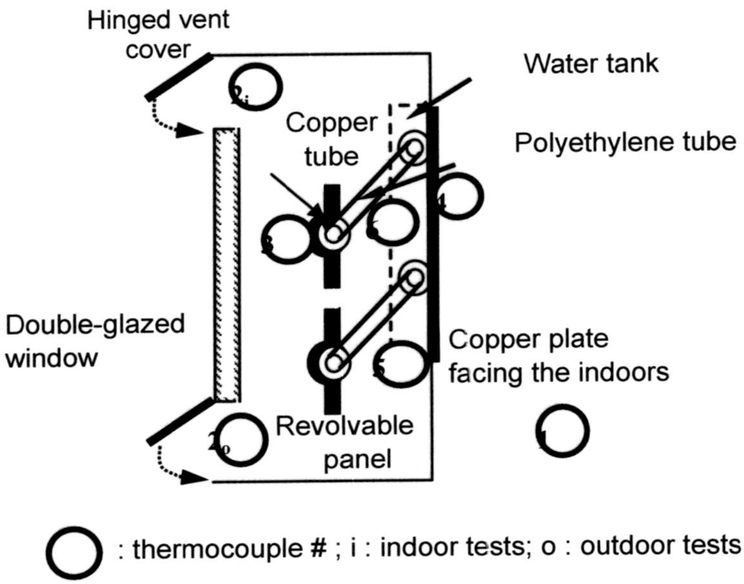

Figure 28. Thermocouple locations of the thermodiode system used in Chun et al.'s indoor tests [Chun and Chen 2002].

Analytical models were developed for heat transfer between different diode components during both the start-up and heat-up phases in this paper. Based on the measured diode temperature variations and the analytical solutions, the diode under forward bias had an effectiveness (heat transfer from the heated copper tube to the radiator divided by conduction heat transfer through the insulating materials and stagnant water in the tubes) of 93 for 600 W/m^2 heat flux. The heat transfer rates between different diode components and their averages during the heat –up phase are plotted in Figure 30 for 600 W/m^2 heat flux on the diode surface under forward bias. These heat transfer rates were normalized by radiation absorbed by the copper sheets, which was about 61 % of the radiation incident on the diode surface. The copper sheets were slightly colder than the ambient air at the beginning of heating. As a result heat loss from the copper sheets to the surroundings ($Q^*_{s\text{-}a}$) was negative during the first few seconds in Figure 30. Subscripts a, r, s, t, w in this plot refer to the surroundings, radiator, copper sheets (the revolvable panels wrapped around the copper tubes in Figure 28), copper tubes, and fluid in the copper tubes and the polyethylene tubes connecting the radiator and absorber plates.

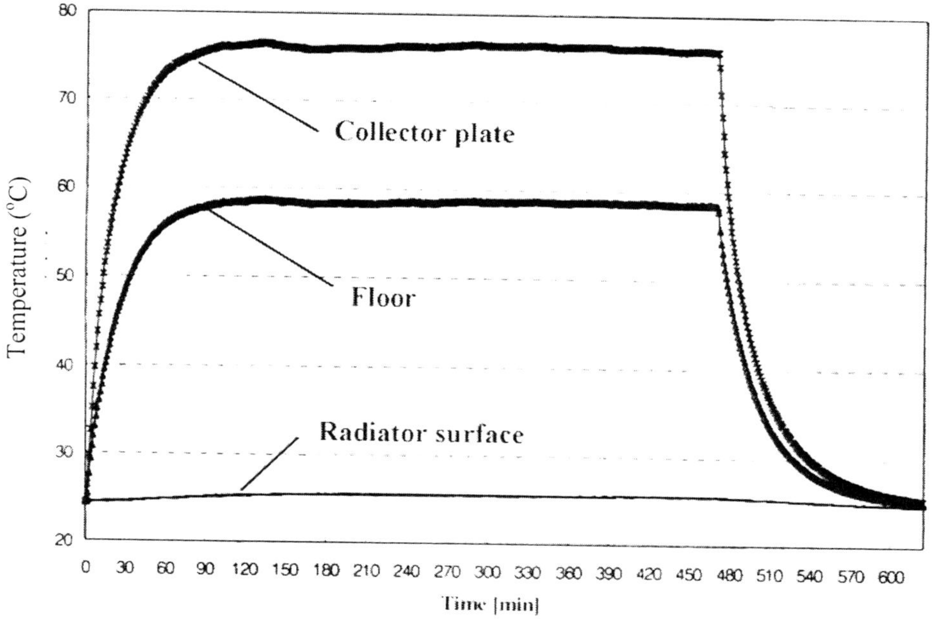

Figure 29. Temperature variations of the plane-type thermodiode system under reverse bias [Chun and Chen 2002].

During the start-up phase (t = 0 to about 20 minutes for 600 W/m^2 incident radiation), the rate of heat transfer from the heated copper tubes to the fluid in the copper and polyethylene tubes (Q^*_{t-w}) first increased faster than the rate of heat loss to the surroundings (Q^*_{s-a}). However, as the temperatures of the copper tubes and the fluid inside become higher and higher, Q^*_{s-a} continuously increased while Q^*_{t-w} reached a maximum value of 50 % before dropping to 33 % at the end of the start-up phase. As heat was still pumped into the fluid in the tubes from the copper tube walls during the second phase of the start-up interval, the temperature of the fluid in the tubes continuously increased until natural convection between the copper tubes and the radiator was induced at about t = 20 minutes.

In the heat- up period, which started from the onset of natural convection throughout the diode and ended when the radiant heater was turned off, about 50% of the energy absorbed by the copper sheets was transferred to the copper tube ($Q^*_{s-t,ave}$), and then to the radiator ($Q^*_{t-r,ave}$) by natural convection throughout the diode. The rest was lost primarily from the copper sheet surfaces to their surrounding ($Q^*_{s-a,ave}$). As heat was transferred to the radiator, its temperature raised continuously during heating, and natural convection from the outer surface of the radiator to the indoor air (Q^*_{r-a}) increased with time. The normalized rate of heat transfer to the indoor air reached 28 % after two hours of heating for a radiation flux 600 W/m^2 on the diode outdoor-facing surface.

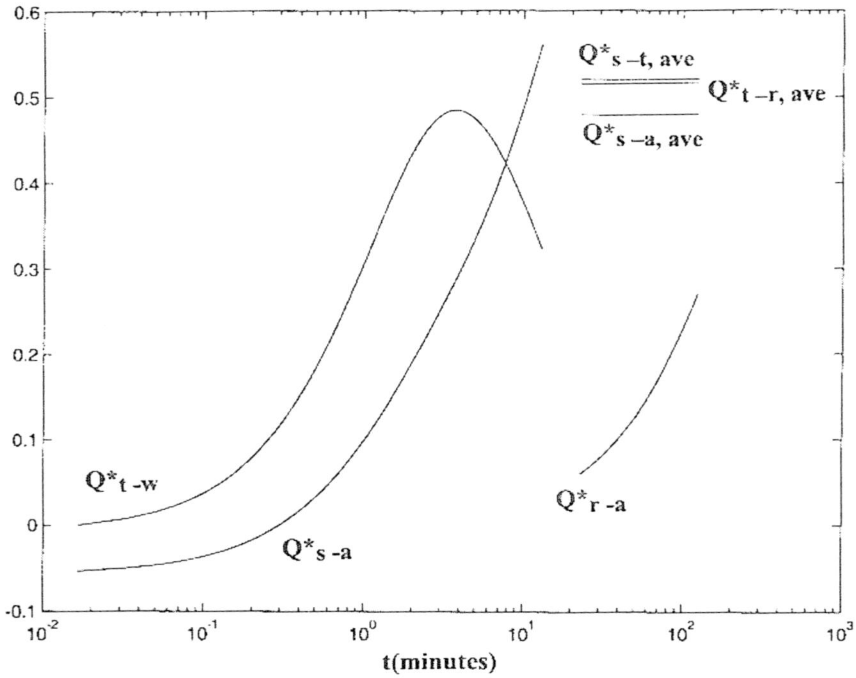

Figure 30. Normalized heat transfer rates during the start-up and heat-up phases for 600 W/m^2 radiation on the diode system [Chun, Chen, and Kim 2002].

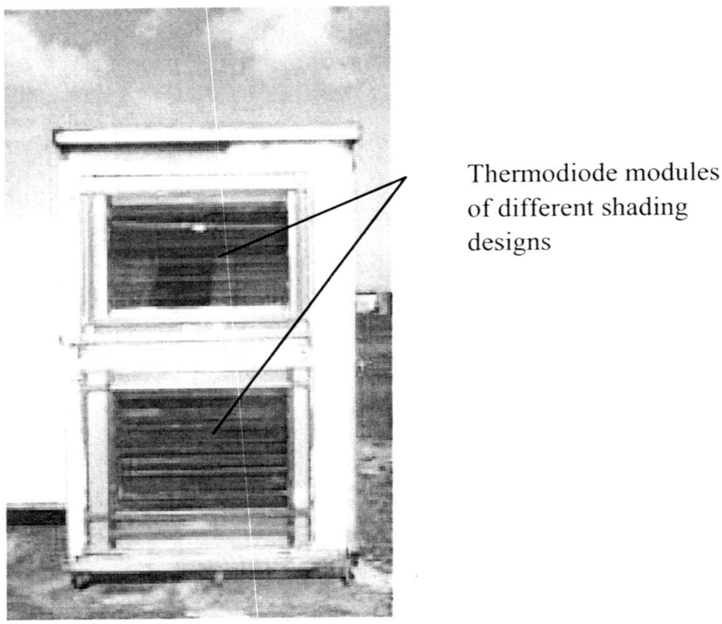

Figure 31. Photo of a cell used in the outdoor test in Cheju, Korea [Chun and Chen 2002].

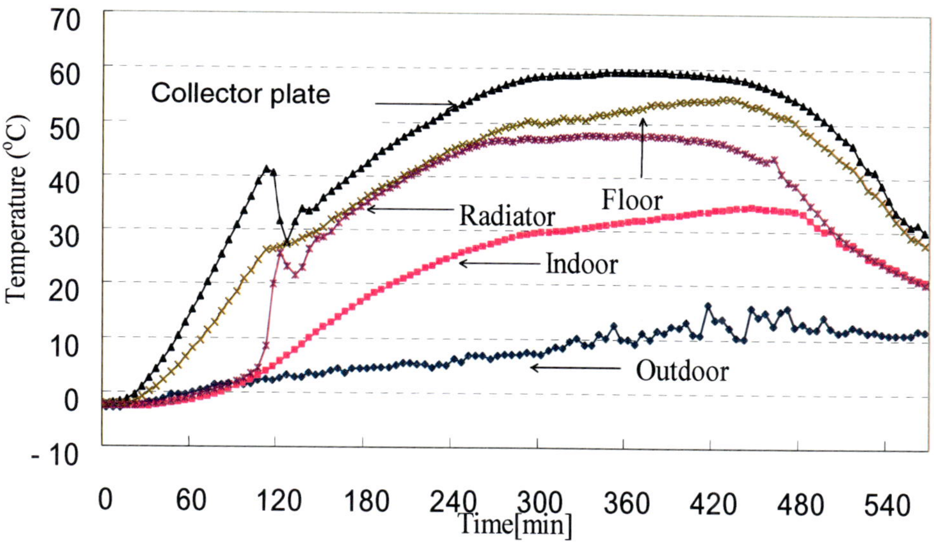

Figure 32. Outdoor test results of the thermodiode system equipped with revolvable panels for solar flux adjustment [Chun and Chen 2002].

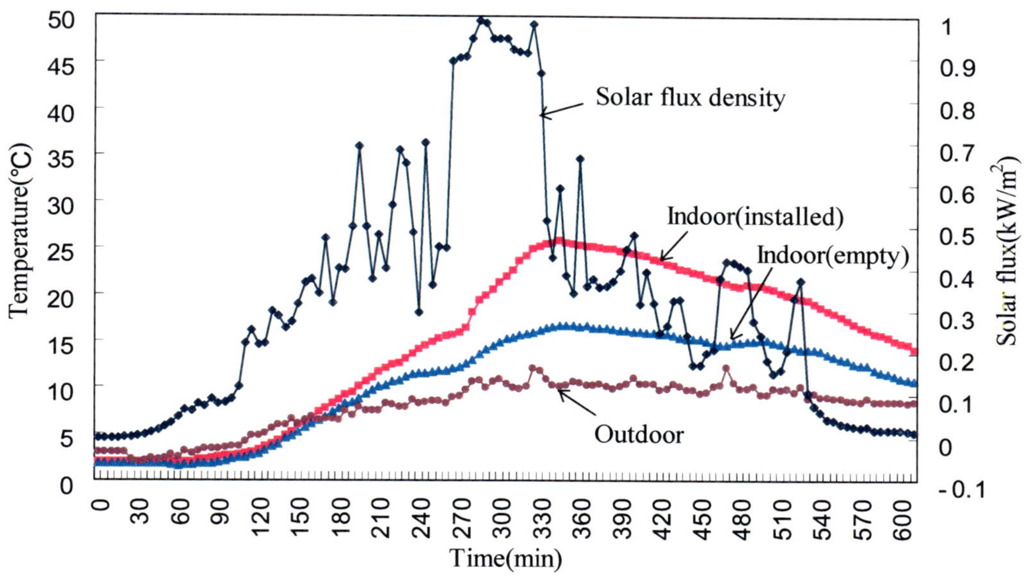

Figure 33. Interior temperatures of test cells with and without thermodiodes installed [Chun and Chen 2002].

Outdoor tests of the plane-type thermodiode were carried out on the campus of the Cheju National University in Korea. Two identical test cells were constructed from polystyrene boards and aluminum sheets. The plane-type thermodiodes were installed on the south façade of one of the cells. Shown in Figure 31 is a test cell with two 1-m tall thermodiode modules.

Each module consisted of two rectangular loops. Presented in Figures 32 and 33 are the outdoor test data collected on a cloudy winter day. Temperature variations of the thermodiode systems (Figure 32) closely resemble those of the indoor tests. Figure 33 was a comparison of the test cell temperatures. The difference between the cell interior temperatures increased very rapidly during the noon time when the sun came out of the cloud, and reached a peak of about 10 °C before the sky turned cloudy again. Analytical solutions similar to those for the bayonet and rectangular-loop thermodiodes were obtained to predict the diode temperature variation with convection coefficients determined from:

$$h = 0.59\ \kappa Ra^{1/4}/\mathrm{H} = 0.59\ \kappa(\mathrm{Gr_H\ Pr})^{1/4}/\mathrm{H} \text{ [Incropera and DeWitt 2002]} \qquad (24)$$

The above equation is the empirical correlation for laminar natural convection on a vertical isothermal surface of height H. A good agreement was obtained when the analytical solution of the average radiator temperature was compared to the indoor test data during heating (Figure 34). The radiator temperature dropped slightly faster than the laminar flow correlation predicted during the cool-down phase.

The bi-directional thermodiodes designed and tested by the authors in the United States and Korea have been proven to be very effective for winter heating applications. The efficiencies of the bayonet thermodiode and the plane-type thermodiodes with and without a storage tank are very close to the highest efficiency of single-phase thermosyphons used for solar energy utilization. The weather tests of the bayonet diode for summer cooling in Utah showed only a slight reduction in daily cooling load, but the diode is still attractive to cooling of office buildings which is only needed during the daytime, and for combined daytime heating/nighttime cooling in early spring and late autumn. The small natural convection coefficients associated with the indoor and outdoor surfaces are one reason for the less effective cooling of the diode fluid at night. The other reason is the small view factor of the vertical diode surface for radiation heat loss to the night. Other systems or devices that utilize the night sky for cooling [e.g., Ezekwe 1990] are almost always laid horizontally to have a non-obstructed view of the sky. A mechanism to tilt the cavity of the bayonet diode to get a less-obstructed view of the sky (see Figure 25) should improve the effectiveness of the bayonet diode for summer cooling applications. The reflective surface attached to the insulation behind the cavity can further enhance the radiation loss to the night sky. The cavity surface area available for natural convection cooling is also doubled in this revised design.

CONCLUSION

The good agreements between the authors' analytical solutions and measured thermodiode temperatures for the various thermosyphons and thermodiodes they studied reveal that the thermal resistance inside a thermosyphon or a thermodiode under forward bias is typically very small in comparison with other thermal resistances along the heat flow path for low-temperature applications of solar energy such as space heating. The dominant thermal resistances for winter heating and summer cooling are the resistances associated with natural convection on the indoor and outdoor surfaces, and radiation to the night sky when the diode is used for summer cooling. Unless these thermal resistances can be considerably reduced through the use of fins, forced convection, and greater night sky view factor, any improvement in diode and thermosyphon heat transfer design would not have a significant impact on the overall thermal resistance and heat transfer rate. Consequently it is unnecessary and not economical to use heat pipes or two-phase thermosyphons in this kind of solar energy systems. A simple single-phase thermosyphon is adequate for these slowly varying, low heat transfer rate processes.

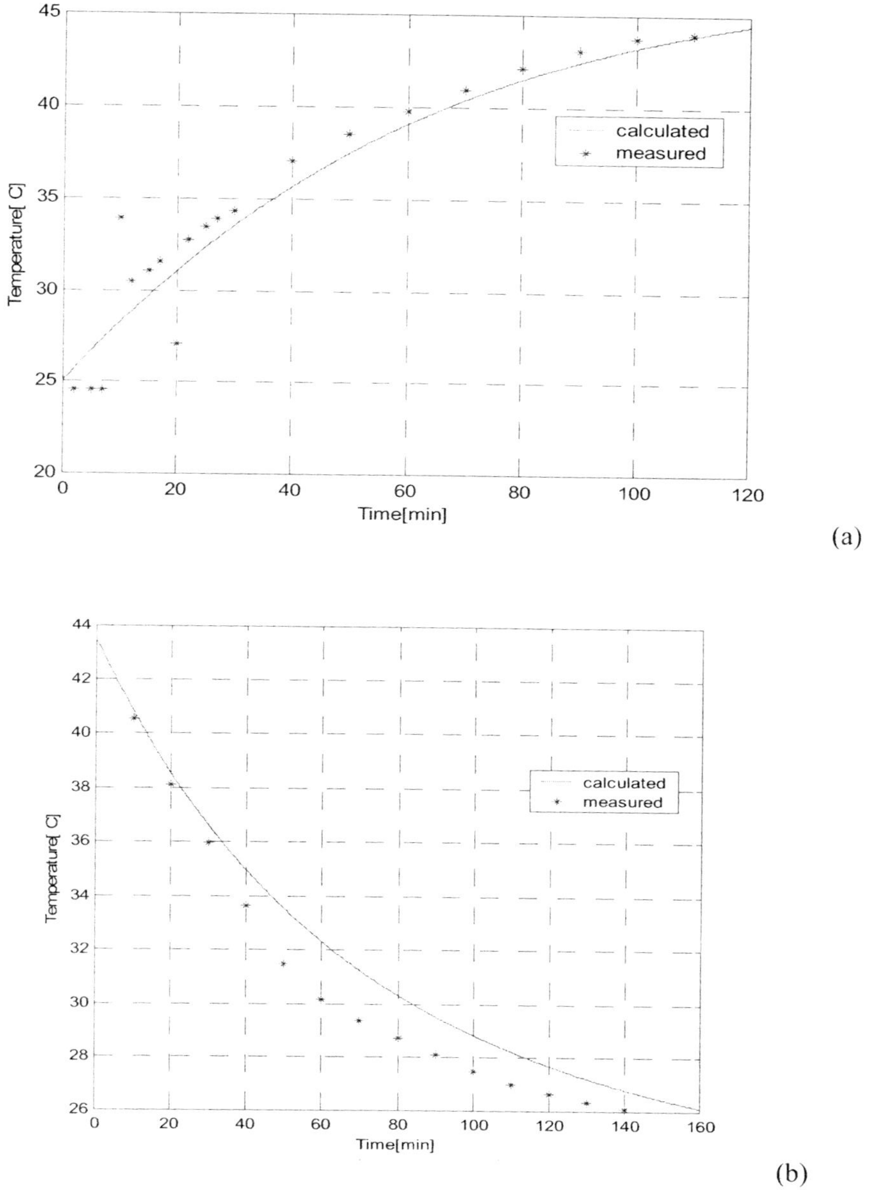

Figure 34. A comparison of the analytical solution and experimental results for 800 W/m^2 radiation on the diode system. (a) Heating phase. (b) Cooling phase [Chun and Chen 2002].

Secondly, the little effect of the change in diode thermal resistance on the overall heat transfer rate suggests the flow channels connecting the diode heating and cooling sections should be long and have small cross-sectional areas. The diode thermal resistance under forward bias will increase, but the overall heat transfer rate between the indoors and the outdoors is not expected to be reduced considerably. The major advantages of long and narrow flow channels between the heating and cooling sections are more insulation can be added and less conduction leakage through the connecting channels when the diode is reverse biased.

REFERENCES

A-STAR (2005) website : www.ime.a-star.edu.sg/02research/ mmc/pdf/mmc_coolingtech.pdf

Bau, H. H., & Torrance, K. E. (1981). On the Stability and Flow Reversal of an Asymmetrically Heated Open Convection Loop. *J. Fluid Mech., 106*, pp. 417-433.

Behnia, M., Morrison, G. L., & Paramasivam, S. (1987). Heat Transfer and Flow in Inclined Open Thermosyphons. *Proc. 1987 ASME/JSME Thermal Engineering Joint Conference*, pp. 7-14.

Chen, K. (1985a). The Optimum Configuration of Natural Convection Loops. *Sol. Energy, 34(4/5)*, pp. 407-416.

Chen, K. (1985b). On the Oscillatory Instability of Closed-Loop Thermosyphons. *ASME J. Heat Transfer, 107(4)*, pp. 826-832.

Chen, K. (1988). Design of a Plane-Type Bidirectional Thermal Diode. *ASME J. Solar Energy Engineering, 110*, pp. 299-305.

Chen, K., Shorthill, R. W., Chu, S. S., Chailapo, P., & Narasimhan, S. (1995). *An Energy-Efficient Construction Module of Variable Direction of Heat Flow, Heat Capacity, and Surface Absorptivity, USAF Report WL-TR-95-3046.*

Chen, K., Chailapo, P., Chun, W. G., Kim, S., & Lee, K. J. (1998). The Dynamic Behavior of a Bayonet-Type Thermal Diode. *Solar Energy, 64(4-6)*, pp. 257-263.

Chu, S. S. (1996). *Design and Performance Evaluation of a Smart Construction Module to Improve Building Energy Efficiency*, Ph.D. dissertation, Mechanical Engineering Department, University of Utah.

Chun, W. G., & Chen, K. (2002). Test Results of a Bi-Directional Thermodiode System for Solar Energy Utilization. *Sol. Energy, 73(4)*, pp. 269-280.

Chun, W. G., Chen, K., & Kim, H. T. (2002). Performance Study of a Bi-Directional Thermodiode Designed for Energy-Efficient Buildings. *ASME J. Sol. Energy Eng. 124*, pp.291-299.

Chun, W. G., Kang, Y. H., Kwak, H. Y., & Lee, Y. S. (1999). An Experimental Study of the Utilization of Heat Pipes for Solar Water Heaters. *Applied Thermal Engineering, 19*, pp. 807-817.

Creveling, H. F., dePaz, J. F., Baladi, J. Y., & Schoenhals, R. J. (1975). Stability Characteristics of a Single Phase Free Convection Loop. *J. Fluid Mech., 67(1)*, pp. 65-84.

Damerell, P. S., & Schoenhals, R. J. (1978). Flow in a Toroidal Thermosyphon with Angular Displacement of Heated and Cooled Sections. *ASME Paper No. 78-HT-44.*

Dunn, P. D., & Reay, D. A. (1994). *Heat Pipes, 4th ed.*, Elsevier Science Ltd, Tarrytown, NY.

Ezekwe, C. L. (1990). Performance of a Heat Pipe Assisted Night Sky Radiative Cooler. *Energy Convers. Mgmt., 30(4),* pp. 403-408.

Grief, R., Zvirin, Y., & Mertol, A. (1979a). The Transient and Stability Behavior of a Natural Convection Loop. *J. Heat Transfer, ASME Series C, 101*, pp. 684-688.

Hellman, S. K., Habetler, G., & Babrov, H. (1956). Use of Numerical Analysis in Transient Solution of Two-Dimensional Heat Transfer Problems with Natural and Forced Convection. *Trans. ASME 78*, pp. 1155-1161.

Hobson, P. A., & Norton, B. (1988). Verified Accurate Performance Simulation Model of Direct Thermosyphon Solar Energy Water Heaters. *ASME J. Solar Energy Engineering, 110*, pp. 282-292.

Huang, B. J., & Hsieh, C. T. (1985). A Simulation Method for a Solar Thermosyphon Collector. *Solar Energy, 35*, pp. 31-43.

Incropera, F. P., & DeWitt, D. P. (2002). *Fundamentals of Heat and Mass Transfer, 5th edition*, John Wiley & Sons, New York, NY, pp. 11, 223, 545.

Ivanov, V. L., & Lapin, Yu. D. (1966). End Heat Exchange in a Cylindrical Passage under Free Convection Conditions. *Thermal Engng. (USSR), 13(6),* pp. 108-112.

Jones, G. F. (1986) Heat Transfer in a Liquid Convective Diode. *ASME J. Sol. Energy Eng., 108*, pp. 163-171.

Kaizerman, S., Wacholder, E., & Bilas, E. (1981). Stability and Transient Behavior of a Vertical Toroidal Thermosyphon. *ASME Paper No. 81-WA/HT-11.*

Kang, Y. H., Kang, M. C., & Chun, W. G. (2003). A Study on Thermal Characteristics of the Solar Collector Made with a Closed Loop Thermosyphon. *Int. Comm. Heat Mass Transf., 30(7),* pp. 955-964.

Keller, J. B. (1966). Periodic Oscillations in a Model of Thermal Convection. *J. Fluid Mech., 26(3)*, pp. 599-606.

Kulhavy, J. T. (1964). Tractor Engine Cooling. *Winter Meeting of the American Society Agricultural Engineers, New Orleans, Louisiana, Paper, No. 64-637.*

Larsen, E. H. (2003). A Capillary Assisted Thermosyphon for Shipboard Electronics Cooling. *Trident Scholar project report A 524914*, Naval Academy Annapolis, MD.

Madejski, J., & Mikielewicz, J. (1971). Liquid Fin-A New Device for Heat Transfer Equipment. *Int. J. Heat Mass Transf., 14*, pp. 357-363.

Mertol, A., Greif, R., & Zvirin, Y. (1981). The Transient Steady State and Stability Behavior of a Thermosyphon with Throughflow. *Int. J. Heat Mass Transf., 24(4),* pp. 621-633.

Morrison, G. L., & Tran, H. N. (1984). Simulation of the Long Term Performance of Thermosyphon Solar Water Heaters. *Solar Energy, 33*, pp. 515-526.

Ong, K. S. (1974). A Finite Difference Method to Evaluate the Performance of a Solar Energy Water Heater. *Solar Energy, 16*, pp. 131-147.

Norton, B., Probert, S. D., & Gidney, J. T. (1987). Diurnal Performance of Thermosyphonic Solar Water-Heater – an Empirical Prediction Method. *Solar Energy, 39*, pp. 257 – 265.

Ong, K. S. (1976). An Improved Computer Program for the Thermal Performance of a Solar Water Heater. *Solar Energy, 18*, pp. 183-191.

Rosenhead, L. (ed.) (1963). *Laminar Boundary Layers*, Oxford University Press, pp. 506-549.

Shitzer, A., Kalmanoviz, D., Zvirin, Y., & Grossman, G. (1978). Experiments with a Flat Plate Solar Water Heating System in Thermosyphonic Flow. *Solar Energy, 22*, pp. 27-35.

Turner, J. S. (1973). *Buoyancy Effects in Fluids*, Cambridge University Press, pp. 207-225.

Welander, P. (1967). On the Oscillatory Instability of Differentially Heated Fluid Loop. *J. Fluid Mech., 29(1),* pp. 17-30.

Williams, R. J. (1978). Investigation of a cryogenic thermal diode. *Paper 78-417, Proc. III Int. Heat Pipe Conference* in Palo Alto, CA, *AIAA Report CP784*, New York.

Zvirin, Y., & Grief, R. (1979b). Transient Behavior of Natural Circulation Loops: Two Vertical Branches with Point Heat Source and Sink. *Int. J. Heat. Mass. Transf, 22*, pp. 499-504.

In: Solar Energy: New Research
Editor: Tom P. Hough, pp.239-254

ISBN 1-59454-630-4

Chapter 7

SOLAR BOOSTED HEAT PUMP (SBHP)

L. Aye, W. W. S. Charters and C. Chaichana
International Technologies Centre, the University of Melbourne, Victoria, Australia

ABSTRACT

Solar boosted heat pumps (SBHPs) can capture energy from low temperature ambient air and solar energy, then upgrading to a higher temperature level for useful applications. For water heating the highest delivered water temperature can be much higher than that provided by regular thermosyphon solar water heaters. Solar boosted heat pump water heaters do not require auxiliary back up heating, they could provide hot water even when the solar energy is not available. There is negligible collector heat loss compared to regular solar collectors since the collector temperature is lower, or about the same as the surrounding ambient air temperature. The collector of a SBHP water heater does not require a cover glazing; this makes the system cost competitive with conventional solar water heaters. A brief history of the development of SBHPs and fundamental principles are presented. The current and potential applications, and recent research directions are also discussed in this chapter.

INTRODUCTION

Brief History

Although most people are familiar with the common household refrigerator the concept of using such a device to provide heating rather than cooling is less widely understood. However the basic operational principles of this machine called a heat pump was laid down in thermodynamic terms by Lord Kelvin the first professor of natural philosophy at The University of Glasgow, Scotland, in the middle of the 19th century. Whereas, the thermodynamic principle of the heat pump was founded at the beginning of the 19th century, it was not realised until 1855 for producing heat, by means of an open-cycle mechanical vapour recompression unit in Ebensee, Austria (Halozan & Rieberer 2004). Then much later the closed vapour compression process was used for generating useful heat. After World War II, heat pump units for air conditioning homes and individual rooms became common.

Fundamentals

There are many types of heat pumps available but in this chapter we will restrict ourselves to consideration of the vapour compression units in which a suitable working fluid is capable of evaporation and condensation as shown in Figure 1. To carry out these processes any practical heat pump uses four separate components:

- Compressor,
- Condenser,
- Expansion device, and
- Evaporator

as illustrated in Figure 2.

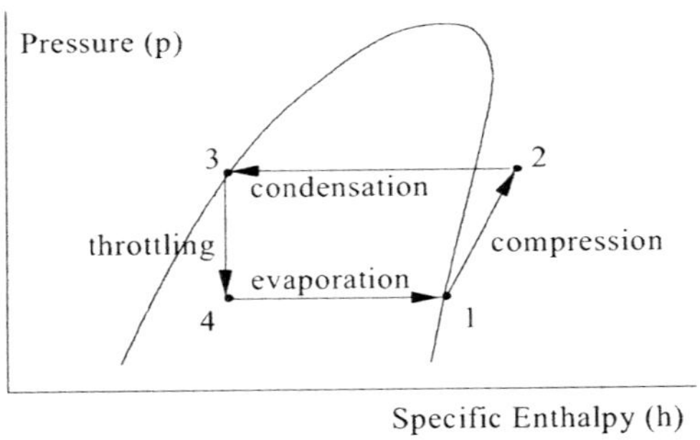

Figure 1. Pressure enthalpy diagram for ideal heat pump cycle

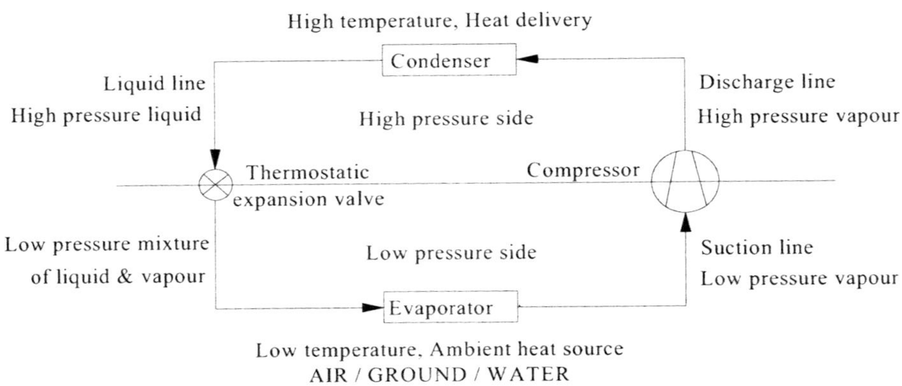

Figure 2. Schematic of heat pump circuit

The heat pump is essentially a device used to take in energy from a low grade source, i.e., one at low temperature, and to upgrade this thermal energy to a more suitable temperature level by providing a mechanical energy input (Figure 3). The heat pump must, by the principle of energy conservation, provide more energy at the high temperature side than is used to drive the mechanics of the compressor. The ratio of output heat to input mechanical work is known as the heat pump heating coefficient of performance (COP) and must always theoretically be greater than unity.

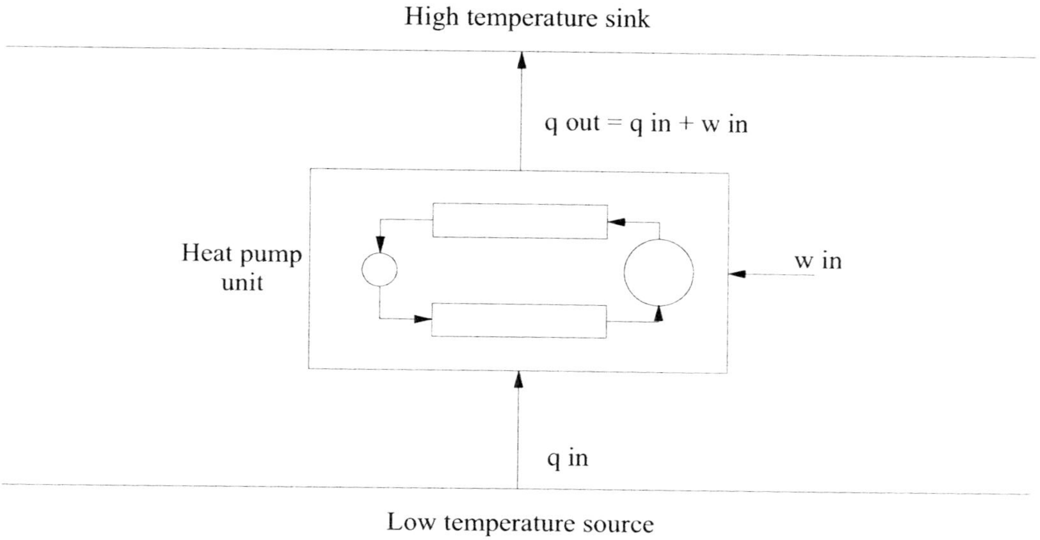

Figure 3. Energy flow from low temperature source to high temperature sink

$$COP_{heating} = \frac{q\ out}{w\ in} = \frac{q\ in + w\ in}{w\ in} = 1 + \frac{q\ out}{w\ in} > 1$$

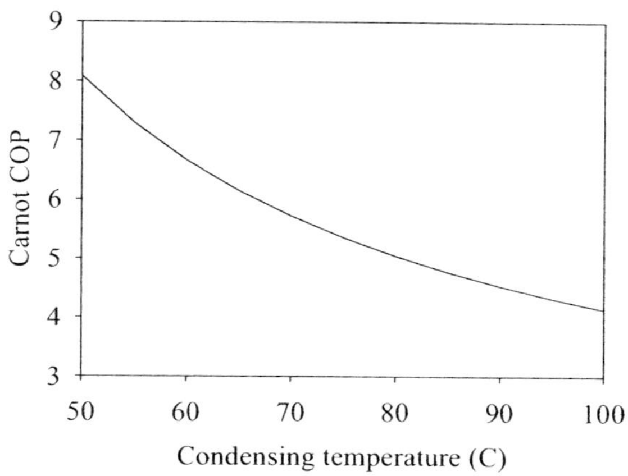

Figure 4. Effect of condensing temperature on ideal COP (te = 10 C)

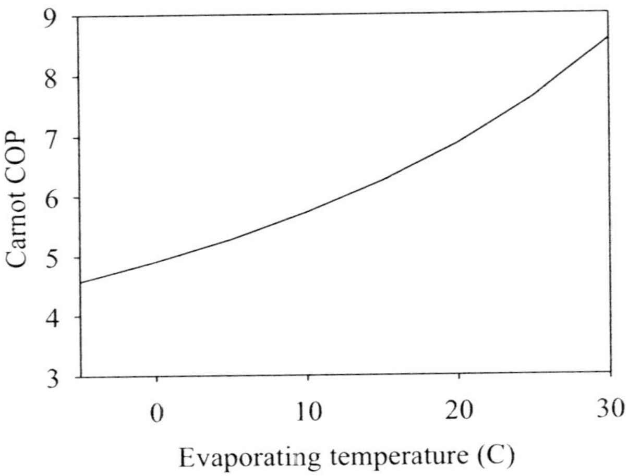

Figure 5. Effect of evaporating temperature on ideal COP (tc = 70 C)

Changes in condensing and evaporating temperatures affect the work term at the compressor and consequently COP. Decreasing condenser or increasing evaporator temperatures will decrease the compressor work and increase the COP. Figure 4 shows the effect of condensing temperature on COP for a fixed evaporator temperature. The effect of evaporating temperature on COP for a fixed condenser load temperature is shown in Figure 5. For simplicity these curves have been obtained assuming operation on the ideal heat pump cycle, i.e., the Carnot cycle of two isothermal and two isentropic (reversible adiabatic) processes not allowing any practical deviations. Although the COP values for real machines will be substantially lower, the trend will be the same.

Practical Heat Pump Cycle

In practice, nearly all vapour compression heat pumps use thermostatic expansion valves (T-X valves) because of the availability of the valve to handle a wide range of operating conditions and the fact that the pressure reduction is essentially isothermal. These valves, which are relatively low-cost devices, control the liquid working fluid flow to the evaporator by sensing the superheated condition of the working fluid leaving the evaporator. Such control ensures that nearly all of the available evaporator surface is covered with a forced convection nucleate boiling film with consequentially excellent heat transfer characteristics in the evaporation process. The factory setting of these valves is usually at 5-6 C of superheat with about 2-3 C change in the exit superheat required to move the valve stem from the 'fully open' to the 'fully closed' position.

Compressors used in practical units are generally of the positive displacement type: either reciprocating or rotary vanes. As the compression process is not reversible, in a thermodynamic sense, the compression is not isentropic, but is often taken to be adiabatic in cycle analysis if the polytropic exponent for the compression is not known.

In each of the evaporation and condensation processes there will be a small fluid flow pressure drop; and there will be entry and exit pressure drops for each of the components in the heat pump circuit. A small degree of sub-cooling often occurs in practice and the degree of superheat is controlled by the T-X valve setting (Figures 6 and 7).

Reversible circuit heat pumps for building cooling and heating are equipped with a four-way exchange valve, which can be automatically or manually operated. This valve enables the normal evaporator to become the condenser and the normal condenser to become the evaporator so that the machine can be made to operate in the reverse fashion. This arrangement is often incorrectly called "reverse cycle" operation, because of the end effect achieved. i.e., the interchange of roles for the cooling and heating parts of the circuit. It would be better to call it "reverse flow" operation, as the basic thermodynamic cycle remains the same for each mode of operation.

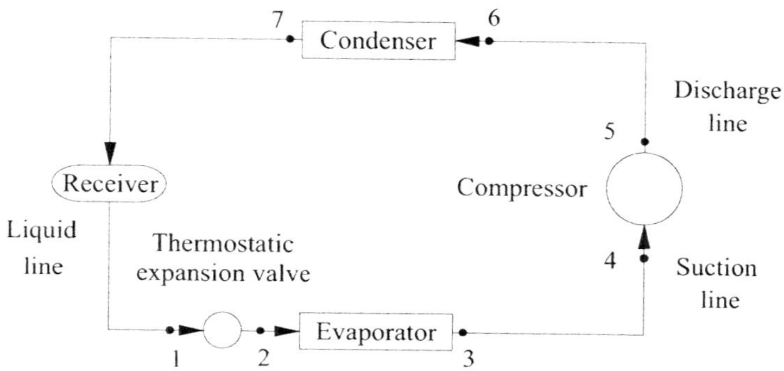

Figure 6. Schematic diagram of a practical heat pump

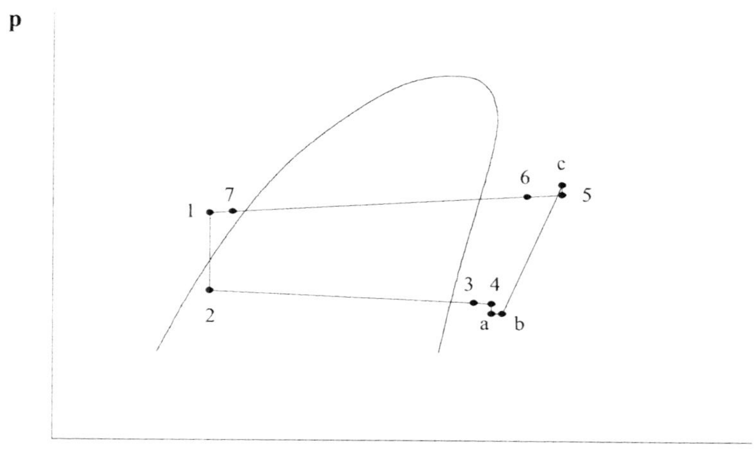

Figure 7 Pressure-enthalpy diagram of a practical heat pump cycle

Processes

1-2 = throttling process	b-c = polytropic compression
2-3 = evaporating process	c-5 = discharge value pressure drop
3-4 = suction line heating	5-6 = discharge line heat loss
4-a = suction valve pressure drop	6-7 = condensing process
a-b = cylinder heating of vapour	7-1 = liquid line heat loss

In general, heat pumps are classified in many different ways, but one method commonly used distinguishes between air, water and ground sources of low temperature energy, as well as between the high temperature energy delivery medium, i.e., air or water. In such a classification the most common types of heat pump units would include air-to-air and water-to-water units.

Solar Energy and Heat Pumps

Although it is fair to say that all heat pumps are solar units as the low temperature energy source whether it be ambient air, river or lake source water, or earth ground effect is always indirectly solar heated, this is like saying that all houses are effectively solar heated. Just as in practice most architects and engineers identify solar houses as one in which particular attention has been paid to maximising the useful solar input so we distinguish here solar heat pump units as ones in which a deliberate attempt has been made to achieve improved performance levels through utilising a solar energy input. If the solar input is used indirectly to raise the evaporator temperature, the unit is termed solar assisted heat pump (SAHP) (Figure 8) and if the evaporator becomes an evaporator/collector unit for direct capture of the solar radiation, we term it solar boosted heat pump (SBHP) (Figure 9). One of the early attempts at harnessing solar energy to improve the overall performance of a heat pump system has been described by Sporn and Ambrose (1955). Charters and Taylor (1976) reported on a SBHP using a flat-plate collector without a cover as the evaporator. This type of system is also named as direct expansion solar collector-heat pump system. Unlike thermosyphon solar water heaters, solar heat pump systems provide a real capability of upgrading low-grade energy sources from the surroundings as well as using solar energy (Charters *et al.* 1980; Lu Aye & Charters 2003).

As explained earlier on the effect of increasing the evaporating temperature for a fixed condensing load temperature is to reduce the source sink temperature difference and to reduce the compression required compression energy. This will inevitably lead to a better COP because of the reduced compressor work requirement. As a side effect, the capacity of the will also effectively increases with the solar effect because of the increased working fluid density at the constant volume nature of the compressor units used. Depending on the nature of the slope of the saturated vapour curve in the pressure-enthalpy diagram of the working fluid there will be changes in specific enthalpy of vaporisation at the higher evaporation temperature. Positive slope working fluids will have a reduction in specific enthalpy of vaporisation as can be seen in Figure 7. For a solar heat pump, which uses R-12 and R-22 as working fluid, the increased mass flow rate effects, more than counterbalances this reduction in specific enthalpy of vaporisation at the higher evaporation pressure and temperature condition.

For all solar heat pump systems some form of thermal energy storage is generally incorporated to obviate the deleterious effects of solar radiation variations on unit performance. For water source systems the storage is often on the lower temperature side of the circuit and this is obviously true of the ground source systems. With air source systems on the other hand, the thermal store is invariably on the high temperature side of the circuit.

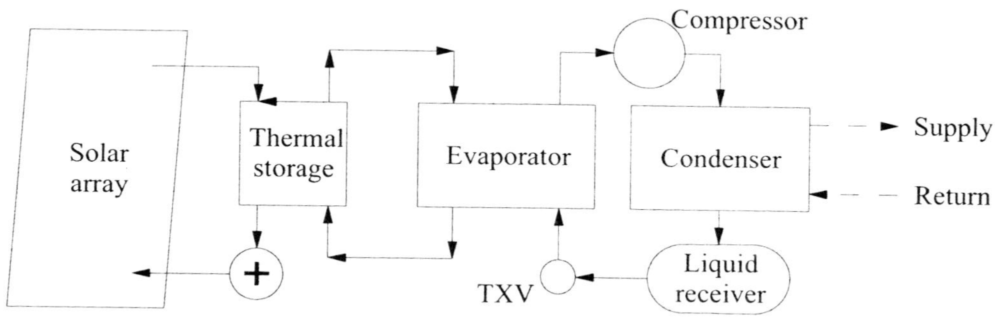

Figure 8. Solar assistant heat pump (SAHP) schematic circuit

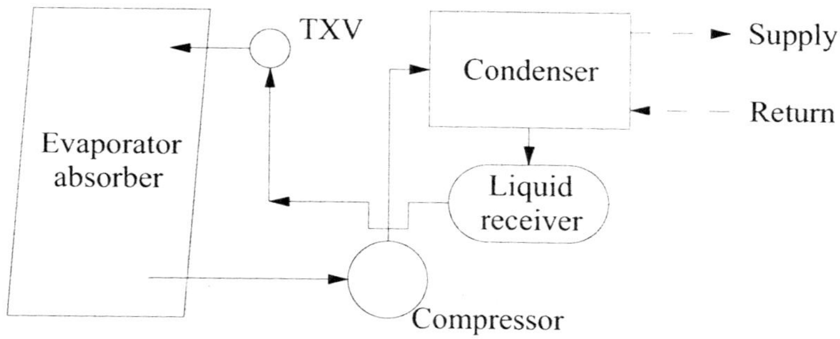

Figure 9. Solar boosted heat pump (SBHP) schematic circuit

The evaporator/collector of a SBHP can collect more solar energy than conventional flat-plate collectors for the same aperture area. The reason is that the evaporator/collector is operating at a lower temperature and consequently has a lower heat loss to the ambient. Because of a lower operating temperature (generally lower than the ambient air) the evaporator/collectors of SBHPs do not requir covers (glazing) to reduce the heat loss. It should be noted that at temperatures lower than ambient air the evaporator/collector can collect solar radiation, sensible heat and latent heat from ambient air. SBHP systems have a more consistent energy delivery capacity than flat-plate solar water heaters due to the ability of the heat pump to operate in air source mode during cloudy periods and to collect heat of condensation during humid conditions. An additional benefit is minimal corrosion problems due to the use of refrigerant. This would increase the life of the collector significantly over that of a water based solar collector; its life is often limited to 10 years (Gorozabel Chanta 2005). Another advantage of using the refrigerant as the collector fluid is that the freezing problem is eliminated due to the very low freezing point of refrigerants. Elimination of the intermediate thermal storage heat exchanger, which is required in the conventional SAHP (Figure 8), also improves the overall performance of SBHP.

The first generation SBHP hot water systems proposed has not achieved commercial success due to the complexity of the combination of heat pump and solar collector components and high installation costs due to the need for plumbing, electrical, and refrigeration connections during installation (Morrison 1994). To minimise system costs and parasitic energy requirements Charters *et al.* (1980) proposed a system that incorporated the heat pump condenser directly in the water storage tank. This configuration removed the need for a separate refrigerant condenser and eliminated the parasitic energy of the pump to heat from the heat pump condenser to the water storage. This system has limited heat transfer by free convection within the tank. A system using a wrap-around condenser on the outside of the storage tank was developed to increase the heat transfer area. This system has achieved significant acceptance in Australia, South East Asia, and the Pacific Islands. To avoid heat-pump circuit connection and working fluid charging at the installation site, a next generation design was developed that incorporated the evaporator as an integral part of the storage tank. The package system has the advantage that all the components are assembled in the factory.

Component Sizing

With the conventional heat pump systems the operational conditions are usually well specified and the designer can choose the components carefully in the initial design process with the full knowledge that the system will run with little variation from the set design conditions. On the other hand, with a solar boosted heat pump there may well be substantial variations in the evaporator temperature as the solar input fluctuates. Extra care should be exercised in the choice of components to allow for the mass flow variations in working fluid and the associated capacity effects.

A suitable condenser can be selected based on the temperature and heating capacity required for the application. The required working fluid mass flow can be estimated. O'Dell *et al.* (1984) reported a general procedure to predict the seasonal performance (heat collected by the solar collector, the condenser heat delivery and COP for various weather conditions). The procedure accounts for variations in collector design and orientation and also for heat pump capacity and efficiency. To provide a tool for assessing the performance of SBHP water heaters, a simulation model based on extension of the TRNSYS program (Klein *et al.* 2004) has been developed by Morrison (1994). Ito *et al.* (1999) also presented a simplified analysis, which can predict heat pump performance under various weather conditions. Effects of solar radiation gained, sky radiation loss, ambient air temperature, wind, speed, collector design, and the temperature of hot water supplied, can be studied with the simplified analysis developed by Ito *et al.* (1999).

To achieve performance better than an air source heat pump (ASHP), a SBHP must operate at high evaporation temperatures for a significant part of its operating time. This requires that the heat collection rate of the solar collector/evaporator and the refrigerant rate through it to the compressor be carefully matched. For example, an oversized compressor connected to a small evaporator plate would draw so much heat from it that the evaporator plate would operate at a very low evaporation temperature. Conversely an undersized compressor, while achieving high evaporation temperatures and hence COPs, would be uneconomical because of its low thermal output for a given plate area. The principal aim of the design is to determine the evaporation temperature, which when known establishes the heat pumping rate and the system COP for any given condenser temperature. A compressor sizing procedure for an air source, based on a direct solar boosted heat pump system is described in the following sections.

Establishing Evaporator Temperature

The steady state balance evaporator temperature is obtained by matching the evaporator heat collection rate and compressor capacity. If the evaporation temperature is T_e and the air temperature is T_a with solar heat input to evaporator $Q_r = IA_e\alpha$, the total heat collected at the evaporator may be written as:

$$Q_e = (UA_e)(T_a - T_e) + IA_e\alpha \quad (1)$$

where A_e = evaporator area (m^2), U = heat transfer coefficient (kW/m^{2}C), I = solar irradiance on the solar collector/evaporator plane (kW/m^2) and α = solar absorptivity of the collector plate. In equation (1) the last term is independent of evaporation temperature, therefore the heat flow rate is proportional (for a given geometry) to the temperature difference $(T_a - T_e)$. It should be noted that large collector heat loss coefficients, U (heat transfer coefficients in equation 1) are necessary to effectively extract ambient energy from air for good performance when the solar radiation level is low. It was reported by O'Dell *et al.* (1984) that uncovered collector heat pump systems have consistently superior performance over covered collector heat pump systems.

The compressor capacity *CC* (kW), defined as the heat input to the vapour in the evaporator necessary to evaporate vapour at the same mass flow rate as vapour is being drawn by the compressor. This is equal to the mass flow rate multiplied by the specific enthalpy change of working fluid passing through the evaporator i.e.,:

$$CC = \dot{m}(h_4 - h_2) \quad (2)$$

The specific enthalpies h_4 and h_2 can be found from the properties of the working fluid used. The working fluid mass flow rate $\dot{m}$ (kg/s) can be estimated from effective volume displacement rate *EVD* (m^3/s) and the specific volume of vapour at the compressor inlet.

$$\dot{m} = \frac{EVD}{v_4} \quad (3)$$

The compressor always has a known displacement rate defined as the swept volume of the space used to draw in refrigerant vapour. The events occurring within the cylinder of a reciprocating compressor are shown in Figure 7. The mean effective valve pressure drops are "4-a" and "c-5". The cylinder heating of vapour is "a-b". A multi cylinder, single acting, reciprocating compressor of stroke *L* (m), bore *D* (m) and rotational speed *N* (r/s) has the following piston displacement rate *PD* (m^3/s).

$$PD = \frac{\pi}{4}D^2 L N n_c \quad (4)$$

where n_c = number of cylinders

Ideally, this displacement rate is also the rate of vapour volume extraction from the evaporator as the evaporator exit is connected to the compressor inlet (or suction line). In practice, the effective volume flow rate is affected by factors such as:

- Re-expansion of the clearance volume,
- Valve pressure drops and
- Cylinder heating effects.

Working fluid mass flow rate driven by the compressor may also be affected by leakage during compression, the presence of other gases with the working fluid and lubrication oil entrainment in the clearance space. The latter factors are generally small, and are neglected in this analysis. So the effective volume flow rate *EVD* (m^3/s) is:

$$EVD = \eta_v PD \tag{5}$$

η_v is the volumetric efficiency which can be estimated by using conventional compressor theory (Kuehn *et al.* 1998).

$$\eta_v = [1 + C - C(\frac{p_c}{p_b})^{1/n}]\frac{v_4}{v_b} \tag{6}$$

where C = clearance factor (dimensionless)
p_c = absolute pressure at state c in Figure 7 (kPa)
p_b = absolute pressure at state b in Figure 7 (kPa)
v_4 = refrigerant specific volume at state 4 in Figure 7 (m^3/kg)
v_b = refrigerant specific volume at state b in Figure 7 (m^3/kg)
n = polytropic compression index (dimensionless)

Hence the compressor capacity *CC* is

$$CC = [1 + C - C(\frac{p_c}{p_b})^{1/n}]\frac{v_4}{v_b} \cdot \frac{\pi}{4v_4} D^2 L N n_c (h_4 - h_2) \tag{7}$$

v_4, h_4 and h_2 are working fluid properties and are functions of the evaporation temperature T_e. For a compressor with known dimensions and rotational speed the compressor capacity (equation 7) can be obtained or computed, and this it is a function of evaporating temperature (for a given condensing temperature). As evaporating temperature T_e increases (and therefore also evaporating pressure) the specific volume v_4 becomes smaller (i.e. vapour is becoming more dense) and the mass flow rate and capacity increase.

It should be noted these:

- Both evaporator heat collection rate Q_e (equation 1) and the compressor capacity CC (equation 7) are functions of evaporating temperature.
- Both evaporator heat collection rate Q_e and the compressor capacity CC must be the same as any steady state operating temperature T_e.
- COP, useful heat delivery rate, compressor work may be determined when the steady state operating temperature T_e is known.

Equations (1) and (7) can be solved for steady state T_e by graphically plotting both curves on a common evaporating temperature axis (Figure 10) or by an iterative numerical method. There are two evaporator heat collection rate curves, one with solar radiation input and the other without solar radiation input, which represents the case of a conventional evaporator. Where the heat collection rate curve and compressor capacity curve intersect a stable operating point exists, and for the given system evaporation will proceed at such a stable operating point. A larger compressor (higher compressor capacity) would yield a lower balance evaporation temperature and conversely a smaller compressor (lower compressor capacity) will clearly give a higher evaporation temperature. The smallest compressor capacity possible is zero i.e., no displacement, and this will intersect the conventional evaporator capacity curve on the T_e axis – that is T_a is the equilibrium evaporator temperature at zero compressor capacity. This represents the highest possible evaporation temperature with the air source for conventional evaporators, which rely upon heat transfer by conduction and convection depending upon the difference in temperature. Balance point with radiation input (1) is clearly higher than that without radiation input (1). Note that balance point with radiation input may be higher than T_a, and this condition is not possible with the conventional system evaporator.

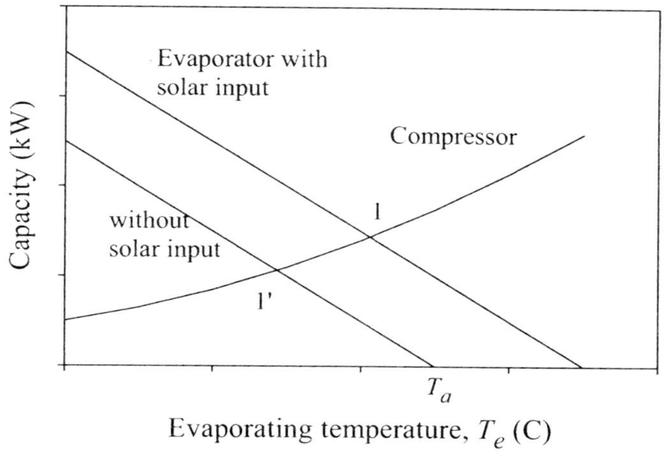

Figure 10. Graphical technique of system balancing

Compressor Work Input

The theoretical work input to a compressor of the type in Figure 7 per unit mass of working fluid, W (kJ/kg) can be calculated by:

$$W = \frac{n}{n-1} p_b \, v_b \, [(\frac{p_c}{p_b})^{(n-1)/n} - 1] \tag{8}$$

By introducing the motor efficiency, η_{mot}, and compressor mechanical efficiency, η_{mech}, the actual power required by the compressor, P (kW), may be estimated as:

$$P = \frac{\dot{m} \, W}{\eta_{mot} \eta_{mech}} \tag{9}$$

A motor efficiency is included to account for friction and electrical losses in the electric motor and a general mechanical efficiency is included to account for friction losses in the drive to the compressor.

Design Considerations

For a given evaporator plate area, evaporation temperature can be obtained as a function of compressor displacement rate. The thermal output and ambient energy extraction can also be obtained as a function of the compressor displacement rate. By suitable matching of the evaporator and the compressor, a design can be arrived at that should provide a reasonable degree of solar boosting on a clear winter day.

For optimal control of a solar heat pump system it would be essential to have a variable speed motor compressor drive unit, but in practice this would be too expansive a choice. In some installations it may well be appropriate to incorporate a two-speed motor drive unit for improved performance.

There have been indications of dynamic instability in the operation of conventional T-X valves under a widely ranging flow and temperature conditions as experienced in practice by the University of Melbourne solar boosted heat pump systems and this effect should be borne in mind when sizing the orifice in the T-X valve.

Renewable Energy Gained by SBHP

SBHP upgrades freely available heat from the environment (solar and ambient air). The renewable heat R gained by the heat pump is the difference between the thermal output Q and the drive energy E.

$$R = Q - E = E \cdot COP - E = E(COP - 1)$$
$$= Q - Q / COP = Q(1 - 1/COP) \qquad (10)$$

Applications

The combination of heat pump and solar technology provides an inherently efficient machine capable of providing processed heat for a variety of end uses. Naturally, the extra complexity of the systems is likely to introduce some extra initial cost margin when compared to the stand alone heat pump or individual solar system, but in general this will be more than offset by the increased efficiency of heat collection and by the built-in capability for operations to provide cooling as well as heating.

Typical applications which have been identified where heating and cooling are required simultaneously include sports/leisure complexes with ice-skating rinks and swimming pools side by side; and large farm dairies where there is a requirement for milk chilling as well as large use of hot water for pasteurisation, sterilisation, and utensil washing. Such applications have been reported in the literature from Europe and from Australia. Another potentially large market for solar heat pump systems lies in those areas where seasonal heating and cooling may be required, such as in building space heating with steady requirements for winter heating and summer cooling to attain year round comfort levels. Applications in this field have been reported worldwide but particularly in the U.S.A.; in Europe; in Japan and in Australia. Domestic SBHP hot water heaters are now available off-the-shelf (Lu Aye *et al.* 2002). Heat pump water heaters, based on air-source evaporators or solar-boosted evaporators, are being adopted in many countries (Morrison *et al.* 2004).

Recent Developments

Phasing out of CFCs due to their harmful effects on the stratospheric ozone layer has led to research on alternative working fluids. During the 1990s, the challenge of finding environmentally-friendly alternative working fluids has become one of the most popular topics in heat pump research (Lorentzen 1994, Lorentzen 1995, Hwong 1995, Halozan 1996, Pearson 1999). Analytical and experimental studies were performed on a SBHP water heating system for the refrigerant R-134a by Hawlader *et al.* (2001). Chaichana *et al.* (2003) explored the possibilities of replacing R-22 with natural working fluids in SBHP water heaters. Six natural working fluids, R-290, R-600, R600a, R-1290, R-744, and R-717, were compared. It was found that R-290 (propane) could be used as direct drop-in replacement for R-22 with only minor modifications to the existing heat pump. On the other hand, R-717 (ammonia) was also identified as one of the suitable replacement working fluids. However, using R-717 requires a totally new heat pump due to the incompatibility with copper and brass in the existing heat pumps. The comprehensive detailed performance of a R-290 SBHP and a R-717 SBHP were reported in Chaichana (2003). Gorazabel Chata *et al.* (2005) performed thermodynamic analysis of a SBHP for a variety of refrigerants (R-12, R-22, R-134a, R-404A, R-407C and R-410A) and recently reported the results.

Huang and Lee (2003) investigated a long-term reliability test of a SBHP water heater. The total accumulated operating time for the prototype is over 20,000 h during 5 years and reported that no machine failure has ever occurred. The study has verified that SBHP is very reliable if it is carefully designed.

Aluminium roll-bond panels with polycrystalline PV modules bound to surfaces were developed to make a cheaper and lighter photovoltaic and thermal hybrid (PV/T) panels for the collector/evaporator of a SBHP (Fujita *et al.* 1998). These PV/T panels were incorporated together with PV modules into a system for electrical generation, hot water supply, and room heating. The experimental results of modified panels with reduced pressure loss were reported by Ito *et al.* (2005). A collector efficiency factor of 0.9 was obtained for the evaporator. It should be noted that for a refrigerant-filled collector the collector heat removal factor (collector efficiency factor) is generally assumed to be unity (O'Dell *et al.* 1984). This is a reasonable assumption since the convection heat exchange coefficient is very high due to a boiling of the refrigerant, which contributes to a very high collector plate efficiency.

Ozgener and Hepbasli (2006) installed and investigated an experimental coupled solar boosting and ground-source system for greenhouse heating (4.19 kW heating load). The experimental results showed that it could be used for greenhouse heating in the Mediterranean and Aegean regions of Turkey.

CONCLUSION

Benefits and fundamental principles of SBHP have been presented. A method for selecting an appropriate compressor for a given collector/evaporator was outlined. Current popular applications of SBHP are domestic hot water heating and swimming pool heating. SBHP is a technology which has not been extensively used because both energy and electricity prices are low. The driving energy for SBHPs is generally fossil fuel powered electricity. Greenhouse gas emissions from the burning of fossil fuels may be a good reason for the increasing market deployment of this technology. SBHP may be one of the key technologies for reducing greenhouse gas emissions.

SBHPs with environmentally friendly working fluids would be available in the immediate future and SBHP hybrid systems (e.g. PV/T) are on the way in the near future. Advancement of computing capabilities, simulation and optimisation techniques would speed up the development of the technology.

REFERENCES

Lu Aye and Charters, W. W. S. (2003) 'Electrical and engine driven heat pumps for effective utilisation of renewable energy resources', *Applied Thermal Engineering*, 23(10): 1295-300.

Lu Aye; Charters, W. W. S. and Chaichana, C. (2002) 'Solar heat pump systems for domestic hot water', *Solar Energy*, 73(3): 169-75.

Chaichana, C. (2003) *Natural Working Fluids for Heat Pumps*, PhD thesis, Department of Civil and Environmental Engineering, The University of Melbourne.

Chaichana, C.; Lu Aye & Charters, W. W. S. (2003) 'Natural working fluids for solar-boosted heat pumps'. *Revue Internationale du Froid, International Journal of Refrigeration*, International Institute of Refrigeration (IIR), 26(6): 637-43.

Charters, W. W. S.; de Forest, L.; Dixon, C. W. S. and Taylor, L. E. (1980) 'Design and performance of solar boosted heat pumps. In Australian and New Zealand Solar Energy Society Annual Conference, Melbourne.

Charters, W. W. S. and Taylor, L. E. (1976) 'Some performance characteristics of a solar boosted heat pump'. In Proceedings of IIR Conference: Towards an ideal refrigerated food chain, 641-48, Melbourne, Australia.

Gorozabel Chata, F. B.; Chaturvedi, S. K. and Almogbel, A. (2005 article in press) 'Analysis of a direct expansion solar assisted heat pump using different refrigerants', *Energy Conversion and Management*.

Halozan, H. (1996) 'Natural refrigerants – an option for heat pumps? paper presented to the IIR Conference: Refrigeration, Climate Control and Energy Conservation, Melbourne, 11-14 February 1996.

Halozan, H. and Rieberer, R. (2004) 'Energy-efficient heating and cooling systems for buildings' IIR Bulletin, vol. LXXXI, no. 2004-6: 6-22.

Hawlader, M. N. A.; Chou, S. K. and Ullah, M. Z. (2001) 'The performance of solar assisted heat pump water heating system'. *Applied Thermal Engineering*, 21: 1049-65.

Hwong, H. (1995) 'Transition of CFCs and HCFCs to alternative refrigerants in the United States'. Proceedings of the International Conference on Energy and Environment, May 1995, 374-80.

Ito, S.; Miura, N. and Takano, Y. (2005) 'Studies of heat pumps using direct expansion type solar collectors. Journal of Solar Energy Engineering, Transactions of the ASME, 127: 60-4.

Ito, S.; Miura, N. and Wang, K. (1999) Performance of a heat pump using direct expansion solar collectors. *Solar Energy*, 65(3): 189-96.

Klein, A. *et al.* (1994) TRNSYS 16 : *A Transient System Simulation Program*, Solar Energy Laboratory, University of Wisconsin, Madison, WI.

Kuehn, T. H.; Ramsey, J. W. and Threlkeld, J. L. (1998) Thermal environmental engineering (revised edition of: Thermal environmental engineering by J. L. Threlkeld. 2nd ed. 1970) , Upper Saddle River, N.J. : Prentice Hall.

Lorentzen, G. (1994) 'Revival of carbon dioxide as a refrigerant'. *Revue Internationale du Froid, International Journal of Refrigeration*, International Institute of Refrigeration (IIR), 17(5): 292-301.

Lorentzen, G. (1995) 'The use of natural refrigerants: A complete solutions to the CFC/HCFC predicament'. *Revue Internationale du Froid, International Journal of Refrigeration*, International Institute of Refrigeration (IIR), 18(3): 190-7.

Morrison, G. L. (1994) 'Simulation of packaged solar heat-pump water heaters'. *Solar Energy*, 53(3): 249-57.

Morrison, G. L.; Anderson, T. and Behnia, M. (2004) 'Seasonal performance rating of heat pump water heaters'. *Solar Energy*, 76(1-3): 147-52.

O'Dell, M. P.; Mitchell, J. W. and Beckman, W. A. (1984) 'Design method and performance of heat pumps with refrigerant-filled solar collectors'. Transactions of ASME, *J. Solar Energy Engineering* 106: 159-64.

Ozgener, O. and Hepbasli, A. (2005 article in press) 'Performance analysis of a solar-assisted ground-source heat pump system for greenhouse heating': an experimental study'. *Building and Environment.*

Pearson, S. F. (1999) 'Natural selection – the struggle for survival in the competitive world of refrigeration equipment'. *20th International Congress of Refrigeration*, International Institute of Refrigeration, Sydney, 19-24 September 1999.

Spon, P. and Ambrose, E. R. (1955) 'The heat pump and solar energy', Proceedings of World Symposium on Applied Solar Energy, Phoenix, Arizona.

In: Solar Energy: New Research
Editor: Tom P. Hough, pp.255-275

ISBN 1-59454-630-4

Chapter 8

SOLAR WATER HEATING SYSTEMS WITH THERMOSYPHONIC FLOW

M. R. Riazi*

Professor of Chemical Engineering, Chemical Engineering Department, Kuwait University, Safat, Kuwait

ABSTRACT

One major source of energy consumption at homes and industry is water heating. Solar water heaters are important because they provide cheapest and clean energy to homeowners especially when cost of energy from fossil fuels is on the rise.

In this chapter we study types and components of solar water heating systems especially those based on thermosyphonic flow. Performance of a solar water heater depends on its orientation toward the equator with respect to the local latitude. Optimization of tilt angles and orientations can significantly improve solar water heater performance. We show how optimum collector slope with the horizon can be obtained based on a set of experimental data obtained using this type of collector in the Middle East.

Design and operational aspects of a thermosyphonic solar water heater made of polypropylene tubes is discussed. Also the rate of heat transfer to the water is measured and correlated through Nusselt, Reynolds and Prandtl numbers. It is also shown how one can estimate amount of increase in water temperature from the collector specifications, water flow rate and sun radiation. Finally a recently proposed method for calculation of thermal efficiency of thermosyphonic solar water heaters is presented.

Key words: Solar water heating system, thermosyphonic flow, design and operational considerations, experiments, polypropylene tubes, collector performance, collector slope, heat transfer rates, system efficiency.

---

* Professor of Chemical Engineering, Chemical Engineering Department, Kuwait University, P.O Box 5969, Safat 13060- Kuwait. Tel: (+965) 4817662, Fax: (+965) 4839498; Email: riazi@kuc01.kuniv.edu.kw

NOMENCLATURE

A_i = internal surface area of pipes, m^2 (or cm^2)
a,b,c = Coefficients in Equation 13
C_p = heat capacity, J/kg-K
d = tube diameter, m (or cm)
h = heat transfer coefficient, W/m^2K
k_w = thermal conductivity of water, W/mK
ℓ = tube length, m (or cm)
$\dot{m}$ = mass flow rate of water, kg/s
Nu = Nusselt number, dimensionless
Pr = Prandtl number, dimensionless
Q_r = heat transfer rate from solar radiation, W
Q_t = the total heat which is collected in the storage tank
Re = Reynold's number, dimensionless
T_a = ambiant air température, °C
$\overline{T}_a$ = the mean ambient temperature
T_{in} = inlet water temperature (city water temperature, °C)
$\overline{T}_{in}$ = the mean cool water temperature into the system
T_o = outlet water temperature, °C.
T_{so} = outside surface temperature of tubes, °C
$\dot{V}$ = volumetric flow rate of water in tubes, mL/s (cm^3/s)

Greek Letters

α_o = the overall solar absorption, dimensionless
ϕ = angle between sun radiation and a plate normal to the collector board, degrees (°).
ρ_w = water density, kg/L
μ_w = water viscosity, kg/m.s
η_S = daily system efficiency, dimensionless
η_R = heat removal efficiency, dimensionless

INTRODUCTION

Water heating accounts for a good portion of energy consumption at homes and industry. In the United States water heating consumes nearly one fifth of residential energy consumption second only to heating and cooling.

In solar water heating systems energy required to heat water comes from sun rather than gas, oil or coal. As the era of cheap oil is almost over, use of solar water heating systems can significantly reduce energy costs. In addition solar water heaters produce no pollution and have minimum operation and maintenance expenses. With impending end of the fossil fuel age and the hard choices ahead renewable energy is the only source of energy for the future. In North America it is expected that the price of natural gas will dramatically increase before this decade is out due to increase in demand. Solar water heating systems are the cheapest and most easily affordable clean energy available to homeowners [1]. Solar water heating systems can supply 90 percent of a family's hot water from May to October, and more than half of what they use throughout the year even in areas such as in Western Oregon [2].

The first commercial solar water heater named Climax was introduced in the 1890's. Californians could save up to $ 9 a year in coal costs by $ 25 investment. In 1920's when natural gas was discovered in southern California and cheap energy became available and market for solar water heaters ended and shifted to Florida. After the second war nearly half the homes in Miami had solar water systems and 80% of new homes were equipped with solar systems [1]. In the 1950's electricity became cheap in Florida and the industry collapsed. The 1973 oil embargo and subsequent tax credits encouraged the people to have solar water heater for their homes and hundreds of manufacturers emerged in the United States. The industry suffered another setback in 1986 when the oil priced collapsed from $ 30 a barrel to $ 10 in which cheap energy from fossil fuel became available. In the past few years the oil price has continued to rise and last year it reached an unprecedented price of $ 70 per barrel. This is another warning to the public that the era of cheap oil is perhaps over. Once again solar water heaters become valuable devices to collect heat from renewable source of energy for home water heating purposes.

In the Canadian province of Manitoba, direct solar radiation received is equal almost to 2 billion terajoules of energy per year. If only 1% of this energy was converted into electricity, it would be equal to almost 200 times the annual output of all the hydroelectric dams in Manitoba [3].

The sun generates huge amount of energy by converting hydrogen to helium which is called solar radiation. This energy reaches the earth as 47% in sunlight, 7% in ultraviolet rays and 46% as infrared radiation or heat. The sunlight and infrared radiation can be absorbed by a solar collector [3]. Photovoltaic (PV) cells produce electricity directly from sunlight. PV cells are also called solar cells or photocells. As a comparison for every $ 20-30 spent on PV cells you can spent $ 1 on solar water heaters to capture the same amount of energy [1].

In general 50 to 65% of annual hot water demand can be obtained through solar water heaters. The heart of a solar heating system is the collector. This absorber is embedded in a thermally insulated box with a transparent cover (usually glass) to minimize thermal loss. A heat conducting liquid (usually a mixture of water and non-environmentally damaging anti-freeze) flows through the absorber and circulates between the collector and the warm water storage tank [4].

With the installation of a thermal solar heating system, one is actively contributing to the lowering of environmentally harmful CO_2 emissions. The sun, however, supplies its energy free of charge.The initial investment cost for solar heaters seem to be high. But from the time of installation of the system on, there are no more operating costs, except very low costs for maintenance. [4]

In general solar water heating systems are efficient, cost effective, reliable and environmentally clean and can preserve natural resources. It is never cost effective to design a system to provide 100% of the load with solar because of the excessive investment in collector area and storage volume. In effective design, solar heaters provide 70% of annual energy load for water heating in a residential home. However, during summer seasons with good sunshine we can cover 100% of hot water demand from solar radiation. Solar water heaters can be easily installed and fit on almost any residential roof with minimum capacity of 100 liters per day [4]. It can deliver hot water at 60 °C for domestic application and 80 °C for commerçial application. For 100 LPD (Liter per day) Capacity Solar Water Heating System, we need 1 solar collector of 2 m^2 area. It requires 4 m^2 shadow free area for installation on the roof [5].

Principle and Types of Solar Water Heaters – Thermosyphonic Flow Heaters

In solar water heaters sunlight strikes heating surface absorber known as collector and heat can be transferred to water directly through tubes that water flows inside them or indirectly through a heat exchanger. In indirect heating systems heat is transferred to a heating fluid such as propylene glycol and then to water through a heat exchanger. The propylene glycol (antifreeze) works in a closed loop system.

Solar water heating system can be classified in two types based on the collector system: a) flat plate and b) evacuated tube collectors. In the first type, solar radiation is absorbed by flat plate collectors which consists of an insulated outer metallic box covered on the top with glass sheet. Inside there are blackened metallic absorber sheets with built in tubes to carry water. The absorber absorbs the solar radiation and transfers the heat to the flowing water [5].

In evacuated-tube collectors, solar collector is made of double layer borosilicate glass tubes evacuated for providing insulation. The outer wall of the inner tube is coated with selective absorbing material. This helps absorption of solar radiation and transfers the heat to the water which flows through the inner tube [5].

Solar water heating systems are also classified as Direct System (Open Loop System) and Indirect System (Closed Loop System).

Open Loop System

The Direct System is the one where the water circulates through the entire system i.e. collector to storage tank and the same hot water is used for various applications. These types of systems are generally not suitable for hard water due to the scaling problems in the risers and headers of the collector.

Closed Loop System

In the Indirect System, the thermic fluid is circulated between the collector and the insulated storage tank with heat exchanger. The heat from the thermic fluid is then transferred to the water through heat exchanger in the storage tank. In this system, the thermic fluids are not in physical contact with the water in storage tank. These systems are generally suitable for hard water [5]. Most commonly used thermic fluid is solution of propylene glycol and water. It should be noticed that ethylene glycol must never be used as it is toxic [6].

Solar water heating system also classified in two types based on the circulation of water.

a) Forced flow or forced circulation system.
b) Thermosyphon system.

Forced Circulation System

In the forced flow system, a pump is used for circulating water between the collectors and the insulated hot water storage tank. The forced flow systems are more efficient as compared to thermosyphon systems due to higher flow rate. Generally, the pumps are operated by differential temperature control (DTC) system, which senses the pre-setting temperature difference between inlet and outlet of the collectors [5].

Thermosyphon System

In the thermosyphon (also spelled thermosiphon) system, water comes from the over head tank to bottom of solar collector by natural circulation and water circulates from the collector to storage tank as long as the absorber keeps absorbing heat from the sun and water gets heated in the collector. The cold water at the bottom of storage tank run into the collector and replaces the hot water, which is then forced inside the insulated hot water storage tank. The process of the circulation stops when there is no solar radiation on the collector. Thermosyphon system is simple and requires less maintenance due to absence of controls and instrumentation [5].

A solar water heater with forced circulation system that electric power is needed to activate pumps is also called active. Solar water heaters with thermosyphonic system is also called passive system where buoyancy (natural convection) rather than electric power is used to circulate the water. Basis of circulation of water in passive systems is natural convection and the flow is called thermosyphonic in which storage tank is located above the solar collector. In thermosyphonic flow, hot water rises and cold water settles because cold water is more dense (heavier) than hot water. For the same reason colder water is at the bottom of a tank [7]. In this chapter we study types and components of solar water heating systems especially those based on thermosyphonic flow. Since beginning of 20th century solar water heating systems with thermosyphonic flow have been in use in the United States (Arizona, California and Florida) as well as in Japan and Australia and later in Israel. General schematic of a solar water heater with thermosyphonic flow is shown in Figure 1.

Materials and components used in solar water heating systems vary depending on the expected operating temperature range. Low temperature systems operate at up to 10 °C above ambient temperature and are often used to heat swimming pools. In these systems tubes are not galvanized. Mid temperature systems heat water 10 to 50 °C above outside temperature and are often used for heating domestic hot water. High temperature systems with operating temperatures more than 50 °C above outside temperature utilize evacuated tubes around the receiver tube to provide high levels of insulation and often use focusing curved mirrors to concentrate sunlight. Cost of such collectors is 3 to 5 times of those for low temperature systems [6].

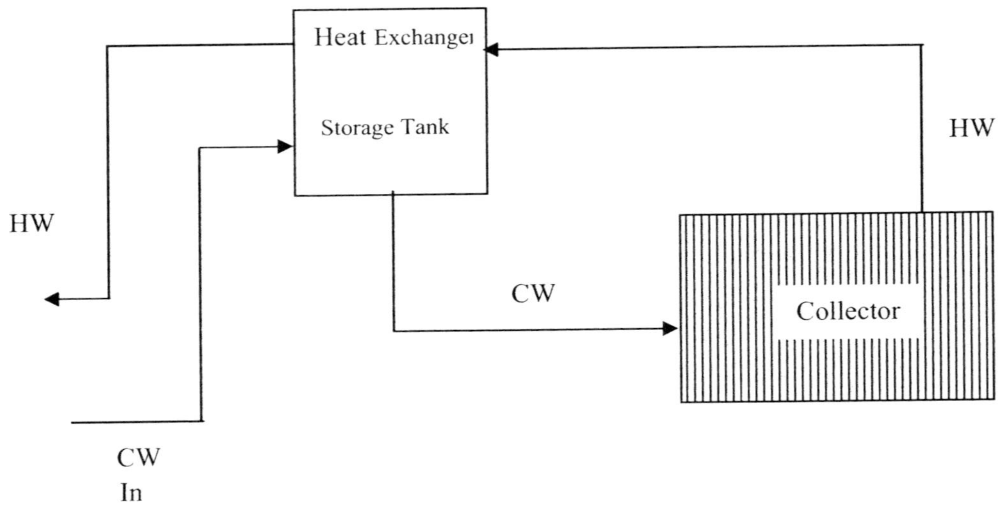

Figure 1. Schematic of solar water heater by thermosyphonic flow.CW = cold water; HW = hot water

Efficiency of a collector depends on the difference between collector temperature and ambient temperature and inversely proportional to the intensity of solar radiation. The inexpensive, unglazed collectors are very efficient at low ambient temperatures, but efficiency drops off very quickly as temperature increases. They offer the best performance for low temperature applications, but glazed collectors are required to efficiently achieve higher temperatures [6]. Later at the end of this chapter it is shown how one can calculate thermal efficiency of solar water heaters in thermosyphon systems.

Performance of a solar water heater also depends on its orientation toward the equator with respect to the local latitude. Optimization of tilt angles and orientations can significantly improve solar water heater performance [8].

Design and Operational Aspects of Solar Water Heaters

In optimum design of a solar water heater that the life-cycle cost is minimum, a system will usually produce about 70% of the annual load. Other design considerations include freeze protection, overheating protection and orientation. The best orientation is facing toward the equator with a tilt up from the horizon equal to the local latitude. Collector performance is maximized when it is placed perpendicular to the sun. Studies show that when tilt angles and orientations vary form this still good performance can be achieved. In the United States, collectors should be rotated within 30° of south. Also, optimize the tilt of the collecting array. Surfaces tilted up from the horizontal at an angle of latitude minus 15° maximize summer solar gains, but reduce winter gains. Surfaces tilted up at latitude plus 15° maximize winter solar gains, and result in a solar delivery that is uniform throughout the year

[6]. For more hot water collector face can be tilted toward the east of the south in the mornings and west of the south in the day and early afternoon [9,10]. A tilt angle equal to the local latitude maximizes year-round solar gains and is usually appropriate for solar water heating [6]. Information on maps and tables of solar resource information in the U.S. are given at the Renewable Resource Data Center webpage [9]. The best location of collector is in an unshaded area where it has access to the sun throughout the year. Obviously, the best place is the roof of buildings and houses. To get better performance the piping between collector and storage should be minimal.

Sizing a solar water heater involves determining the total collector area and the storage volume required to provide total hot water requirement during summer. Storage tank volumes may vary from 50 to 120 gallons or 200 to 450 liters. The large size of 120 gallons may be sufficient for a family of 4 to 6 people. As a rule of thumb a collector area of 1 m^2 is required for every 57 liters tank capacity which means for a family of 2 persons collector area of 2 m^2 is needed. In colder and less sunny areas (northern U.S.) additional collector area of 1.1 to 1.3 m^2 is needed for each person [11].

Experiments with Solar Polypropylene Collectors in Thermosyphon Systems

Many investigators have studied thermosyphonic solar water heating systems both experimentally [12,13] and analytically [14,15]. Most of these studies were designated to study the overall system performance only. In the experiments conducted by these investigators, flat-plate solar water heating systems were used. Riazi and Razavi [8, 16,17] made experimental study of solar water heating systems with thermosyphonic flow and polypropylene finned-tubes to obtain an optimum condition at which best performance can be achieved.

Various components of the system are shown in Figure 2. The unique feature of this system is use of funned tubes made from polypropylene. These tubes had an average surface area twice that of normal tubes. Use of polypropylene tubes has the following advantages over metallic tubes:

1) Easy to handle and to carry because of tubes' lightness.
2) Low cost of materials.
3) Higher heat transfer rates compared to normal collectors.
4) Higher resistance to corrosion.
5) Good resistance to temperature variation and the sun.
6) No need for electrical equipment in the system.

The components of the system were: 180 liters storage tank, 36 parallel polypropylene finned tubes of 19 mm inside diameter and 1 m long, valve and fittings, a black-painted board, two thermometers, and a water level floater. Finned tubes were fixed on a black board which could be tilted relative to the horizon. These tubes were connected to the 180-liter tank. The tank and all fittings were completely insulated to prevent heat loss. However, the collector was not glazed [8]. Water to the tank was supplied from the city main water pipeline by a level floater. Two thermometers were installed on the system to measure temperatures of the inlet and outlet water to the system. A general view of experimental setup that was set up on is shown in Figure 3. The collector is facing south and solar radiation was perpendicular to the surface. The collector was installed on the roof of chemical engineering department building located in Tehran at the latitude of 36° north of the equator and 52° east of prime meridian.

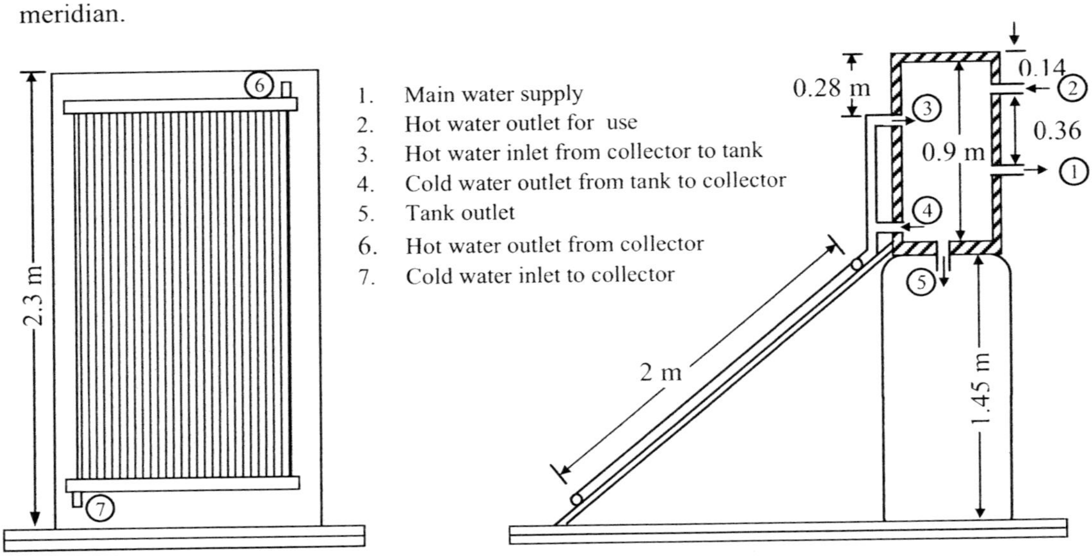

Figure 2. Components of experimental setup using polypropylene finned tubes with thermosyphonic flow (taken with permission from Ref. 17).

Figure 3. General picture of experimental setup using polypropylene finned tubes with thermosyphonic flow

At the first step, an appropriate location for connecting inlet and outlet tubes to the tank was found in order to obtain a natural water flow rate. In the second stage, the best slope of the collector relative to the horizon to gain the maximum energy transfer for this system was studied. In the third stage a comparison was made between efficiency of polypropylene tube and steel tube collectors. In the final stage of the experimental studies rate of heat transfer were measured and a general correlation was obtained to estimate water outlet temperature.

The first set of experiments were conducted in September-November 1992 when the ambient temperature varied from 20 to 30 °C during day times. Total of 12 experiments were conducted at this stage and the temperature readings during one set of experiment is given in Table 1 [8]. To determine the best collector slope, water outlet temperature was measured at various collector slopes. These are shown in Figure 4 where it clearly shows that slope of 36° gives the highest performance which is the same as the latitude of the location.

Table 1. Temperature readings during one experiment

Time	Inlet Temperature, °C	Outlet Temperature, °C
09:40	24	21
12:20	24	25
10:40	24	31
11:00	25	33
11:40	29	36
12:20	30	37
13:00	29	39
13:40	25	40
14:00	25	40
15:00	23	40

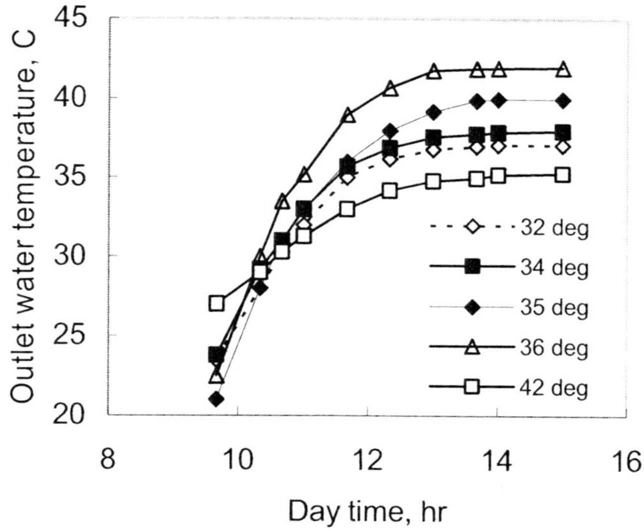

Figure 4. Outlet water temperature versus collector slope relative to the horizon.

The second set of experiments were performed during July and August of 1994 where the air temperature varied from 29 to 36 °C. In this set of experiments performance of polypropylene tubes were compared with those of steel tubes. Experiments began at 9:00 am and ended at 5:00 pm. In each experiment the storage tank was filled with fresh water. No water was drawn off from the system throughout the experiment. In each experiment, ambient, inlet and outlet water temperatures were measured and recorded every 60 min. At first, similar to the results presented in Figure 4, best collector slope was obtained for the steel tubes with three sets of experiments at three different slopes of 35, 36 and 37°. Results shown in Figure 5 indicate again that the best collector slope is 36°. This indicates that the best collector slope is independent of the type of tubes used in the system and mainly depends on the location latitude and direction of the system.

To compare performance of two types of tubes total of 28 experiments were conducted. In this series of experiments, 14 tests were conducted for polypropylene tubes and 14 experiments for steel tubes [16]. Attempts were made to set the conditions so that for every experiment for the polypropylene tube system a similar experiment under similar conditions was conducted for the steel tube system. Experiments for two tubes were conducted at two different dates but the ambient air temperature and the initial temperature of water were the same. In all experiments the collector slope was fixed at the optimum slope of 36° [16]. Results for one pair of the experiment are given in Table 2.

Table 2. Sample of one pair of experimental data for comparison of two types of tubes used in collector.

Daily Reading Time	Polypropylene tube collector					Steel tube collector				
	COT °C	CIT °C	VBT °C	ATS °C	VFR °C	COT °C	CIT °C	VBT °C	ATS °C	VFR °C
09:00	36.2	26.2	25.4	29.0	7.87	33.6	26.0	25.4	29.0	7.55
10:00	45.0	26.8	27.0	32.8	8.99	39.5	26.8	26.6	29.8	7.87
11:00	50.2	30.2	38.2	33.3	9.44	44.0	44.0	27.4	30.2	30.6
12:00	56.0	36.4	45.0	33.8	10.50	46.8	28.8	38.6	31.5	8.58
13:00	60.6	44.0	50.6	34.6	10.50	51.0	34.0	42.0	33.0	8.99
14:00	63.7	46.8	53.2	35.6	9.94	54.4	40.2	45.5	34.2	9.44
15:00	65.2	50.8	57.6	35.6	9.94	56.5	43.4	48.6	34.8	8.99
16:00	66.0	53.0	59.0	36.2	9.44	57.8	47.0	51.9	35.2	8.21
17:00	64.2	53.8	59.2	36.2	8.21	55.4	48.6	52.5	35.8	7.55

Data source: Ref. [16]

Initial water temperature was: 25 C. Dates of experiments: July 16 and August 11, 1994, for propylene and steel tubes, respectively. COT = collector outlet temperature; CIT = collector inlet temperature; VBT = vessel bulk temperature; ATS = ambient temperature at shade; VFR = volumetric flow rate, cm^3/s.

Figure 6 shows a graphical comparison between the water outlet temperature for the polypropylene and steel systems for which experimental data are given in Table 2. Results for other experiments show similar behavior. This indicates that under similar conditions, polypropylene tubes show better performance over steel when they are used in solar water heating systems with thermosyphonic flow.

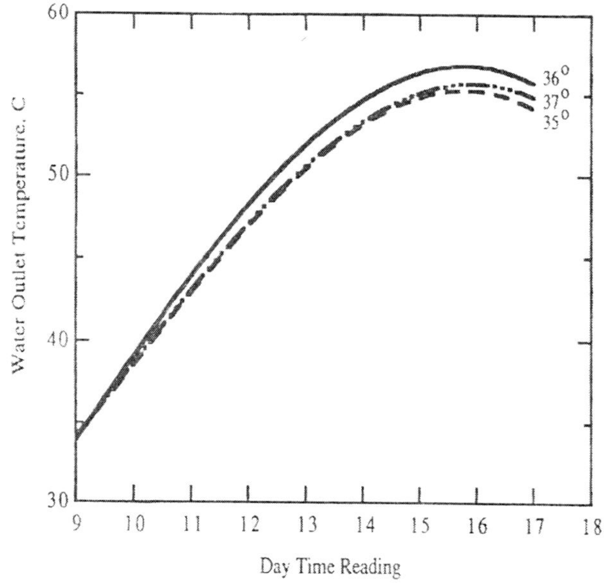

Figure 5. Water outlet temperature versus collector slope for steel tubes. [Taken with permission from Ref. 16]

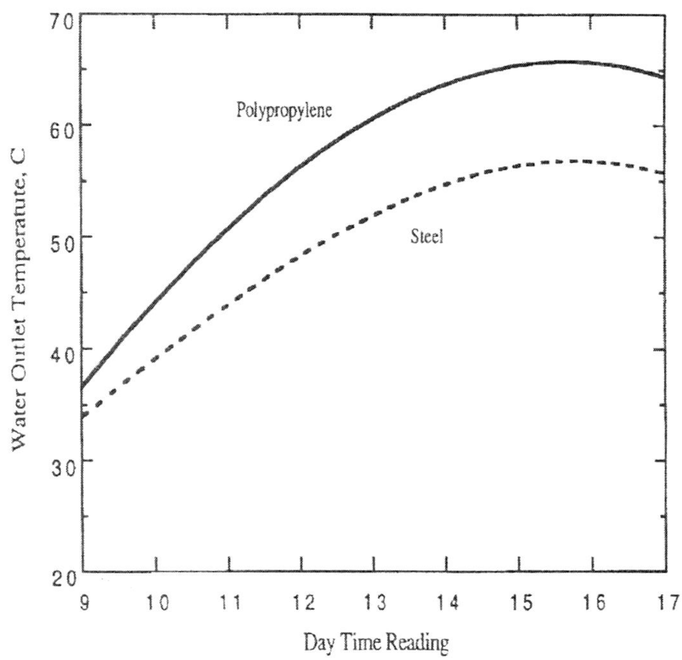

Figure 6. Comparison of water outlet temperature for polypropylene and steel tube collectors. Initial water temperature is 25 °C. [Taken with permission from Ref. 16]

To simplify analysis of data, a summary of all experimental data is shown in Table 3. Generally, for polypropylene tubes, water temperature was increased by 40 °C from the initial water temperature and by 30 °C from the ambient temperature, while for the steel tubes these increases were 30 °C and 20 °C, respectively. In other words, by using polypropylene tubes, we can increase water temperature 10 °C more than that for steel tubes under similar conditions. The average water flow rate in the collector for steel tubes was 8.4 and for the polypropylene tubes was 9.4 cm^3/s.

Table 3. A comparison between polypropylene and steel tubes and summary of experimental results.

Type of Tube	No. of experiments	Ambient temp. °C	Initial water temp., °C	Maximum water outlet temp., °C	Water flow rate, cm^3/s
Polypropylene	14	33	25	65	9.4
Steel	14	33	25	55	8.4

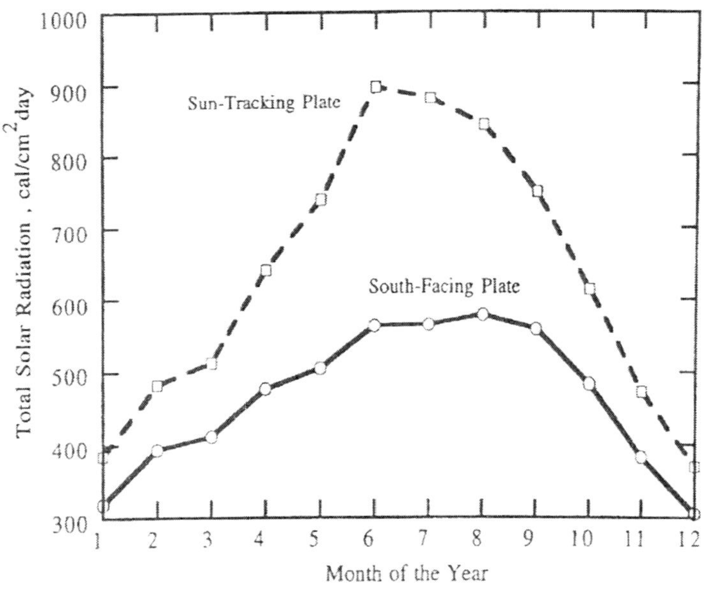

Figure 7. Total solar radiation during the year in the city of Tehran with latitude of 36° north of the equator and 52° east of prime meridian. [Taken with permission from Ref. 16]

For the experiments shown here, the collector blackboard was facing toward the south, and its slope with respect to the horizon was fixed at 36°. It is possible to get temperatures well above 65 °C, which was obtained for the polypropylene collector if a sun-tracking collector system was used. Based on the methods discussed by Reddy [18,19] and Sabbagh et al. [20], total solar radiation on a south-facing plate inclined at latitude angle and on a sun-tracking plate were estimated during various months of the year for the City of Tehran, where the experiments for this study were performed. Results of the calculations are shown in Figure 7. This figure shows the difference between total daily solar radiation between a sun-tracking plate and south-facing plate in various months of the year. In order to calculate the mean monthly total solar radiation, empirical relationships were employed that use meteorological data, such as bright sunshine hours, maximum air temperature, and relative humidity. Results shown in Figure 7 indicates that in the months of June, July and August (numbered 6, 7 and 8 in the figure) maximum solar radiation in Tehran can be obtained. In addition, by using sun tracking systems, it would be possible to further increase the performance of polypropylene tubes by some 50%.

Rate of Heat Transfer in Polypropylene Tubes with Thermosyphonic Flow

In the final stage of this study a heat transfer rate was determined for polypropylene tubes in solar water heaters. Heat transfer rates can be calculated through heat transfer coefficients. In this study the same experimental set-up as in previous studies (Figure 2) was used to determine heat transfer in polypropylene tubes used for solar water heating systems. Heat transfer coefficient is then correlated to flow rate and water properties through a generalized correlation in terms of Nusselt, Reynold and Prandtl numbers. The correlation can be used to predict the rate of heat transfer in such solar water heaters [17].

All the experiments were conducted during the summer season (July 2002) when ambient temperatures varied from 34 to 37 °C. In each set of experiments, temperatures of the inlet and outlet water from the tubes were measured. The angle of solar radiation with respect to the plane normal to the black board (ϕ) and the air temperature were measured every hour. Surface temperature of polypropylene tubes was measured by a thermistor. The city water used in all experiments had an average temperature of 19°C during the period at which experiments were conducted. In each experiment the following parameters have been measured:

T_{in} = inlet water temperature (city water temperature, °C)
T_o = outlet water temperature, °C.
ϕ = angle between sun radiation and a plate normal to the collector board, degrees (°).
$\dot{V}$ = volumetric flow rate of water in tubes, mL/s (cm^3/s)
T_{air} = air temperature, °C
T_{so} = outside surface temperature of tubes, °C

During each experiment T_{in} was constant, however, T_o is the average of measured temperatures at the beginning, middle and end of each set of experiment. T_{so} the outside surface temperature is the average temperature of several locations to represent all area. Further details of experimental measurements are given in our previous publications [8,16]. Measured values for these parameters are given in Table 4 as reported in Reference [17].

Table 4. A Set of Experimental Data for Heat Transfer Rate

Experiment	$\dot{V}$ mL/s	ϕ	T_{in} °C	T_o °C	T_{air} °C	T_{so} °C
1	80	44.8	19	25.7	34	38
2	70	39.6	19	27.4	35	37
3	33	35.3	19	31.1	36	55.7
4	42	31.6	19	32.0	36.3	44.5
5	22	27.8	19	38.2	37.5	59.7
6	15	23.8	19	44.4	36.6	63.5
7	9.4	21	19	50	37	72.6

Analysis and Correlation of Data

Amount of heat transfer due to solar radiation (Q_r) depends on the angle of radiation (ϕ) as:

$$Q_r = Q_{max} \quad \text{Cos}\ \phi \tag{1}$$

where Q_{max} is the maximum rate of heat transfer from the sun when the radiation angle is zero (normal radiation). Q_{max} can be calculated from maximum water flow rate and temperature difference. The apparatus could handle a maximum flow rate of $\dot{V}_{max}$ = 500 mL/s. Before the system reaches steady state when there is no water outlet flow ($\dot{V}$ = 0), the outside surface temperature of tubes reaches a value called maximum water temperature and is shown by $T_{so(max)}$. As mentioned earlier this outlet water temperature is the average of three measurements at the beginning, middle and end of the experiment.

$$Q_w = \dot{m} C_p \Delta T_w \tag{2}$$

$$\Delta T_w = T_o - T_{in} \tag{3}$$

$$\dot{m} = \rho_w \dot{V} \tag{4}$$

where ρ_w is density of water and C_p is heat capacity of water which may be taken as 4.18 J/g °C.

The energy loss, Q_{loss}, can be calculated form the following relation:

$$Q_{loss} = h_o A_o (T_{so} - T_{air}) \tag{5}$$

where h_o is the heat transfer coefficient for the external surface. Razavi and Riazi [17] have shown that h_o can be calculated from

$$h_o = \frac{Q_{max} \text{Cos}\phi - \dot{m} C_p (T_o - T_{in})}{A_o (T_{so} - T_{air})} \quad (6)$$

Based on the thickness and thermal conductivity of polypropylene tubes the temperature difference between outside (T_{so}) and inside (T_{si}) surface temperatures is about 3°C. As the thickness of the tube increases this temperature difference also increases.

$$T_{si} = T_{so} - 3 \quad (7)$$

Now we define inside heat transfer coefficient, h_i in the following form:

$$Q_w = h_i A_i \Delta T \quad (8)$$

where ΔT is defined as :

$$\Delta T = T_{si} - T_{water}$$

$$T_{water} = \frac{T_{in} + T_o}{2} \quad (9)$$

A_i is the total internal surface area of all tubes and it is given as:

$$A_i = n\pi\ell d \quad (10)$$

where n is the number of tubes, ℓ and d are the length and inside diameter of tubes. In our calculations n = 36, d = 1.9 cm and ℓ = 200 cm (A_i = 42955 cm^2). Reynolds and Nusselt numbers are calculated as:

$$Re = \frac{4\rho_w \dot{V}}{\pi d \mu_w} \quad (11)$$

$$Nu = \frac{h_i d}{k_w} \quad (12)$$

where μ_w and k_w are the viscosity and thermal conductivity of water, respectively. Physical properties of water were taken from Holman [21]. Calculated values of Q_r, Q_w, h_i, Re and Nu are given in Table 5. The general relation between Nu and Re is in the following form:

$$Nu = a\ Re^{b}\ Pr^{c} \tag{13}$$

Table 5. Calculation of Heat Transfer Rate and Nusselt Number

Experiment	Q_r, W	Q_w, W	h_i, W/m^2°C	Re	Nu
1	2350	2239	41.2	5600	1.05
2	2551	2457	53.0	5400	1.28
3	2700	1668	14.0	2500	0.39
4	2693	2264	21.2	3230	0.51
5	2928	1765	14.6	1900	0.41
6	3028	1613	13.1	1400	0.38
7	3088	1217	8.1	800	0.23

where Pr is the Prandtl number (C_p μ/k). Value of c when Re < 5000 is given as 1/3 by Holman [21]. In this study Pr is constant since the fluid is water and the effect of temperature on Pr is neglected.

Using values of Nu and Re in Table 5, constants a and b have been determined through linear regression analysis. The result was

$$Nu = 0.0015\ Re^{0.75}\ Pr^{1/3} \tag{14}$$

The correlation parameter in the regression was 0.95. Through use of these equations one can predict the outlet water temperature (T_o) through Equations (2), (8) and (9). This is demonstrated in the following example.

Example 1: A polypropylene tube collector is installed with the slope of 36 ° at the latitude of 36° north of the equator. The collector is made of 30 parallel tubes of length 1 m, inside diameter of 10 mm and thickness of 1 mm. The water flow rate is measured at the rate of 25 cm^3/s. The air temperature is 35 °C and average surface temperature of tubes is measured as 55 °C. If the water inlet temperature is at 25 °C calculate the increase in water temperature after leaving the collector.

Solution: Based on the information available we have n = 30, d = 10 cm and ℓ = 100 cm and from Equation (10): A_i = 9424.8 cm^2. The physical properties of water at 30 °C are given by Rowley et al. [22] as: ρ_w=944 kg/m^3, μ_w=0.82×10^{-3} kg/m.s, $C^L{}_P$=4.18×10^3 J/kg.K, k_w=0.613 W/m.K. Pr=(4.18×10^3×0.82×10^{-3})/0.613=5.59. From Equation (11) we have: Re=(4×944×25×10^{-6})/(3.14×1.0×10^{-2}×0.82×10^{-3})=3666.3. From Equation (14) we calculate: Nu=0.0015×(3666.3)$^{0.75}$×(5.59)$^{0.333}$=1.25311. From Equation (12): h_i=(1.06×0.613)/ 1.0×10^{-2})=64.95 W/m^{2}K. $\Delta T=T_{si}-T_{water}$ where T_{si}=55-3=52 °C. Assuming 10 °C increase in water temperature, from Equation (9): T_{water}=(35+25)/2=30 °C. ΔT = 52-30=22 °C. Equation (8) should be used to calculate the rate of heat transfer to water: Q_w=64.95×9424.8×10^{-4}×22=1346.7 W. From Equation (4) $\dot{m}$ =944×25×10^{-6}=0.0236 kg/s. From Equation (2) the rise in temperature of water can be calculated as: ΔT_w =1346.7/(0.0236×4.18×10^3)= 13.7 °C. In the next round of calculations we assume the temperature rise is 12.8 °C which gives T_{water}=31.4 °C and Q_w==1261 W. New calculated water temperature rise is ΔT_w = 12.8 °C. There is no need for another round of calculation and the answer is 12.8 °C.♦

EFFICIENCY OF THERMOSYPHONIC SOLAR HEATING SYSTEMS

The most recent discussion on the performance of thermosyphon solar water heaters is given by Chang et al. [23]. They suggested two efficiencies: 1) daily system efficiency denoted by η_S and 2) heat removal efficiency denoted by η_R. These two efficiencies that may be used to indicate performance of thermosyphonic system are discussed here.

Daily efficiency, η_S is defined as:

$$\eta_S = \alpha_o - U_S \frac{T_i - \overline{T}_a}{H_t} \tag{15}$$

where α_o is the overall solar absorption, U_s is the coefficient of overall system heat loss rate in the energy collecting phase, H_t is the daily total radiation upon collector which should be greater than 7 MJ/(m^2.day). T_i is the initial mean tank temperature and $\overline{T}_a$ is the mean ambient temperature. α_o is in fact equivalent to efficiency η_S^* at $M/A_S = 75$ where M is the total mass of water in the system and A_S the collector area. Figure 8 shows that value of η_S at $\frac{T_i - \overline{T}_a}{H_t} \rightarrow 0$ is almost the same as α_o for one set of experimental data reported by Chang et al. [23]. Data are represented by straight line of Eq. (15) with correlation coefficient of 0.991. For this system $\eta_S^* = 0.543$ and $\alpha_o = 0.544$.

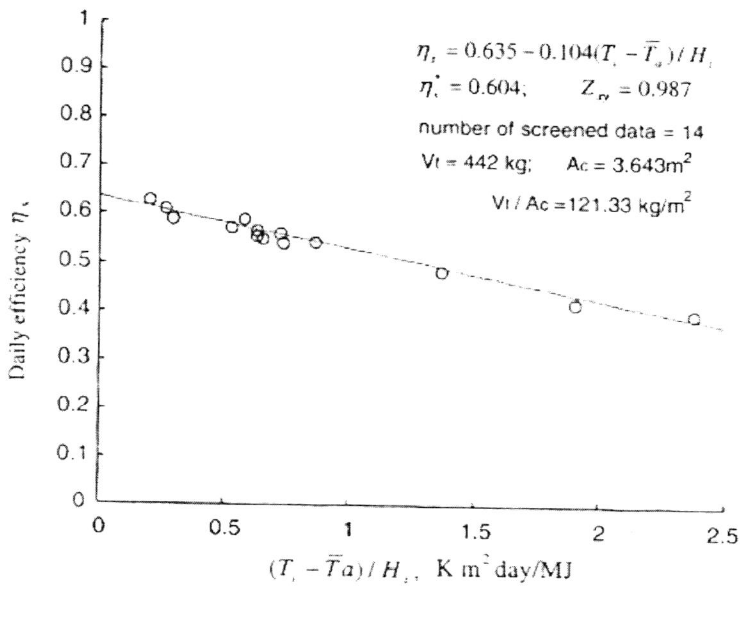

Figure 8. Daily efficiency test results for a thermosyphonic solar water heater (Taken with permission from reference [23])

The heat removal efficiency η_R is defined as:

$$\eta_R = \frac{Q_u}{Q_t} \tag{16}$$

where Q_u is the useful heat which user obtain from the system and Q_t is the total heat which is collected in the storage tank. Q_u is always less than Q_t. Values of Q_u, Q_t and finally η_R can be calculated from the following relations.

$$Q_u = \int_0^{t_f} \dot{V}\rho_w C_P \left[T_o(t) - T_{in}(t)\right]dt \tag{17}$$

where $\dot{V}$ is the volumetric flow rate and $T_o(t)$ is discharge water temperature versus time, t. $T_{in}(t)$ is the cool charge water temperature versus time. Q_t is calculated as:

$$Q_t = \rho_w V_t C_P (T_i - \overline{T}_{in}) \tag{18}$$

where V_t is the tank volume and $\overline{T}_{in}$ is the mean cool water temperature into the system and is defined as:

$$\overline{T}_{in} = \frac{\int_0^{t_f} T_{in}(t)dt}{t_f} \tag{19}$$

t_f can be determined from the following relation:

$$t_f = \frac{V_t}{\dot{V}} \tag{20}$$

Therefore from Eq. (16) η_R is calculated as:

$$\eta_R = \frac{\int_0^{t_f} \left[T_o(t) - T_{in}(t)\right]dt}{\int_0^{t_f} \left[T_i(t) - T_w(t)\right]dt} \tag{21}$$

Figure 9 shows variation of η_R with $(T_i - \overline{T}_{in})/\dot{m}_d$ where $\dot{m}_d$ is the same as $\dot{V}$ in Eq. (20) and it is in the unit of L/min. The limiting value of η_R shown by η_R^* is 0.846 where represents value of η_R at $(T_i - \overline{T}_{in})/\dot{m}_d = 2$. A normal collector should have $\eta_R^* \geq 0.8$. The relation between η_R and $(T_i - \overline{T}_{in})/\dot{m}_d$ is nearly linear as shown in Figure 9. Operational range of $(T_i - \overline{T}_{in})/\dot{m}_d$ is between 0 and 6 but normally varies from 1 to 3 with mean value of 2.0 at which $\eta_R = \eta_R^*$.

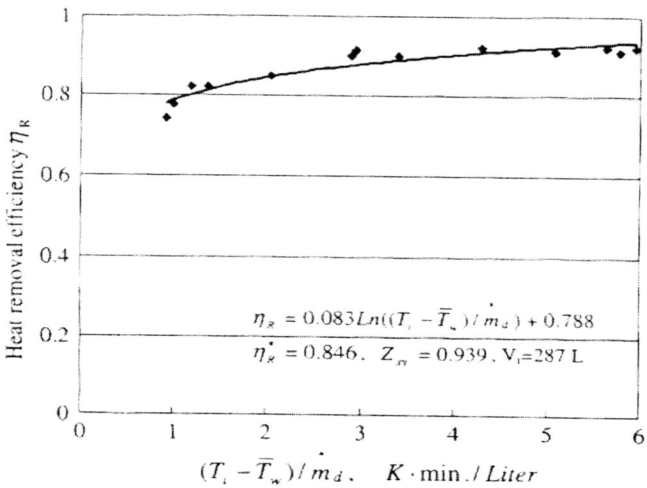

Figure 9. Heat removal efficiency test results for a thermosyphonic solar water heater (Taken with permission from reference [23])

Conclusion

In this study the history, principle and operational as well as design aspects of solar water heaters with thermosyphonic system were discussed. Determination of optimum collector slope, effect of type of tubes and advantages of polypropylene tubes have been discussed. It has been shown that these tubes can be used to build solar water heaters with thermosyphonic system with higher efficiency and lower cost.

In this study heat transfer rate in polypropylene tubes used in solar water heaters with thermosyphonic flow has also been determined from experimental measurements for the Reynolds number range of 800 – 5600. A correlation has been proposed which may be used to calculate heat transfer coefficient and subsequent heat transfer rate in polypropylene tubes under similar conditions. Methods of calculation of efficiencies of solar water heaters with thermosyphonic flow are also presented with case studies that show two types of efficiencies defined for such systems. Information presented in this chapter may be used to build and operate cost effective solar water heaters for homes and buildings.

REFERENCES

[1] Solar Energy Systems Inc., "A Brief History of the American Solar Water Heating Industry," 2004, November 9, Available from URL: *http://www.ecs-solar.com/Solar_Articles/briefHistory.pdf*

[2] Energy Guide, "Solar Water Heating," 2003, Available from URL: http://energyguide.com/ and *http://library.energyguide.com/EnergyLibraryTopic.asp?bid=minnpower&prd=10&TID=13280&SubjectID=7734*

[3] Manitoba Energy, Science & Technology, "Energy Initiatives – Solar Energy," Energy Development Initiative © 2003, Available from URL: *http://www.gov.mb.ca/est/energy/initiatives/solar.html*

[4] Solar Energy Ireland, Initiative © 2003, How Solar Heating Works, 2003, Available from URL: *http://www.solarenergyireland.com/index.html*

[5] Indian Renewable Energy Development Agency Limited (IREDA), "Frequently Asked Questions about Solar Thermal Water Heating Systems," 2004, December 20, Available from URL: *http://www.iredaltd.com/solarthermal.htm*

[6] Walker, A., "Solar Water Heating," 2003, July 14, Available from URL: *http://www.wbdg.org/design/resource.php?cn=0&cx=0&rp=23*

[7] *ASHREA Applications Handbook*, Chapter 33, Solar Energy Use, Water Heating, 33.13-33.16, American Society of Heating, Refrigerating and Air-Conditioning Engineers, Atlanta, GA, 2003.

[8] Razavi, J; Riazi, M.R. Experiments with polypropylene finned-tube solar water heating system in thermosyphonic flow, *Energy Sources*, 1994. Vol. 16, No. 1, 129-133.

[9] National Renewable Energy Laboratory "Solar Radiation Resource Information" 2004, February 13, Available from URL*: http://rredc.nrel.gov/solar/*

[10] Arizona Solar Center, 2005, January 3, Available from URL: *http://www.azsolarcenter.com/arizona/modules/hotwater.html*

[11] ToolBase Services, NAHB Research Center, 2004 Available from URL: *http://www.toolbase.org/tertiaryT.asp?TrackID=&DocumentID=3216&CategoryID=949*

[12] Chinnappa, J. C. V., Gnanalingam, K., 1973, Performance at Colombo, Cylon, of a pressurized solar water heater of the combined collector and storage type. *Solar Energy*, Vol. 15, 195-204.

[13] Shitzer, A., Kalmanoviz, D., Zviring, Y., and G. Grossman. 1979. Experiments with a flat plate solar water heating system, *Solar Energy*, Vol. 22, 27-35.

[14] Close, D.J. The performance of solar water heaters with natural circulation. *Solar Energy* , 1962, Vol. 6, 33-40.

[15] Ong, K.S. 1976 An improved computer program for the thermal performance of a solar water heater. *Solar Energy* , 1976, 18, 183-191.

[16] Riazi, M.R.; Razavi. J. Performance of polypropylene and steel tubes in solar water heaters with natural circulation. *Energy Sources* , 1997, Vol. 19, No. 2, 147-152.

[17] Razavi, J.; Riazi, M. R.; Mahmoodi, M. Rate of Heat Transfer in Polypropylene Finned Tube Solar Water Heating System, *Solar Energy*, 2003, Vol. 74, No. 6, 441-445.

[18] Reddy, S.J. An empirical method for the estimation of total solar radiation. *Solar Energy*, 1971, Vol. 13, 289-290.

[19] Reddy, S.J. An empirical method for the estimation of net radiation intensity. *Solar Energy*, 1971, Vol.13, 290-291.

[20] Sabbagh, J.A., Sayigh, A.A.M., El-Sale E.M.A., Estimation of the total solar radiation from meteorological data. *Solar Energy*, 1977, Vol. 19, 307-311.

[21] Holman, J.P. (1997) *Heat Transfer,* 8th edn, McGraw-Hill, New York.

[22] Rowley, R. L., Wilding, W. V., Oscarson, J. L., Zundel, N. A., Marshall, T. L., Daubert, T. E., Danner, R. P. *"DIPPR Data Compilation of Pure Compound Properties,"* Design Institute for Physical Properties (DIPPR), Taylor and Francis Publishing Co., New York, NY, 2002. (Internet site: http://dippr.byu.edu/).

[23] Chang, J. M., Leu, J. S., Shen, M. C., Huang, B. J. A proposed modified efficiency for thermosyphonic solar heating systems, *Solar Energy*, 2004, Vol. 76, 693-701.

In: Solar Energy: New Research
Editor: Tom P. Hough, pp.277-307

ISBN 1-59454-630-4

Chapter 9

NEW METHOD FOR ESTIMATING THE TIME SERIES FRACTAL DIMENSION: APPLICATION TO SOLAR IRRADIANCES SIGNALS

S. Harrouni *

Solar Instrumentation & Modeling Group / LINS – Faculty of Electronics and Computer
University of Science and Technology H. Boumediene (USTHB)
P.O Box 32 - El-Alia - 16111 Algiers – Algeria

A. Guessoum **

Signal Processing and Imagery Laboratory, University of Blida
P. O Box 720 – Blida – Algeria

ABSTRACT

To electrify remote areas, the use of solar energy is the best economical and technological solution. The choice of the sites for the installation of photovoltaic systems and analysis of their performance requires the knowledge of the solar irradiation source. The evolution of the solar irradiation obeys to seasonal variations and random fluctuations. These latter considerably affects the behavior of the PV systems, therefore it is necessary to quantify them. The concept of fractals may then be used for modeling the solar irradiances fluctuations. For daily solar irradiances, the fractal dimension ranges from 1 to 2. D close to 1 describes a clear sky state without clouds while a value of D close to 2 reveals a perturbed sky state with clouds. Indeed, a straight line has a fractal dimension of one, just like its Euclidean dimension, once the line shows curls, its fractal dimension increases. The curling line will fill the plane more and more, and once the plane is filled up, it has a fractal dimension of two. Thus, the fractal dimension of a temporal signal has a fractal dimension between 1 and 2.

* sharrouni@yahoo.fr

** aguessoum@caramail.com

To measure the fractal dimension of time series, several methods and algorithms based on various coverings has been elaborated. In order to improve the complexity and the precision of the fractal dimension estimation of the discreet time series, we developed a simple method based on a multi-scale covering using the rectangle as a structuring element of covering. An optimization technique has been associated to this method in order to determine the optimal time interval $\Delta\tau_{max}$ through which the line log-log is fitted whose slope represents the fractal dimension. This optimization technique is permitted to improve the precision and the time computing of the method.
In order to measure the performance and the robustness of the proposed method, we applied it to fractal parametric signals whose theoretical fractal dimension is known, namely: the Weierstrass function (WF) and the fractional Brownian motion (FBM). Experimental results show that the proposed method presents a good precision since the estimation error averaged over 180 tests for the two types of signals is 3.7 %.
The proposed method is then applied to estimate the fractal dimension of solar irradiances of two sites: Boulder and Golden located at Colorado. A good concordance has been found between the fractal dimension estimated and the shapes of the solar irradiances curves.

INTRODUCTION

Fractals are objects presenting a high degree of geometrical complexity, their description and modeling is carried out using a powerful index called *fractal dimension*. Indeed, the fractal dimension is an important characteristic of fractals that contains information about their geometrical irregularities over multiple scales. The fractal dimension of a curve, for instance, will lie between 1 and 2, depending on how much area it fills. The fractal dimension can than be used to compare the complexity of two curves [1]. In a solar field, the fractal dimension is directly related to the roughness of the irradiance signals. We can then quantify the irradiance fluctuations in order to establish a classification according to the atmospheric state [2], [3], [4], [5].

This chapter is devoted to this mathematical tool that has important practical applications. We start in Section II with generalities on the concept of dimension. Then, we deal in the Section III with the problem of the theoretical estimation of this index used for the Euclidean objects or self-similar objects. Section IV focuses on experimental estimation of the fractal dimension used in the case of natural fractals; a survey of existing methods is presented. Thereafter, we will be interested in the Section V in the fractal dimension estimation of curves or one-dimensional profiles e.g. temporal signals, which quantifies the degree of their fluctuations. The fractal dimension can thus be used to compare the complexity of two curves, and therein lies its importance for applications. Also, we present a new method to evaluate the fractal dimension of discrete temporal signals or curves with an optimization technique: *the rectangular covering method*. To evaluate its accuracy, the proposed method is applied to fractal signals whose theoretical fractal dimensions are known: Weierstrass function (WF) and fractional Brownian motion (FBM).The application of the proposed method to solar irradiance signals is presented in the Section VI. Finally, in Section VII, we give a conclusion and discuss experimental results.

PRELIMINARIES

Mathematically, any metric space has a characteristic number associated with it called *dimension*, the most frequently used is the so-called *topological* or *Euclidean* dimension. The usual geometrical figures have integer Euclidean dimensions. Thus, points, segments, surfaces and volumes have dimensions 0, 1, 2 and 3, respectively.

But what for the fractals objects, it is more complicated. The coastline is an extremely irregular line in such a way that it would seem to have a surface, it is thus not really a line with a dimension 1, nor completely a surface with dimension 2 but, an object whose dimension is between 1 and 2. In the same way, we can meet fractals whose dimension ranges between 0 and 1 (Like the cantor set which will be seen later) and between 2 and 3 (surface which tends to fill out a volume), etc. So, fractals have dimensions which are not integers but fractional numbers, called *fractal dimension*.

In the classical geometry, an important characteristic of objects whose dimensions are integers is that any curve generated by these elements contours has finite length. Indeed, if we have to measure a straight line of 1 m long with a rule of 20 cm, the number of times that one can apply the rule to the line is the number 5. If a rule of 10 cm is used, the number of applications of the rule will be 10 times, for a rule of 5 cm, the number will be 20 times and so on. If we multiply the rule length used by the number of its utilizations we will find the value 1 m for any rule used.

This result, if it is true for the traditional geometry objects, it is not valid for the fractals objects. Indeed, let us use the same method to measure a fractal curve, with a rule of 20 cm, the measured length will be underestimated but with a rule of 10 cm, the result will be more exact. More the rule used is short more the measure will be precise. Thus, the length of a fractal curve depends on the rule used for the measurement: the smaller it is, the larger the length is found.

It is the conclusion reached by Mandelbrot when he tried to measure the length of the coast of Britain. He found that the measured length depends on the details taken into account. Indeed, the coast can be traversed by car or with foot to better follow the turnings of it, or with the step of ants in order to circumvent all its sand grains. With each time one will obtain a larger length, according to the inverse of the step used for the course.

With the question "What it is the length of the coast of Britain?", Mandelbrot answered, "It is infinite length" since the length of the coast grows as the step of course decreases.

Thus, fractal shapes cannot be measured with a single characteristic length, because of the repeated pattern we continuously discover at different scale levels.

This growth of the length follows a power law found empirically by Richardson

$$L(\eta) \propto \eta^{-\alpha} \tag{1}$$

Where L is the length of the coast, η is the length of the step used and the exponent α represents the fractal dimension of the coast.

Another main property of fractals is the self-similarity. This characteristic means that an object is composed of sub-units and sub-sub-units on multiple levels that resemble the structure of the whole object. So fractal shapes do not change even when observed under a different scale, this nature is also called scale-invariance. Mathematically, this property should hold on all scales. However, in the real world the self-similarity is only observed over some scales, the objects are then statistically self-similar or self-affine.

THE THEORETICAL DETERMINATION OF THE FRACTAL DIMENSION

Fractal dimension being a measurement in the way in which the fractal occupies space, to determine it we have to draw up the relationship between this way of occupation of space and its variation of scale.

Let N denote the number of the measurement stallion of n length needed to cover an object of L length and D the fractal dimension of the considered object. We have

$$N = \left(\frac{L}{n}\right)^D \tag{2}$$

From where: $$D = \frac{\ln(N)}{\ln\left(\frac{L}{n}\right)} \tag{3}$$

For the self-similar fractals, L/n represents the magnification factor and n/L the reduction factor, n corresponds for these objects to the number of objects similar to the original which are contained in the initial object.

This dimension is often called the Hausdorff–Besicovitch dimension. This is true in the case of the lines and usual surfaces where fractal dimensions and Hausdorff–Besicovitch dimensions are equal. But in the general case of the fractal objects, the dimension of Hausdorff–Besicovitch has a more complicated expression.

Let X denote the set of sequences of subsets $\{X_i,\ i=1,\ 2,\ 3,\ldots\}$, such that $X \subseteq \bigcup_{i=1}^{\infty} X_i$ and $0 < diam(X_i) \leq \varepsilon$ for all i, where $diam(X_i)$ is the largest distance between any two points of X_i.

For each $p \in [\ 0,\ \infty]$ the Hausdorff p–dimensional measure of X is defined by

$$H_p(X) = \lim_{\varepsilon \to 0}\left(\inf_{H(\varepsilon)}\left\{\sum_i [diam(X_i)]^p\right\}\right) \tag{4}$$

There is a critical real number $D_H \geq 0$ such that $H_p(x) = 0$ for $P > D_H$ whereas $H_p(x) = \infty$ for $p < D_H$. This critical D_H is the Hausdorff dimension of X also called Hausdorff–Besicovitch dimension [6], [7].

We note through this complex definition that the Hausdorff –Besicovitch dimension is very abstract. Although mathematically it is the most studied, and in a general way the best known, due to the fact that it is applicable to any element, it is often too difficult to compute.

For traditional geometry objects, the Hausdorff–Besicovitch dimension is equal to its topological dimension which is not the case for the fractals objects. This allows Mandelbrot to state that:

An object is fractal if its Hausdorff–Besicovitch dimension strictly exceeds its topological dimension.

For better understanding of this, we will compute the fractal dimension of some Euclidean and fractal objects.

Determination of the Fractal Dimension of the Usual Euclidean Objects

We mean by Euclidean object any usual object of the traditional geometry which is not fractal. This object can be a segment, a plane or a volume.

In this section we will apply the formula (3) to calculate the fractal dimension of these Euclidean objects and to check that their fractal dimension is equal to their topological dimension.

A straight line

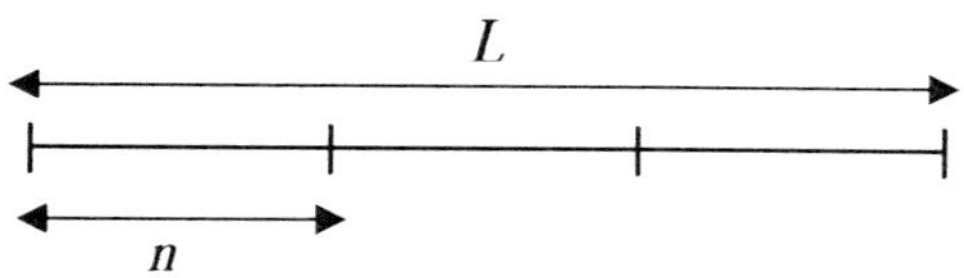

Figure 1. Determination of a straight line fractal dimension

Let's consider a segment of a straight line of length L (Figure 1), if one uses a segment of length $n=L/3$ to measure the length L, it will be necessary for us to use this segment 3 times ($N = 3$). According to formula (3) the dimension of this object will be

$$D = \frac{\ln(N)}{\ln\left(\frac{L}{n}\right)} = \frac{\ln 3}{\ln 3} = 1 \qquad (5)$$

This dimension is the topological dimension of a straight line.

For a square of side L (Figure 2), we need to use 9 squares of side $n=L/3$ to cover the initial square of side L, from where $N = 9$. Hence

$$D = \frac{\ln(N)}{\ln\left(\frac{L}{n}\right)} = \frac{\ln 9}{\ln 3} = 2 \qquad (6)$$

This is nothing other than the topological dimension of the plan or a surface.

A square

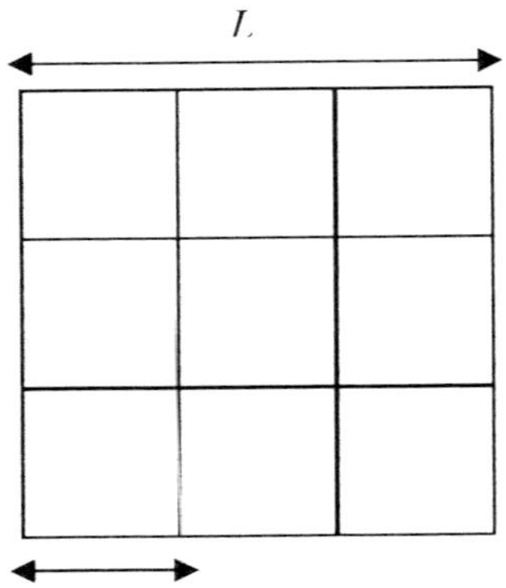

Figure 2. Determination of a square fractal dimension

A cube

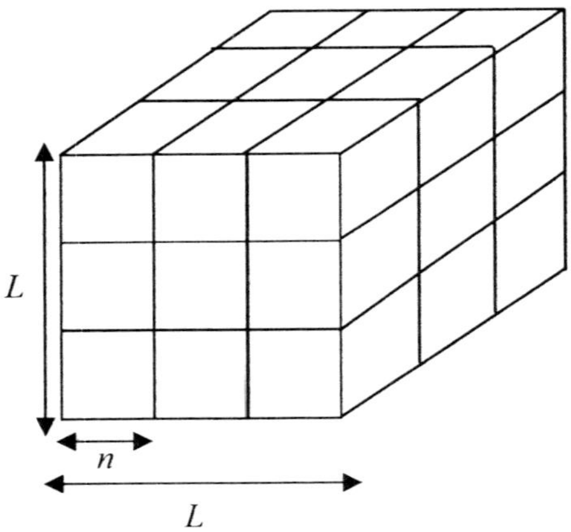

Figure 3. Determination of a cube fractal dimension

This cube of side L contains 27 small cubes of side $n=L/3$, from where the number of cubes needed to cover the initial cube is $N = 27$. It follows

$$D = \frac{\ln(N)}{\ln\left(\frac{L}{n}\right)} = \frac{\ln 27}{\ln 3} = 3 \quad (7)$$

This is the topological dimension of a volume.

In these three examples, we considered a scale (e.g. magnification factor) $L/n = 3$, but one finds the same results with any other magnification factor.

Determination of the Fractal Dimension of the Common Fractal Objects

In this section we present some examples of known fractals and their methods of construction. We then calculate their fractal dimensions by using formula (3).

The Triadic CANTOR set

One of the simplest and best known fractals is the so-called triadic Cantor set. Its construction is demonstrated in Figure 4. This object is built by removing the central segment of the object made up at the beginning of 3 segments to obtain 2 similar segments with a reduction factor of 1/3, from where the number of segments obtained is $N = 2$ and the reduction factor $n/L = 1/3$. Next, this rule is applied to the two newly created segments, and the procedure is repeated ad infinitum [8]. To calculate its fractal dimension we can use Eq. (3). Hence

$$D = \frac{\ln(N)}{\ln\left(\frac{L}{n}\right)} = \frac{\ln 2}{\ln 3} = 0.63 \quad (8)$$

The dimension of the triadic Cantor set is thus an irrational number less than 1, which characterizes dust, which is discontinuous sets of points and which have a dimension between 0 and 1.

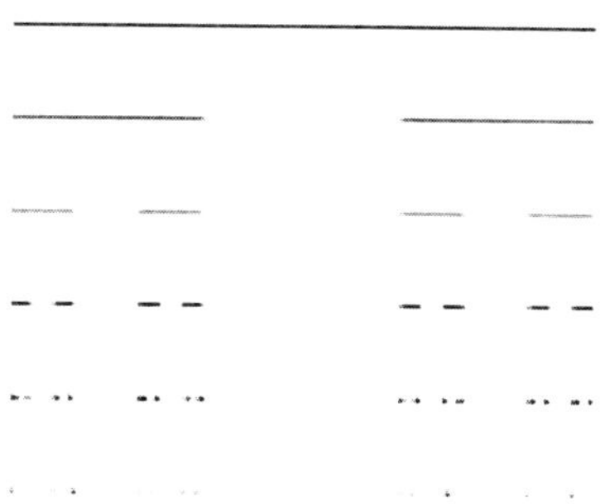

Figure 4. Construction of the triadic Cantor set

The Von Koch Snowflake

Figure 5 shows the construction of the Von Koch snowflake. At every step, as shown on this figure, the middle third of every side of the polygon is replaced with two linear segments at angles 60° and 120°.

Starting from an equilateral triangle, the two first steps lead to the star-like curves plotted on the right of the triangle. If one goes on long enough one finally gets the curve below.

Thus, at each stage from every side of the triangle 4 new segments are created with a reduction factor 1/3, from where $N = 4$ and $L/n = 3$, the fractal dimension of this curve is thus

$$D = \frac{\ln(N)}{\ln\left(\frac{L}{n}\right)} = \frac{\ln 4}{\ln 3} = 1.26 \qquad (9)$$

This is an example of figures whose Euclidean dimension is 1 (it is broken line) but whose fractal dimension is irrational and higher than 1. This is the characteristic of the fractal curves which have a topological dimension equal to 1 and a fractal dimension between 1 and 2.

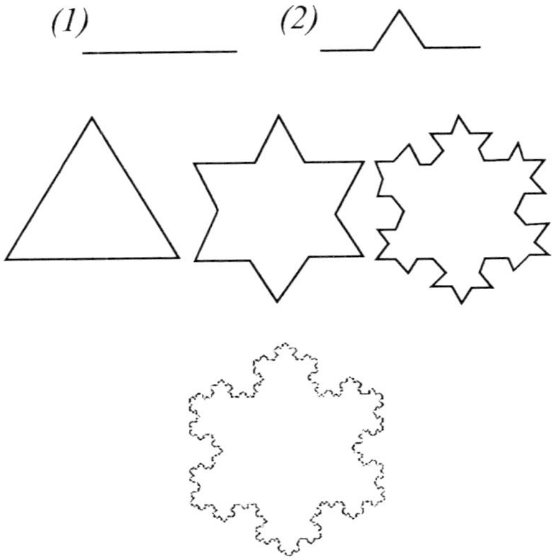

Figure 5. Construction of the Von Koch snowflake

Figure 6. Construction of the Sierpinski gasket

The Sierpinski Gasket

As shown in Figure 6, this object is created by removing at each iteration the central triangle to obtain 3 new triangles similar to the initial one with a reduction factor of 1/2, it follows that $N = 3$ and $L/n = 2$. From where

$$D = \frac{\ln(N)}{\ln\left(\frac{L}{n}\right)} = \frac{\ln 3}{\ln 2} = 1.585 \tag{10}$$

The Sierpinski gasket thus belongs to the fractal curves since its dimension is between 1 and 2.

The Experimental Determination of the Fractal Dimension of Natural Objects

We saw in the previous section that fractals have irrational dimensions. We apply eq. 3 to calculate the fractal dimension of some common fractals and if we reported to their manner of construction we would find that these fractals are self-similar.

Nevertheless, when one tries to determine fractal dimension of natural objects, one is often confronted with the fact that the direct application of eq. (3) is ineffective.

In fact, as we already mentioned, the majority of the natural fractal objects existing in our real world are not self-similar, but rather self-affine. The magnification factor and the reduction factor are thus difficult to obtain since there is not an exact self-similarity. Other methods are then necessary to estimate the fractal dimensions of these objects.

In practice, to measure a fractal dimension, several methods exist, some of which are general, whereas others are applicable only to special classes of fractals.

This section, focuses on the more commonly used methods namely, Box–counting dimension and Minkowski-Bouligand dimension, which are based on the great works of Mandelbrot [9] and Bouligand [10] and from which derive several other algorithms.

Box–Counting Dimension

This method is based upon a quantization of the space in which the object is imbedded. By a grid of squares of side ε. The number $N(\varepsilon)$ of squares that intersect the fractal object is then counted. The Box-counting dimension is then defined by

$$D_B = \lim_{\varepsilon \to 0} \frac{\ln[N(\varepsilon)]}{\ln\left(\frac{1}{\varepsilon}\right)} \tag{11}$$

When for various values of ε, we plot $\ln(N(\varepsilon))$ versus $\ln(1/\varepsilon)$, we obtain a number of points $\{\ln(N(\varepsilon)), \ln(1/\varepsilon)\}$. The slope of the straight line passing through these points gives the estimate of the Box-counting dimension D_B.

(a)

(b)

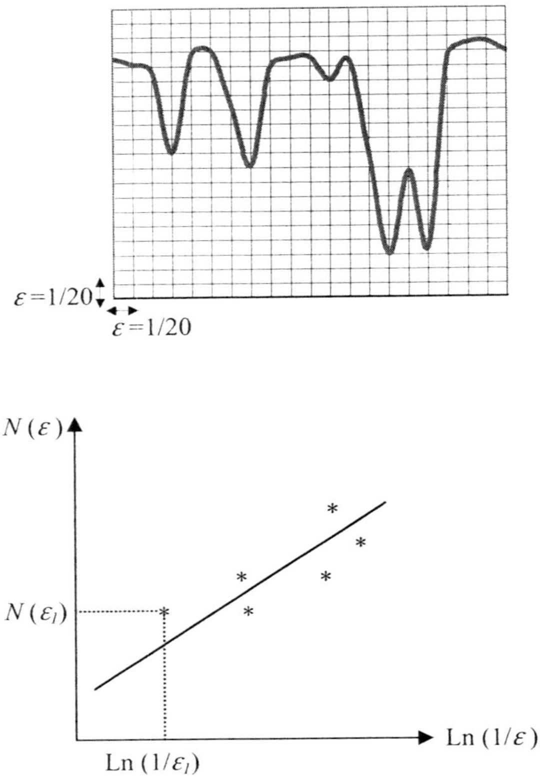

Figure 7. Example illustrating the Box –counting method
(a) Covering the curve by a grid of squares
(b) The log –log plots

Figure 7 gives an example illustrating this method. The object E (a curve) is covered by a grid of squares of side $\varepsilon = 1/20$, the total number of squares contained in the grid is $20^2 = 400$ and the number of squares intersecting the curve E is 84 (Figure 7-a).

In the Figure 7-b, the slope of the straight line fitted by a linear regression constitutes the fractal dimension of the curve E.

Minkowski –Bouligand Dimension

This method is based on Minkowski's idea of dilating the object which one wants to calculate the fractal dimension with disks of radius ε and centered at all points of E. The union of these disks thus create a *Minkowski cover*.

Let $S(\varepsilon)$ be the surface of the object dilated or covered and D_M the Minkowski –Bouligand dimension. Bouligand defined the dimension D_M as follows

$$D_M = 2 - \lambda(S) \tag{12}$$

$\lambda(S)$ is called the similarity factor and it represents the infinitesimal order of $S(\varepsilon)$. It is defined by

$$\lambda(S) = \lim_{\varepsilon \to 0} \frac{\ln[S(\varepsilon)]}{\ln(\varepsilon)} \tag{13}$$

Inserting eq.(13) in the relation (12) we obtain

$$D = \lim_{\varepsilon \to 0} \left[\frac{2 - \ln[S(\varepsilon)]}{\ln(\varepsilon)} \right] \tag{14}$$

The properties of the logarithm permit us to rewrite the relation (14) in the following form

$$D = \lim_{\varepsilon \to 0} \frac{\ln\left[\frac{S(\varepsilon)}{\varepsilon}\right]}{\ln\left(\frac{1}{\varepsilon}\right)} \tag{15}$$

or, rearranged

$$\ln\left(\frac{S(\varepsilon)}{\varepsilon}\right) = D \ln\left(\frac{1}{\varepsilon}\right) + \text{constante} \tag{16}$$

$$\text{avec } \varepsilon \to 0$$

The fractal dimension can then be estimated by the slope of the log-log plot: $\ln\left(\frac{S(\varepsilon)}{\varepsilon}\right) = f\left(\ln\left(\frac{1}{\varepsilon}\right)\right)$ fitted by the least squares method.

Figure 8-a shows the Minkowski covering $E(\varepsilon)$ composed of the union of disks of radius ε.

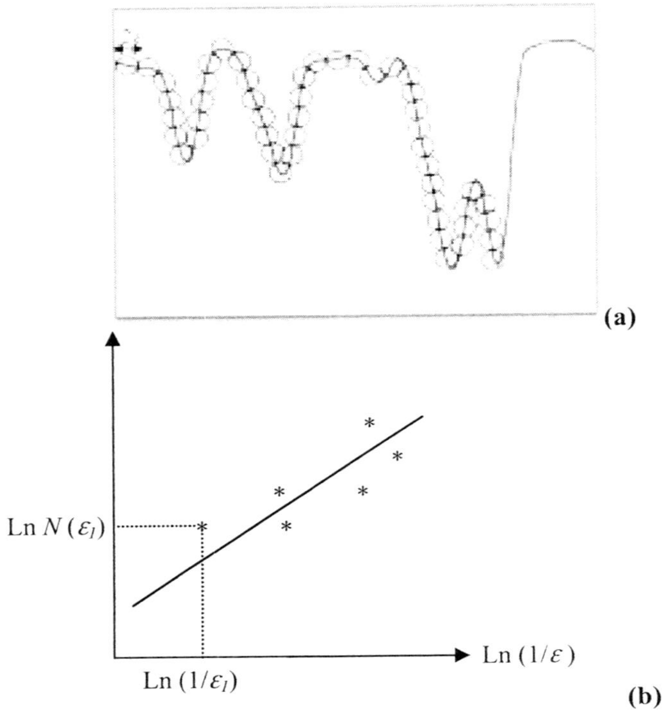

Figure 8. Example illustrating the Minkowski–Bouligand method
a) The Minkowski covering
b) The log–log plots

Discussion of the Box –counting and the Minkowski – Bouligand Methods

According to the analysis of Dubuc et *al.* [1], the Box-counting dimension and the Minkowski –Bouligand dimension are mathematically equivalent in limit thus, $D_B = D_M$. However, they are completely different in practice because of the way that limits are taken, and the manner in which they approach zero.

Experimental results published in the literature [1], [6], [11] showed that these two methods suffer from inaccuracy and uncertainty. Indeed, according to [11] the precision of these estimators are mainly related to the following aspects:

Real Value of the Fractal Dimension D

With big values of *D*, the estimation error is always very high. This can be explained by the effect of resolution [12]. When the value of *D* increases, its estimates can not reflect the roughness of the object and higher resolution is then needed.

Resolution

In the case of the temporal curves, the resolution consists of observation size of the curve (minute, hour, day…). According to Tricot et *al.* [13] estimated fractal dimension decreases with the step of observation. This is due to the fact that a curve tends to become a horizontal line segment and appears more regular. Inversely, estimated fractal dimension tends to increase with the increase of the step of observation because, a curve tends to be a series of vertical line segments and appears then more irregular.

Effect of Theoretical Approximations

Imprecision of the Box-counting and the Minkowski –Bouligand methods is also related to constraints occurring in theoretical approximations of these estimators. For example, the Box-counting dimension causes Jumps on the log –log plots [1] which generates dispersion of the points of the log –log plots with respect to the straight line obtained by linear regression. Moreover, the value of $N(\varepsilon)$ must be integer in this method. The inaccuracy of the method of Minkowski –Bouligand is due to the fact that the Minkowski covering is too thick. In this case the log –log plane results in a concave curve instead of a straight line [1].

Choice of the Interval [$\varepsilon_0, \varepsilon_{max}$]

The value ε_{max} and in general the interval [ε_0, ε_{max}] through which the line of the log-log plots are adjusted is a very significant parameter. Indeed, the precision of the estimators is influenced greatly by the choice of this interval. When ε_0 is too large, the curve is covered per few elements (limp or balls). Conversely, when the value ε_{max} is too small, the number of elements which cover the curve is too large and each element covers few points or pixels.

Some researchers tried to choose this "optimal" interval in order to minimize the error in estimation [12], [1]. For example, Liebovitch et al. [14] proposed a method for determining this interval, Maragos et al. [6] used an empirical rule to determine ε_{max} for temporal signals. In practice, these optimal intervals considerably improve the precision of the fractal dimension estimate for special cases, but not in all cases.

MEASURING THE FRACTAL DIMENSION OF SIGNALS

A Survey of Existing Methods

Many natural processes described by time series (e.g., noises, economical and demographic data, electric signals... etc.) are also fractals in the sense that their graph is a fractal set [6]. Thus, modeling fractal signals is of great interest in signal processing.

Let us recall that the most significant characteristic of fractals is their fractal dimension. This latter which contains information about their geometrical structure is as a powerful tool for measuring the degree of their irregularity over multiple scales.

For a curve, fractal dimension is between 1 and 2, it approaches 2 if it is extremely irregular and tends towards 1 if it is more regular. Fractal dimension can then be used to compare the geometrical complexity of two curves.

Considering the importance of this index and the impact of its use in practice, the precision of its estimate is necessary. Methods of Box –counting and of Minkowski –Bouligand prove then ineffective due to the fact that they suffer from inaccuracy, as we already mentioned. Inspired by the Minkowski –Bouligand method, a class of approaches to compute the fractal dimension of signal curves or one-dimensional profiles called "covering methods" is then proposed by several researchers.

These methods consist of creating multiscale covers around the signal's graph. Indeed, each covering is formed by the union of specified structuring elements. In the method of Box –counting, the structuring element used is the square or limp, that of Minkowski –Bouligand uses the disk.

Dubuc et al. [1], [13] proposed a new method called "Variation method". This one criticizes the standard methods of fractal dimension estimation namely: Box –counting and Minkowski –Bouligand. Indeed, "Variation method" applied to various fractal curves showed a high degree of accuracy and robustness.

Maragos and Sun [6] generalized the method of Minkowski –Bouligand by proposing the "Morphological covering method" which uses multiscale morphological operations with varying structuring elements. Thus, this method unifies and improves other covering methods. Experimentally, "Morphological covering method" demonstrated a good performance, since it has experimentally been found to yield average estimation errors of about 2%-4% or less for discrete fractal signals whose fractal dimension is theoretically known [6]. For deterministic fractal signals (these signals will be detailed further in this chapter) Maragos and Sun developed an optimization method which showed an excellent performance, since the estimation error was found between 0 % and 0.07 %.

Zeng et al. [11] proposed a new method for estimating the fractal dimension of one-dimensional profiles. Their method is a correction procedure based on the Box –counting method and using fuzzy techniques. It permits the reduction of two types of estimation errors: stochastic and systematic errors. It showed very satisfactory results in experiments.

New Method for Estimating the Fractal Dimension of Discrete Temporal Signals

In order to contribute in improving the accuracy of fractal dimension estimation of the discrete temporal signals we developed a simple method based on a covering by rectangles.

Rectangular Covering Method

Presentation of the Method

The method based on Minkowski –Bouligand approach consists of covering the curve for which we want to estimate fractal dimension by rectangles. The choice of this type of structuring element is due to the discrete character of the studied signals. Thus, the rectangle allows combining every point on the x-axis with the corresponding point on the y-axis, thus achieving the covering of the signal without information loss.

From the mathematical point of view, the use of the rectangle as structuring element for the covering is justified. Indeed, Bouligand [3] showed that D_M (Minkowski –Bouligand dimension) can be obtained by also replacing the disks in the previous covers with any other arbitrarily shaped compact sets that posses a nonzero minimum and maximum distance from their center to their boundary.

Thus, as shown in Figure 9, for different time intervals Δτ the area S(Δτ) of this covered curve is calculated by using the following relation

$$S(\Delta\tau) = \sum_{n=0}^{N-1} \Delta\tau \left| f(t_n + \Delta\tau) - f(t_n) \right| \quad (17)$$

N denotes the signal length (number of samples of the considered signal), $f(t_n)$ is the value of the function representing the signal at the time t_n and $| f(t_n+\Delta\tau) - f(t_n) |$ is the function variation related to the interval $\Delta\tau$.

The fractal dimension is then deduced from equation (16) where ε is replaced by the time interval $\Delta\tau$. Hence

$$\ln\left(\frac{S(\Delta\tau)}{\Delta\tau}\right) = D\ln\left(1/\Delta\tau\right) + \text{constante} \quad (18)$$

$$\text{avec } \Delta\tau \to 0$$

Thus, to determine the fractal dimension *D* which represents the slope of the straight line (18), it is necessary to use various time scales $\Delta\tau$ and to measure the corresponding area $S(\Delta\tau)$. We then obtain several points ($\Delta\tau_i$, $S(\Delta\tau_i)$) constituting the line (18).

A good estimation of the fractal dimension *D* requires a good fitting of the log-log plot defined by eq.(18). Therefore, the number of points constituting the plot is important. This number is fixed by $\Delta\tau_{max}$ which is the maximum scale over which we attempt to fit the log-log plot.

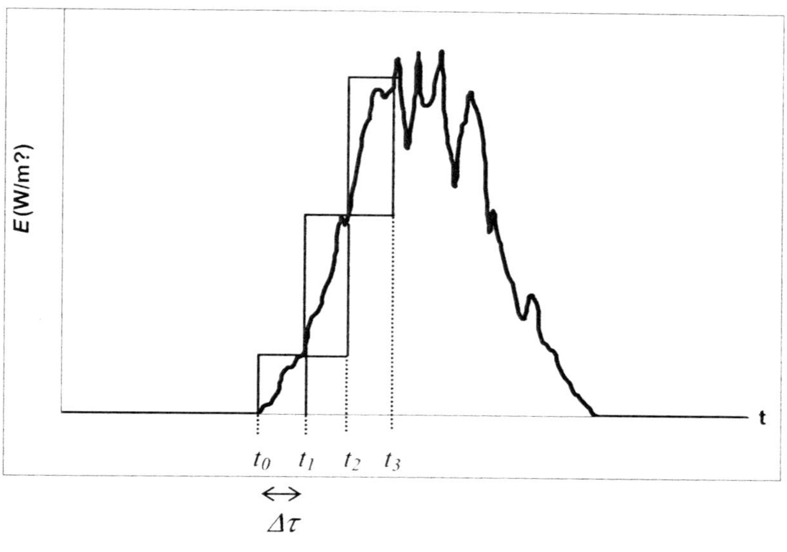

Figure 9. An example of temporal curve covered by rectangles

As mentioned above, to estimate the fractal dimension most methods determine $\Delta\tau_{max}$ experimentally. This procedure requires much time and suffers from precision. Also, we developed an optimization technique to estimate $\Delta\tau_{max}$.

Optimization Technique [15]

Experience shows that $\Delta\tau_{max}$ required for a good estimation of D depends on several parameters, especially the time length of the signal N. $\Delta\tau$ should not be too weak, in order not to skew the fitting of the line, and it must not exceed $N/2$. $\Delta\tau$ must also satisfy the condition of linearity of the line.

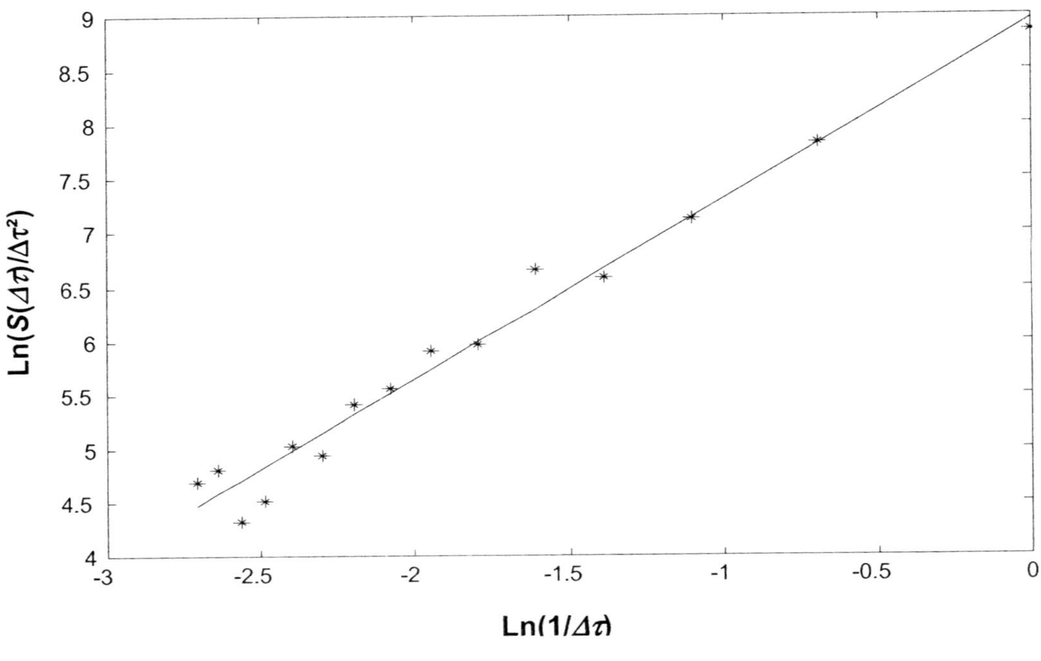

Figure 10. An example of log –log plots fitted by the least squares estimation

Figure 10 represents an example of log-log plot. Our optimization technique consists first in taking a $\Delta\tau_{max}$ minimal about 10, because the number of points constituting the plot should not be very small as signaled above, then, $\Delta\tau_{max}$ is incremented progressively as far as reaching $N/2$. We hence obtain several straight log-log lines which are fitted using the least squares estimation. The $\Delta\tau_{max}$ optimal is the one corresponding to the most linear log-log straight line. The linearity criterion used is the minimum of the quadratic error.

The quadratic error used is defined by the following formula

$$E_{quad} = \frac{\sum_{i=1}^{n} d_i}{n} \qquad (19)$$

In this relation, N denotes the number of points used for the straight log-log line fitting, d_i represents the distance between the points $\left(Ln\left(\frac{1}{\Delta\tau}\right), Ln\left(\frac{S(\Delta\tau)}{\Delta\tau^2}\right)\right)$ and the fitted straight log-log line as showed in Figure 11. In this figure $Ln\left(\frac{1}{\Delta\tau}\right)$ is denoted X and $Ln\left(\frac{S(\Delta\tau)}{\Delta\tau^2}\right)$ denoted Y.

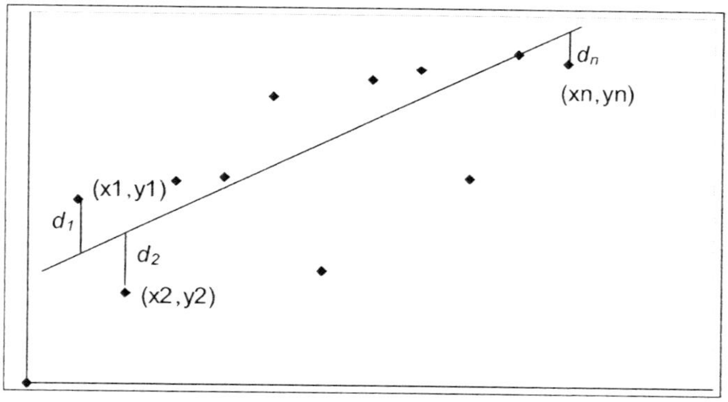

Figure 11. Calculation of the distance d_i of real points constituting the log-log plots to the fitted straight line

Validation of the Method

In order to test the validity and the accuracy of the rectangular covering method, we applied it to two different types of parametric fractal signals whose theoretical fractal dimension is known, these test signals are the Weirstrass function (FW) which is a deterministic signal and the random signal of the fractional Brownian motion (FBM). These fractal signals that will be briefly defined below are often used in various applications.

The Weierstrass Function (FW)

It is defined as [16-18]

$$W_H(t) = \sum_{k=0}^{\infty} \gamma^{-kH} \cos(2\pi\gamma^k t) \qquad (20)$$

$$0 < H < 1$$

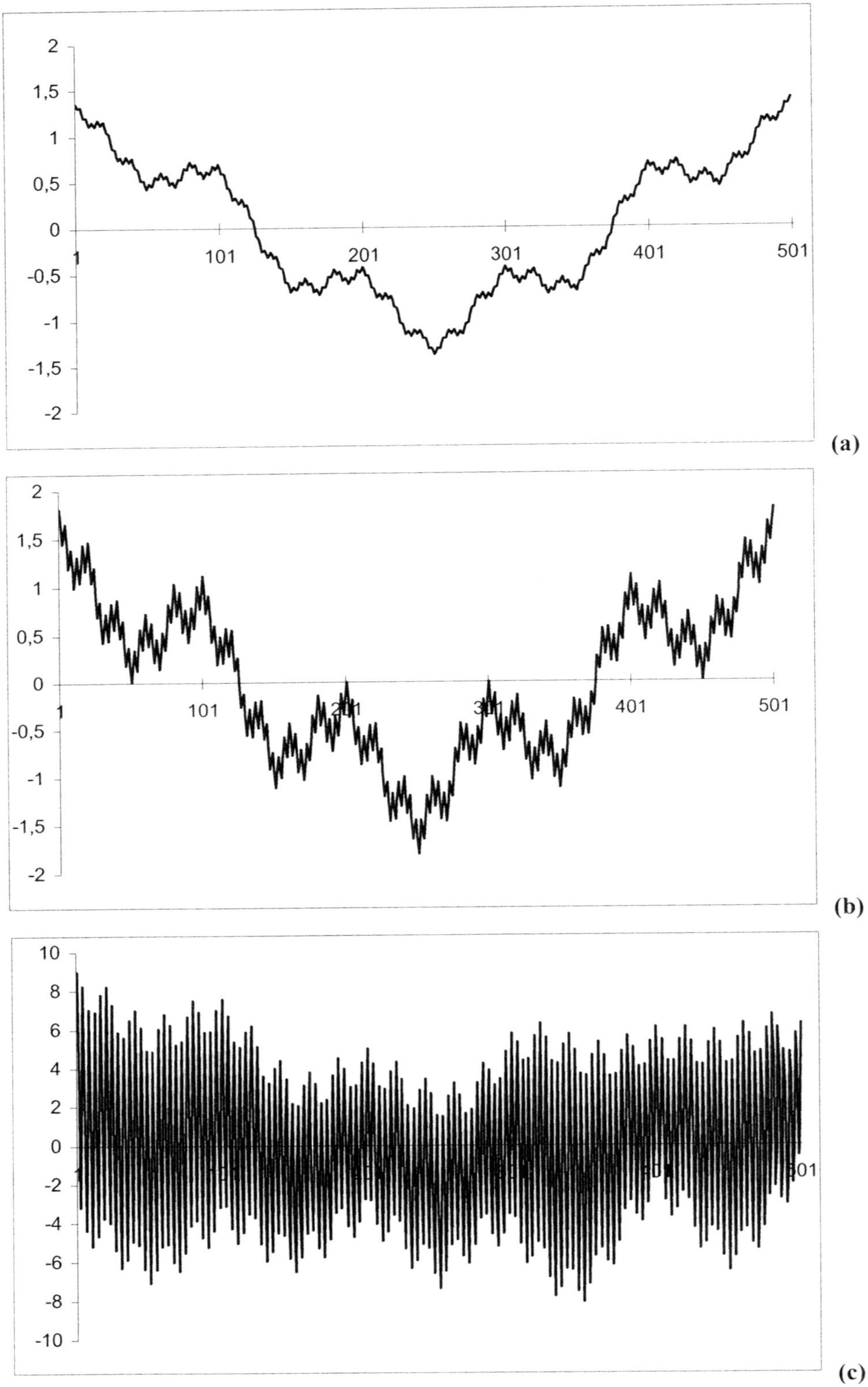

Figure 12. Examples of synthesized FW signals a) D=1.2 b) D=1.5 c) D=1.8

This function is continuous, but nowhere differentiable, if γ is the integer, then it is periodic with period one [ ], $\gamma>1$ is fixed by the experimenter, the fractal dimension of this function is $D = 2 - H$.

In our experiments, we synthesized discrete time signals from WF's by sampling $t \in [0,1]$ at $N+1$ equidistant points, using $\gamma=2.1$ and truncating the infinite series so that the summation is done only for $0 \le k \le k_{max}$. The k_{max} is determined by the inequality $2\pi\gamma^k \le 10^{12}$ [6].

Figure 12 shows three FW resulting signals whose irregularity increases with their fractal dimension D.

Fractional Brownian Motion (FBM)

It is one of the most used mathematical models used to describe self-affine fractals existing in nature.

The Brownian motion discovered by the English botanist R. Brown is a random process that defines the evolution of the microscopic particles of a fluid. Mandelbrot and Wallis proposed an extension of this motion: the fractional Brownian motion.

The function of the Brownian fractional motion $B_H(t)$ with parameter $0 < H < 1$ is a time varying random function with stationary, Gaussian distributed, and statically self-affine increments. So

$$\left\langle [B_H(t) - B_H(t_0)]^2 \right\rangle = 2D(t-t_0)^{2H} \qquad 0 < H < 1 \tag{21}$$

The fractal dimension D of $B_H(t)$ is $D = 2 - H$.

To synthesized FBM signals, several methods exist [19], [20], [21] the most known are: Choleski decomposition method, Durbin–Levinson algorithm, FFT method and circulant matrix method.

In our experiments we synthesized FBM signals via the Durbin –Levinson method. Figure 13 shows synthesized FBM signals of size length $N = 500$ and of varying D. The larger D (the smaller H), the more fragmented these fractal signals look.

Results on Test Signals for the Rectangular Covering Method

In order to validate our rectangular covering method we applied it to the synthesized test signals. The results are shown in Tables 1 and 2. We can see the estimated fractal dimension D^* and the estimation error $|D - D^*|$ for different values of D. $\Delta\tau_{max}$ optimal used for obtaining D^* are also showed in these tables.

The experimental results from tables 1 and 2 indicate that, for the two fractal signals FW and FBM, the rectangular covering method performs well in estimating dimensions $D \in [1.1, 1.9]$, since the estimation error is less than or equal 6 % for the WF signals and 7 % for FBM signals.

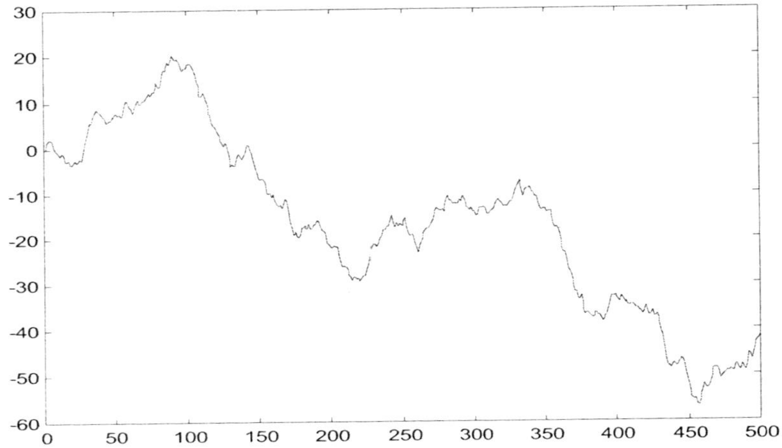

(a)

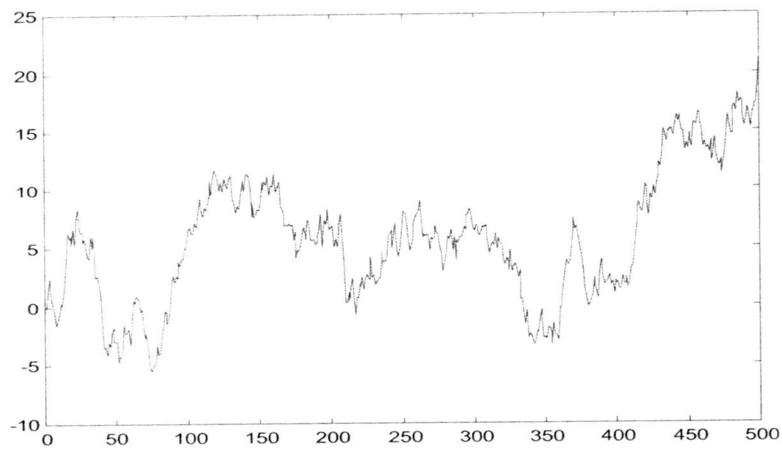

(b)

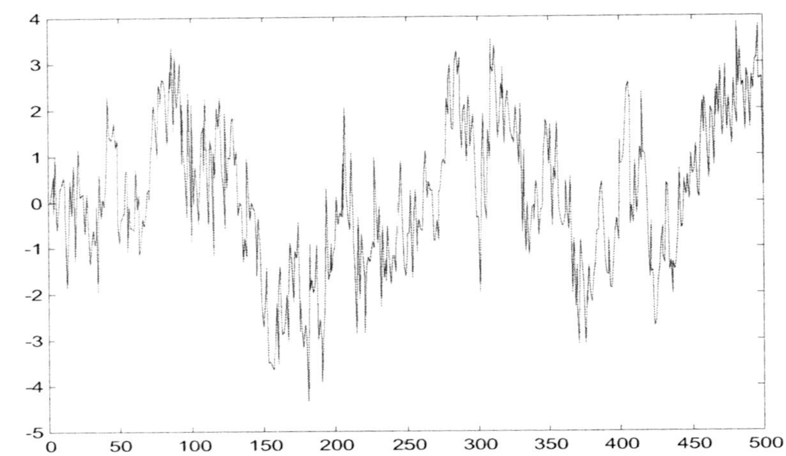

(c)

Figure 13. Examples of synthesized FBM signals a) D=1.2 b) D=1.5 c) D=1.8

Table 1. Rectangular covering method on WF's

True D	D estimated (D*)	Error ((D-D*).100)	$\Delta\tau_{max}$ optimal
1.1	1.16	6 %	13
1.2	1.22	2 %	13
1.3	1.30	0 %	11
1.4	1.39	1 %	13
1.5	1.48	2 %	13
1.6	1.59	1 %	14
1.7	1.69	1 %	14
1.8	1.78	2 %	14
1.9	1.86	4 %	13

Table 2. Rectangular covering method on FBM's

True D (D)	D estimated (D*)	Error ((D-D*).100)	$\Delta\tau_{max}$ optimal
1.1	1.17	7 %	10
1.2	1.28	8 %	10
1.3	1.27	3 %	11
1.4	1.41	1 %	10
1.5	1.51	1 %	13
1.6	1.62	2 %	11
1.7	1.75	5 %	11
1.8	1.75	5 %	14
1.9	1.92	2 %	12

By varying the signals' length $N \in [100, 1000]$ with a step of 100 we have also observed similar performance of this method. Over 99 different combinations of (D, N) the average estimation error of the rectangular covering method was 4 % for both WF's and FBM's.

Examples of fractal dimensions and their estimation errors found by varying the signals length N are given in tables 3 and 4. We observe that for FW signals the estimation error of D decreases with the increase in N. However, this is not observed in all cases for the FBM signals.

Table 3. Results of Rectangular covering method on FW with various lengths N

D \ N	100		500		1000	
D	D*	Er	D*	Er	D*	Er
1.1	1.23	13 %	1.16	6 %	1.16	6 %
1.2	1.30	10 %	1.22	2 %	1.22	2 %
1.3	1.38	8 %	1.30	0 %	1.30	0 %
1.4	1.46	6 %	1.39	1 %	1.40	0 %
1.5	1.55	5 %	1.48	2 %	1.50	0 %
1.6	1.62	2 %	1.59	1 %	1.61	1 %
1.7	1.68	2 %	1.69	1 %	1.71	1 %
1.8	1.73	7 %	1.78	2 %	1.82	2 %
1.9	1.70	20 %	1.86	4 %	1.91	1 %

Table 4. Results of Rectangular covering method on FBM with various lengths *N*

D \ N	100		500		1000	
D	D*	Er	D*	Er	D*	Er
1.1	1.35	25 %	1.17	7 %	1.09	1 %
1.2	1.36	16 %	1.28	8 %	1.2	0%
1.3	1.3	0 %	1.27	3 %	1.36	6 %
1.4	1.47	7 %	1.41	1 %	1.43	3 %
1.5	1.42	8 %	1.51	1 %	1.47	3 %
1.6	1.56	4 %	1.62	2 %	1.6	0 %
1.7	1.7	0 %	1.75	5 %	1.68	2 %
1.8	1.82	2 %	1.75	5 %	1.78	2 %
1.9	1.8	10 %	1.92	2 %	1.93	3 %

Generally, one can clearly note from the results of tables 2 and 3 that the rectangular covering method performs well on tested signals in estimating dimensions $D \in [1.3, 1.8]$ for any signal length. For $D = 1.1$, $D = 1.2$ and $D = 1.9$ e.g. when the signal becomes very regular or very irregular a good estimate of the fractal dimension is obtained for high signal length. This can be explained by the fact that for signals with small length, some degree of fragmentation is irreversibly lost during sampling, since the tested signals are sampled versions.

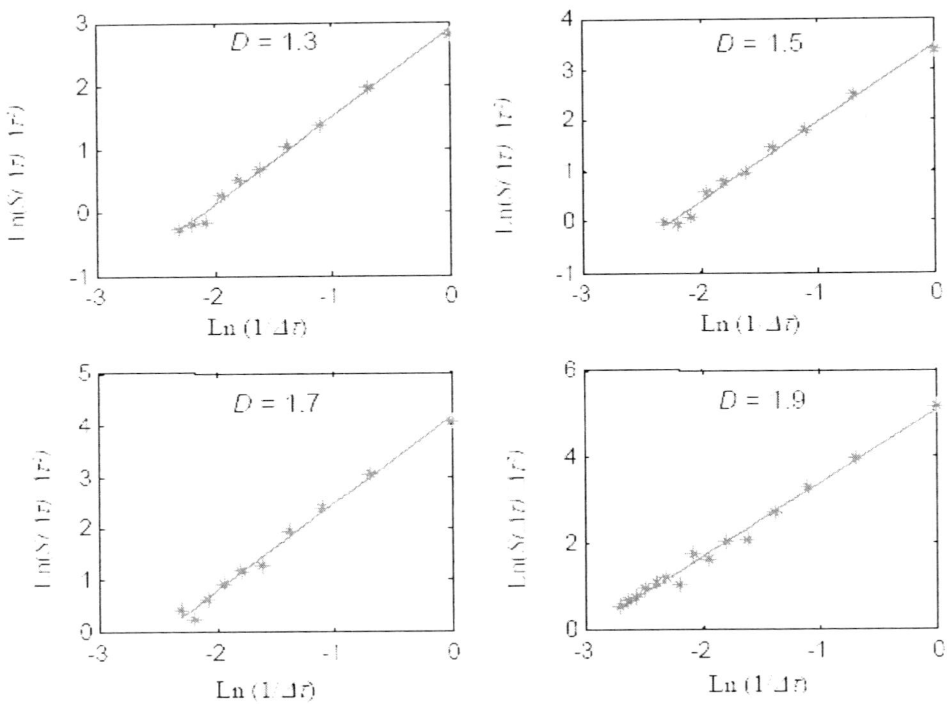

Figure 14. Examples of log-log plots for WF signals with $N = 100$

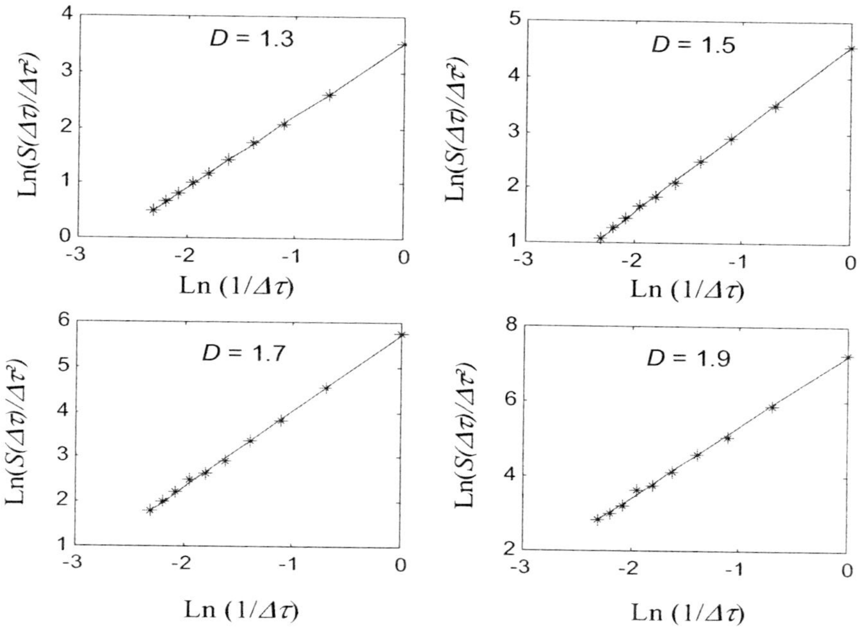

Figure 15. Examples of log-log plots for WF signals with $N = 1000$

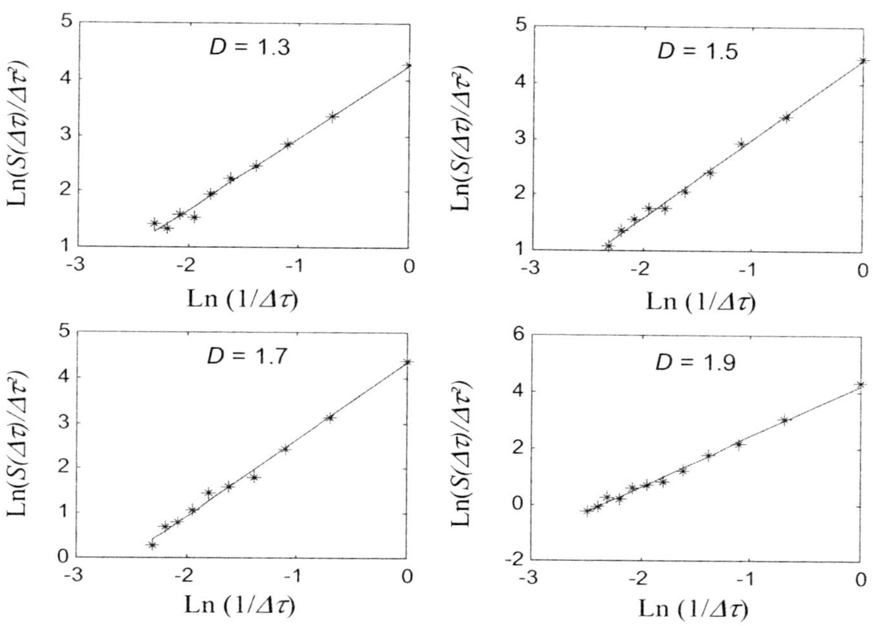

Figure 16. Examples of log-log plots for FBM signals with $N = 100$

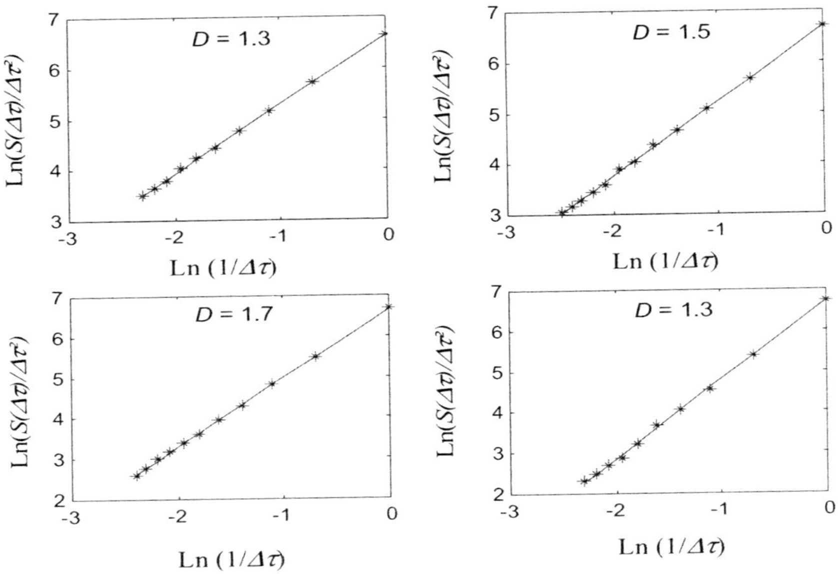

Figure 17. Examples of log-log plots for FBM signals with $N = 1000$

To confirm these results log-log plots of some examples of the tested signals for $N = 100$ and $N = 1000$ are presented in Figures 14, 15, 16 and 17. We notice that generally the log-log plots lie on perfectly straight lines for $N = 1000$ which shows a good performance of the method. However, we observe few concavity in plots for $N = 100$ which demonstrates that the higher signal length the better the estimation of its fractal dimension.

MEASURING THE FRACTAL DIMENSION OF SOLAR IRRADIANCE SIGNALS

To electrify remote areas, the use of solar energy is the best economical and technological solution. The choice of sites for the installation of photovoltaic systems and analysis of their performance requires the knowledge of the solar irradiation source. The evolution of the solar irradiation obeys to seasonal variations and random fluctuations. The latter of these considerably affect the behavior of the PV systems, therefore it is necessary to quantify them. If the seasonal tendencies can be described by deterministic models, the random fluctuations, due to the atmospheric fluctuations are more difficult to model.

The fractal dimension D measures the amount of daily solar irradiance fluctuations. For daily solar irradiances, the fractal dimension ranges from 1 to 2. D close to 1 describes a clear sky state without clouds while a value of D close to 2 reveals a perturbed sky state with clouds.

In this section, we propose to estimate the fractal dimension of solar irradiances using the rectangular covering method for two sites: Boulder and Golden located at Colorado.

Experimental Data

The experimental data bank contains global irradiance data measured at two sites of Colorado: Golden (latitude = 39°74'N, longitude = 105°18'W and altitude = 1829m) and Boulder (latitude = 39°91'N, longitude = 105°25'W and altitude = 1855m). The irradiance data have been recorded during the year 2003 on a horizontal surface by the Precision Spectral Pyranometer (PSP).

Data sample rate is 1-sec, with 1-min averages for Golden site and 2-sec, with 1-min outputs for Boulder site. From the 1 minute data, we extracted the 10 minute data by taking the irradiances measured every 10 minutes.

Experimental Results

In order to estimate the fractal dimensions of the daily irradiances at Golden and Boulder, we used both the one and the ten minute data. Figure 18 presents the log-log lines permitting the estimation of the fractal dimension of irradiance curves with the two time steps.

The Figure shows that the log-log plots lie on straight lines. The fractal dimensions obtained from the log-log lines are illustrated by Figure 19 which represents, for the two sites, two solar irradiance curves with different characteristics and their fractal dimensions $\hat{D}_{1m}$ and $\hat{D}_{10m}$ corresponding to the 1 minute and the 10 minute data.

Figure 19 shows that there is concordance between the shape of the signals and the corresponding fractal dimension. It also shows that the fractal dimensions $\hat{D}_{1m}$ and $\hat{D}_{10m}$ are nearly equal.

Anyway, the closeness of $\hat{D}_{10m}$ to $\hat{D}_{1m}$, shows that the rectangular covering does not require a large number of data. This is very important for the practice where irradiances data measured at a step of 1 minute are rarely available [4], [5], [22].

Consequently, to analyze the obtained results we shall henceforth consider the dimension $\hat{D} = \hat{D}_{10m}$. The fractal dimensions obtained for four months of the year (January, April, July and October) and both sites are shown in Figure 20.

$\hat{D}$ varies from 1.00 to 1.94 for Golden and from 1.00 to 1.93 for Boulder corresponding to the existence of regular and fluctuating daily irradiances. Figure 21 gives the annual variation of the monthly means of $\hat{D}$ for the two sites.

This Figure shows clearly that $\hat{D}$ is fluctuating, the standard deviation of $\hat{D}$ is 7.7 % for Golden and 8.2 % for Boulder. These standard deviations and the annual mean of the fractal dimension $<\hat{D}>$ suggest that the solar irradiances of Golden and Boulder exhibit the same fluctuations ($<\hat{D}>=1.37$ for Golden and 1.38 for Boulder). However, the analysis of $\hat{D}$ month per month leads to the conclusion that the irradiance at Boulder fluctuates slightly more than at Golden. This analysis also permits the detection of the months where the fluctuations of the irradiances are most intense – May and June for both sites – and those where these irradiances are very regular – September and October for both sites. This information allows refining the sizing of photovoltaic system.

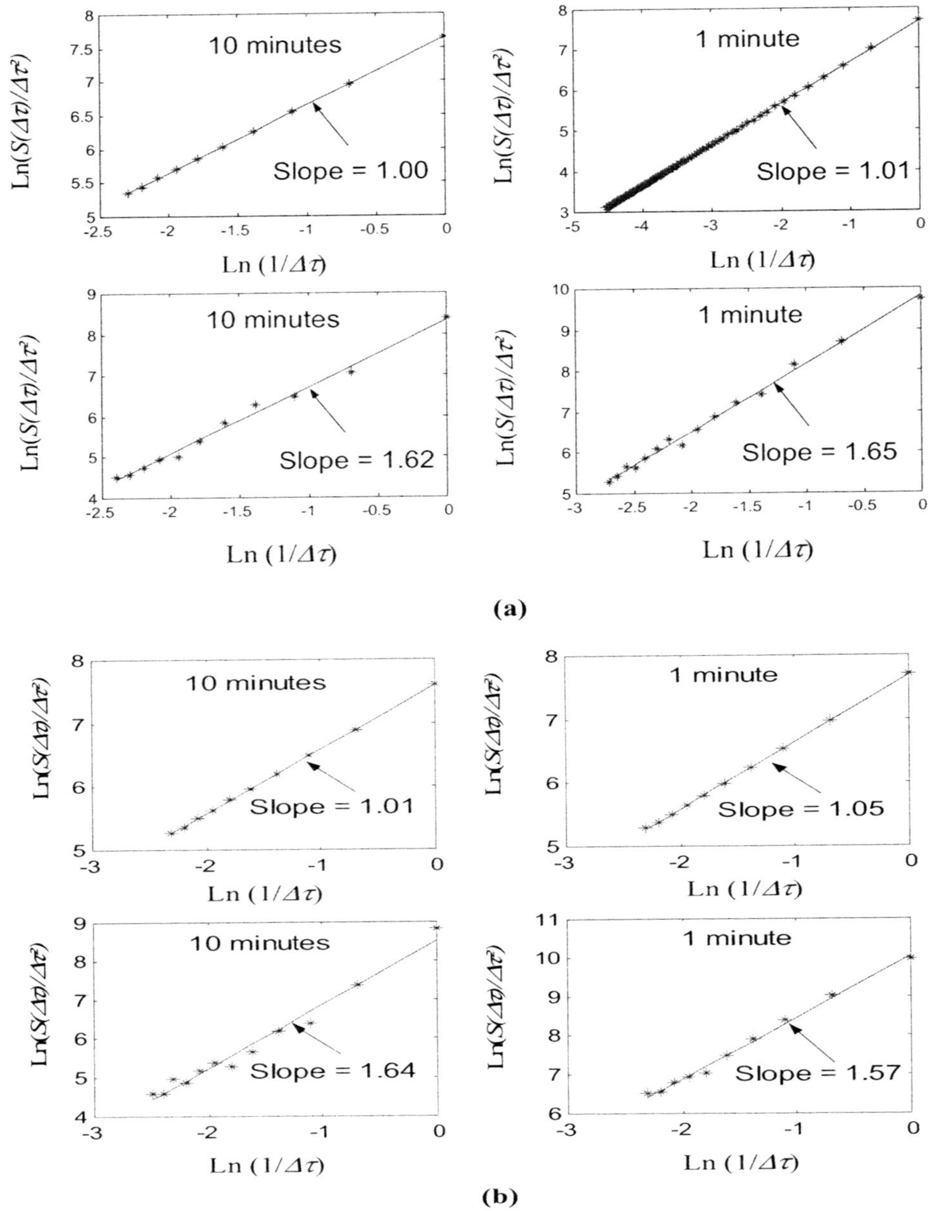

Figure 18. Two Examples of log-log plots fitted by the least-squares estimation with their slopes which represent an estimated fractal dimension $\hat{D}_{1m}$ and $\hat{D}_{10m}$ obtained at the two scales: 1 minute and 10 minute for both sites.
Golden site, upper: day of July 09, lower: day of December 21
Boulder site, upper: day of August 15, lower: day of February 22

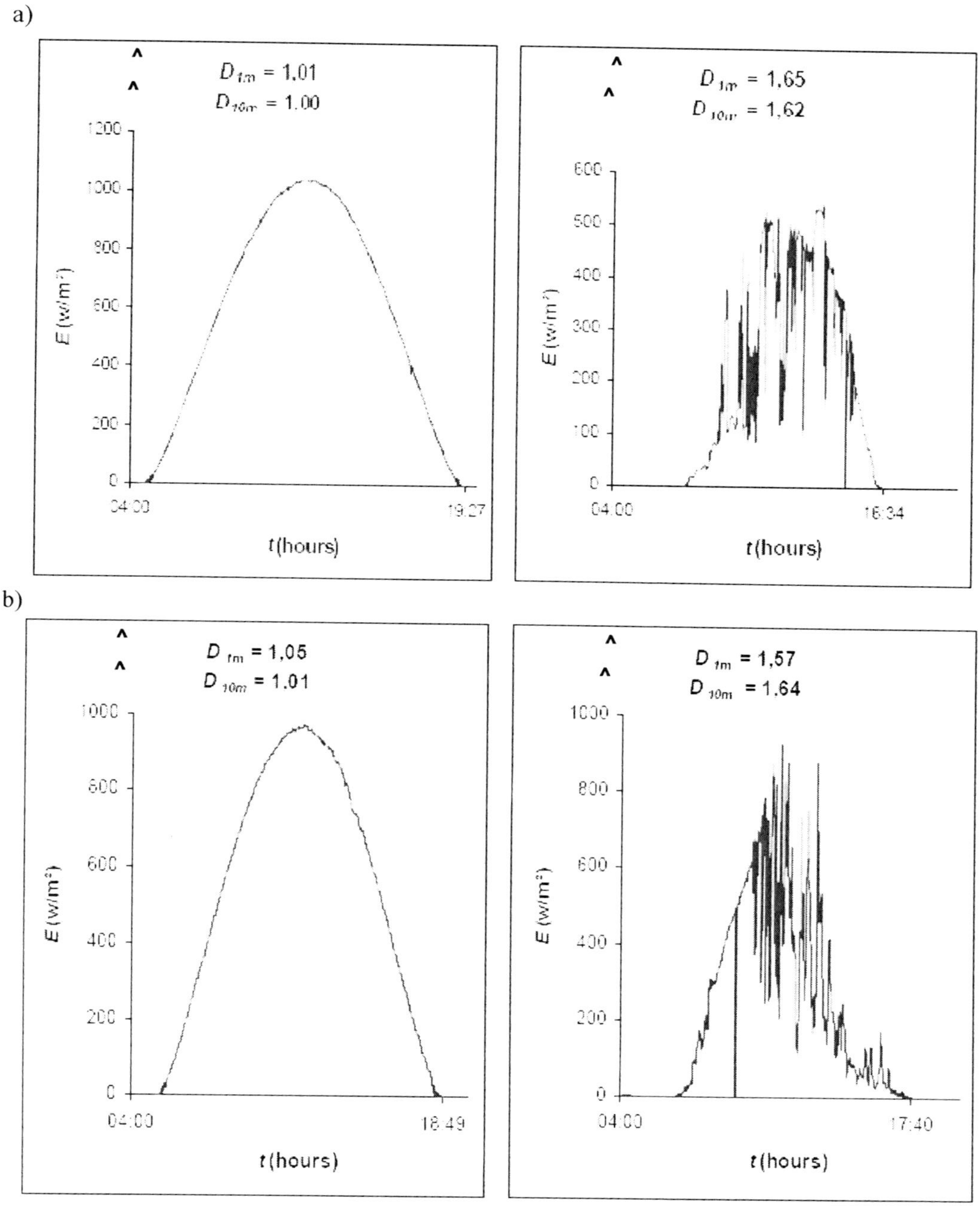

Figure 19. Different irradiance signals with their fractal dimensions $\hat{D}_{1m}$ and $\hat{D}_{10m}$

a) Golden site, the left column: day of July 09 and the right column: day of December 21

b) Boulder site, the left column: day of August 15 and the right column: day of February 23

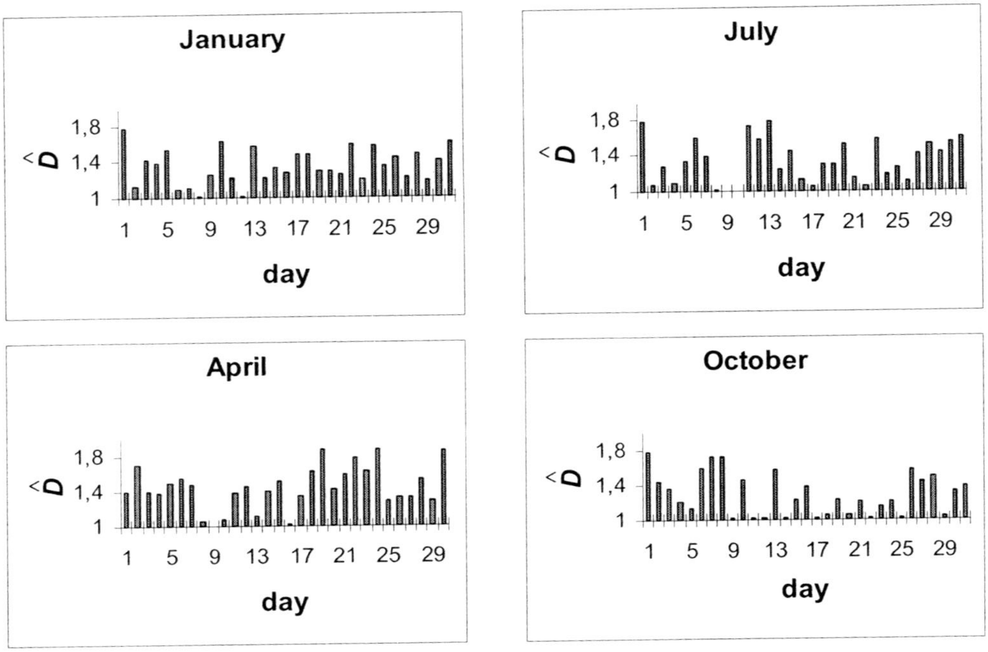

(a)

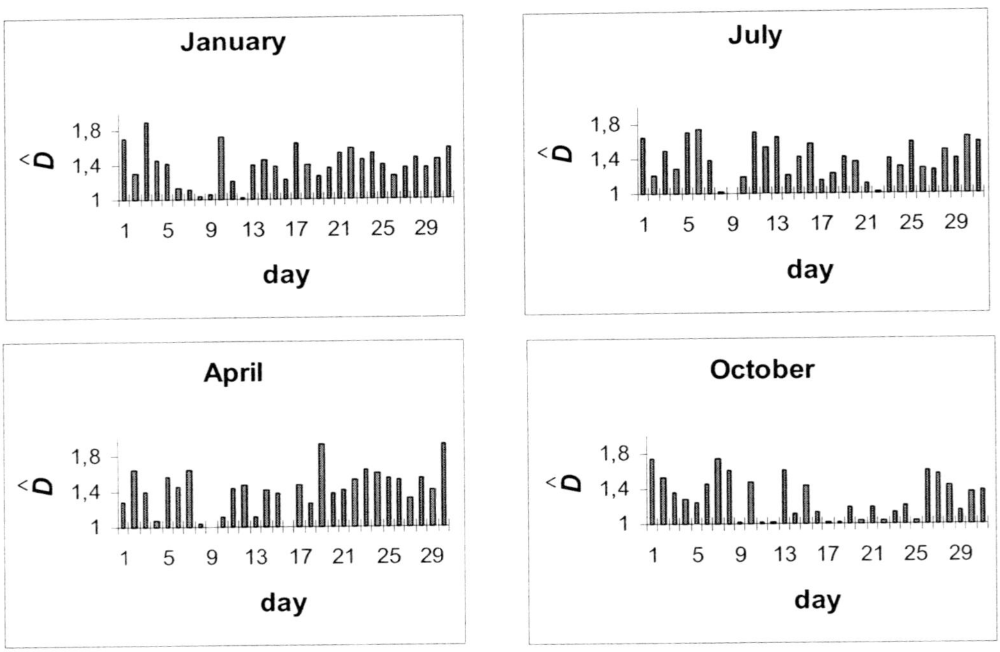

(b)

Figure 20. Fractal dimensions of daily irradiances for four representative months of the year: January, April, July and October.

a) Golden

b) Boulder

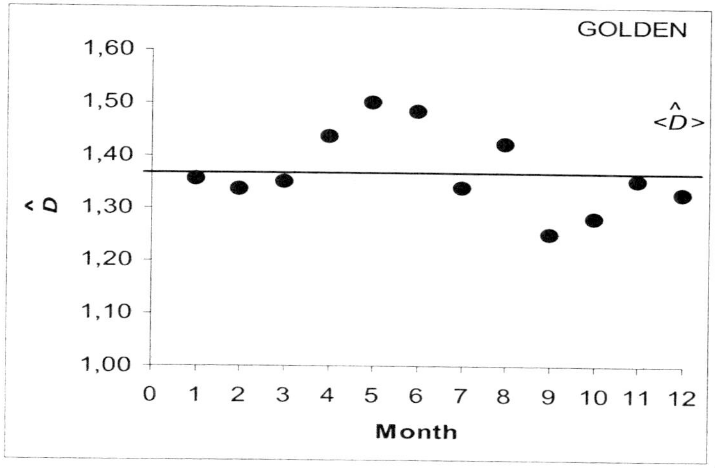

a)

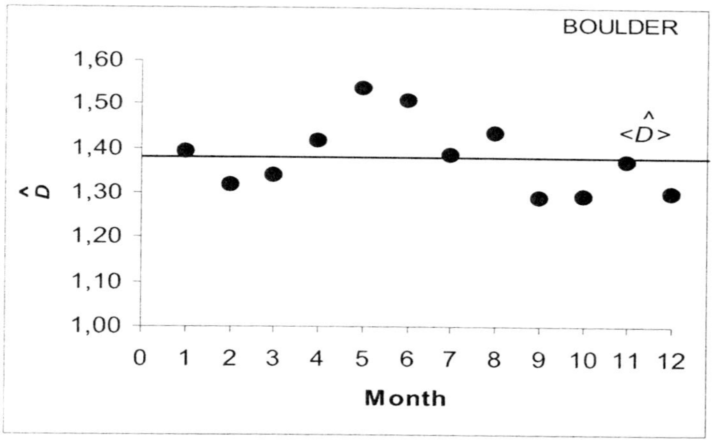

b)

Figure 21. Annual variation of the monthly means of the estimated fractal dimension $\hat{D}$, the straight solid line represents the annual mean of the fractal dimension $<\hat{D}>$

a) Golden, $<\hat{D}>=1.37$

b) Boulder, $<\hat{D}>=1.38$

Conclusion

Fractal dimension is an important parameter that measures signal shape irregularity. It is a powerful index of the degree of their irregularity. For the time-series signals, fractal dimension quantifies the degree of their fluctuation which allows their comparison, classification and prediction.

Several methods and algorithms for measuring the fractal dimension of time series based on various coverings have been elaborated. The difference between these algorithms resides in the covering of the curve.

In order to improve the complexity and the precision of the fractal dimension estimation of the discreet time series, we developed a simple method based on a covering multi-scale using the rectangle as a structuring element of covering.

Our method showed a good performance since it has experimentally been found to yield average estimation errors averaged over 180 tests of about 3.7% for fractal signals synthesized from Weierstrass functions and fractal Brownian motion.

One of the various applications of the fractal dimension is its use to quantify the random fluctuations of the solar irradiance signals. This is very useful for their prediction, simulation and classification thereafter.

Also, in this chapter the rectangular covering method has been applied to estimate the fractal dimension of solar irradiances of two sites: Boulder and Golden located at Colorado. We have shown that our method performs well on irradiance signals in estimating their fractal dimensions since a good concordance has been found between the fractal dimensions estimated and the shapes of the solar irradiances curves

On the other hand, we have demonstrated that the rectangular covering method does not require a large number of data. Indeed, fractal dimensions found for the irradiance signals measured with a time step of 10 minutes are not much different from those of the irradiance signals measured with a time step of 1 minute. This is very advantageous, since data with a short time step (1 minute) are rarely measured due to the costs involved by such measurements.

The fractal dimension quantifies the fluctuations of the daily global irradiances. These fluctuations are closely related to the atmospheric conditions and, therefore, to the state of the sky. A fractal dimension value close to one characterizes a clear sky without haziness, whereas a dimension value close to 2 corresponds to a variable sky with cloud passages. Consequently, the fractal dimension can be used to achieve a classification of the irradiances according to the state of the sky.

REFERENCES

[1] Dubuc, B.; Quiniou, F.; Roques-Carmes, C.; Tricot, C. & Zucker, S. W. Evaluating the fractal dimension of profiles, *Phys. Rev. A*, 1989, Vol. 39, pp. 1500-1512.

[2] Maafi, A. & Harrouni, S. Measuring the fractal dimension of solar irradiance in view of PV systems performance analysis, *Proc.6th WREC*, Brighton (UK), 1-7 July, 2000, Vol. IV, pp. 2032-2035.

[3] Harrouni, S. & Guessoum, A. Fractal classification of solar Irradiances into typical days using a cumulative distribution function, *International Conference on Renewable Energies and Power Quality ICREPQ'03*, Vigo (Spain), 9-11 April 2003.

[4] Harrouni, S. et Maafi A. Classification des éclairements solaires à l'aide de l'analyse fractale,*Revue Internationale des énergies renouvelables (CDER),* 2002, Vol. 5 (2), pp. 107-122.

[5] Maafi, A. & Harrouni, S. Preliminary results of fractal classification of daily solar irradiance, *Solar Energy*, 2003, Vol. 75 (1), pp. 53-61.

[6] Maragos, P. & Sun, F.K. Measuring the fractal dimension of signals: morphological covers and iterative optimization, *IEEE Transaction on Signal Processing*, 1993, Vol. 41 (1), pp. 108-121.

[7] Barnsley, M. *Fractals Everywhere*; Boston, MA : Academic Press, 1988.

[8] Vicsek,T. In *Fractal Growth Phenomena;* Editor, World scientific publishing co. pta. Ltd; 2nd ed.

[9] Bouligand G. Ensembles impropres et nombre dimensionnel, *Bull. Sci. Math.,* 1928, Vol. II-52, pp. 320-344, 361-376.

[10] Minkowski, H. Uber die Bgriffe Lange, Oberflache und Volumen, *Jahresber, Deutch. Mathematikerverein*, 1901, Vol. 9, pp. 115-121.

[11] Zeng, X.; Koehl, L.; Vasseur, C.Design and implementation of an estimator of fractal dimension using fuzzy techniques, *Pattern Recognition*, 2001, Vol. 34, pp. 151-169.

[12] Huang, Q.; Lorch, J.R., & Dubes, R.C. Can the fractal dimension of images be measured?, *Pattern Recognition,*1994, 27 (3), pp. 339-349.

[13] Tricot, C. ; Quiniou, J.F. ; Wehbi, D. ; Roques-Carmes, C. & Dubuc, B. Evaluation de la dimension fractale d'un graphe, *Rev. Phys.* 1988, Appl. 23, pp. 111-124.

[14] Liebovitch, L.S. & Toth. A fast algorithm to determine fractal dimension by box counting, *Phys. Lett. A* 141, 1989, pp. 386-390.

[15] Harrouni, S. ; Guessoum, A. et Maafi, A. Optimisation de la mesure de la dimension fractale des signaux : Application aux éclairements solaires. *Séminaire Nationale sur l'Automatique et les Signaux (SNAS'02)*, Université de Annaba, 27-28 Octobre 2002.

[16] Hardy, G.H. Weierstrass's nondifferentiable function, *Trans. amer. Math. Soc.*, 1916, Vol. 17, pp. 322-323.

[17] Mandelbrot, B.B. *The fractal geometry of nature*, New York: Freeman, 1982/1983.

[18] Berry, M.V. & Lewis, Z.V., On the Weierstrass-Mandelbrot fractal function, *Proc. roy. Soc., Ser. a*, 1980, Vol. 370, pp. 459-484.

[19] Mandelbrot, B.B. & Wallis, J.R. Computer experiments with fractional Brownian motion *Water Resources Res,* 1969, Parts 1-3, Vol. 5, pp. 228-267.

[20] Voss, R.F., *Fractals in nature: From characterization to simulation*; in The science of fractal Images, H-O. Peitgen and D. Saupe, Eds. New York: Springer-Verlag, 1988.

[21] Lundahl, T.; Ohley, W.J.; Kay, S.M. & Siffert, R., Fractional Brownian motion: A maximum likelihood estimator and its application to image texture, *IEEE Trans. Med. Imaging*, 1986, Vol. MI-5, pp. 152-160.

[22] Harrouni, S.; Guessoum, A. & Maafi, A. Classification of daily solar irradiation by fractional analysis of 10-min-means of solar irradiance, *Theoretical and applied climatology,* 2005, Vol. 80 (1), pp. 27-36.

In: Solar Energy: New Research
Editor: Tom P. Hough, pp.309-320

ISBN 1-59454-630-4

Chapter 10

TIME SERIES HOLDS PROSPECTS FOR REAL-TIME FORECASTING OF THE MAXIMUM POWER POINT FOR A PHOTOVOLTAIC ARRAY

S. Shahnawaz Ahmed,* Mahbubul Hasan,
Md. Enamul Haque Khondker, A.K.M. Moniruzzaman,
Bangladesh University of Engineering and Technology, Dhaka, Bangladesh
Azhar B. Khairuddin
Universiti Teknologi Malaysia, Skudai, Johor Bahru, Malaysia

ABSTRACT

A PV systems is by nature a non linear power sources which delivers the maximum power output with the highest efficiency only for a particular load current and voltage termed the maximum power point (MPP). This point depends on the PV array's current-voltage (I-V) characteristics that in its turn varies instant to instant mainly due to random changes in insolation (incoming solar radiation). Moreover, the load connected to the PV array also changes as per consumers' needs and does not correspond to the MPP. A prior knowledge of the MPP will be useful for operating a PV array at that point, irrespective of the connected load magnitude, through on-line control of the firing angle or any other controllable parameter of the power electronic interfaces (e.g. inverter or chopper) between the load and the PV array. MPP information will also be useful for a grid connected PV array for the purpose of load dispatching i.e. buying deficit power from the grid or selling surplus power to the grid. Hence there is a need for prior determination of the maximum power point in real time i.e. within a short interval of time.

It appears from a review of the literature that mainly different variants of artificial neural network (ANN) have been applied to find the MPP of a PV array. However, the methods suffer from one or more of the drawbacks such as requiring sophistication in the choice of various parameters which influence the success of ANN, use of difficult-to-measure

* Corresponding author's e-mail: ssahmed@eee.buet.ac.bd; Fax:+880-2-8613046

input variables, comparison of the output with those obtained through conventional modeling based calculations but not with the measured values, and prediction of MPP on hourly basis whereas solar insolation changes every few minutes.
In this chapter an algorithm based on autoregressive (AR) third order time series has been proposed to forecast a photovoltaic (PV) array's short circuit current and open circuit voltage at a given instant from directly and easily measurable load currents and voltages in shortly spaced three preceding intervals of time. The maximum power point (MPP) of the PV array's current-voltage characteristic curve corresponding to that instant can then be obtained from the forecasted short circuit current and open circuit voltage. This chapter has emphasized on laying the theoretical foundation, achieving preliminary results, their comparison with measured values and identifying further investigations needed for the proposed real-time application of time series which has so far been used mainly for forecasting long term events. The methodology presented in this chapter does not lack in generality and can easily be extended to include refinements. It is applicable for a PV array installed at any geographical location either in stand-alone or in grid connected mode.

INTRODUCTION

Photovoltaic (PV) electricity is one of the major renewable energy sources and requires only the first (capital) cost while zero fuel cost and very little maintenance. With continued research on the line of solar cell materials, power conditioning devices (DC to AC conversion) and battery technology, the initial expenses of installing a PV array is expected to be affordable by the common people in near future. Its application horizon [1] is now not only confined within small-scale and stand-alone use by the isolated and off-grid rural or coastal communities rather it is also making inroad to the grid connected urban people. Under the emerging competitive market environment or deregulation [2] the non-utility generation (NUG) entities have started operating PV plants which are interfaced with the grid systems as distributed or embedded generation sources.

A PV systems is by nature a non linear power sources which delivers the maximum power output with the highest efficiency only for a particular load current and voltage termed the maximum power point (MPP). This point depends on the PV array's current-voltage (I-V) characteristics which in its turn varies instant to instant mainly due to random changes in insolation (sunshine–intensity). Moreover, the load connected to the PV array also changes as per consumers' needs and does not correspond to the MPP. A prior knowledge of the MPP will be useful for operating a PV array at this point, irrespective of the connected load magnitude, through on-line control [3] of the firing angle or any other controllable parameter of the power electronic interfaces (e.g. inverter or chopper) between the load and the PV array. MPP information will also be useful for a grid connected [4] PV array for the purpose of load dispatching i.e. buying deficit power from the grid or selling surplus power to the grid. Hence there is a need for predetermination of the maximum power point in real time i.e. within a short interval of time.

It appears from a review of the literature that mainly different variants of artificial neural network (ANN) have been applied [5,6] to find the MPP of a PV array. However, the methods suffer from one or more of the drawbacks such as requiring sophistication in the choice of various parameters which influence the success of ANN, use of difficult-to-measure input variables, comparison of the output with those obtained through conventional modeling based calculations but not with the measured values, and prediction of MPP on hourly basis whereas solar insolation changes every few minutes.

This chapter proposes determining the MPP at an instant from the immediate past basic data using a time series e.g. a third order Autoregressive (AR) model [7]. The basic data for the time series are the directly measurable load currents and voltages in closely spaced three preceding intervals of time. This work has considered this interval as 10 minutes. Of course this interval can even be made less. The load currents and voltages will be transformed into corresponding interval's open circuit voltage (V_{oc}) and short circuit current (I_{sc}). Then the next interval's I_{sc} and V_{oc} will be forecasted from these transformed ones. The forecasted I_{sc} and V_{oc} will then be used to compute the PV array's maximum power output in that interval.

It is noteworthy that AR model has widely been applied for other long term events such as forecasting the hourly, weekly, monthly or annual demand [8] of a power utility. But the potentials of AR or any time series have not yet been investigated into for forecasting the real-time events like determining MPP. Therefore, this paper has presented the theoretical basis, reported the results of a preliminary analysis and compared those with the physical data collected from a live PV system in stand alone mode, and discussed further refinements needed for the proposed forecasting algorithm which is perhaps the first of its kind.

Maximum Power of PV Array

An elementary solar cell is basically a p-n junction diode usually made of semiconductors like silicon or gallium arsenide. To have an appreciable current output and a practical open circuit voltage, large number of elementary cells are assembled in series-parallel blocks and each such assembly is called a photovoltaic panel. Several panels are electrically connected in parallel to make a PV array.

The refined, simplified and final [3] equivalent circuit model for a single PV cell is shown in Figure 1.

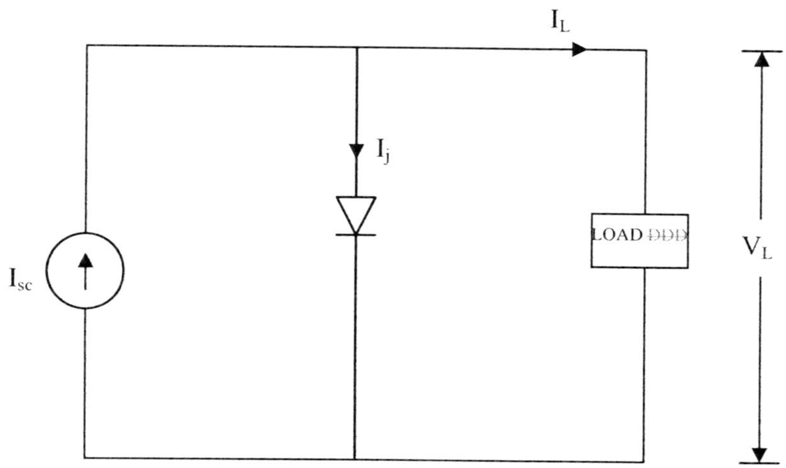

Figure 1: Equivalent circuit of a PV cell

The equation governing the relationship of I_L (the load current) with I_{sc} (the light generated diode current without bias i.e. short cirtcuit) and I_j (the diode current with bias in the dark i.e. absence of light) is given as follows.

$$I_L = I_{sc} - I_j = I_{sc} - I_o\left(e^{\frac{qV_L}{\gamma kT}} - 1\right) \quad (1)$$

where,

I_j is an exponential function of the diode reverse saturation current I_o depending upon the load voltage V_L

q = charge of an electron = 1.5992×10^{-19} coulombs
γ = a factor having a value usually between 1.0 and 3.0.
k = Boltzman's constant = 1.38066×10^{-23} Joules/°K
T = (ambient temperature, t_a°C +273) °K

Eqn. 1 leads to an I-V characteristics of an elementary PV cell as shown in Figure 2. The characteristic curve of a complete PV array is similar in shape to that in Figure 2 with the only exception that the complete array current and voltages are just the current and voltage of a single cell multiplied by appropriate factors considering respectively the number of parallel and series paths in the array of panels.

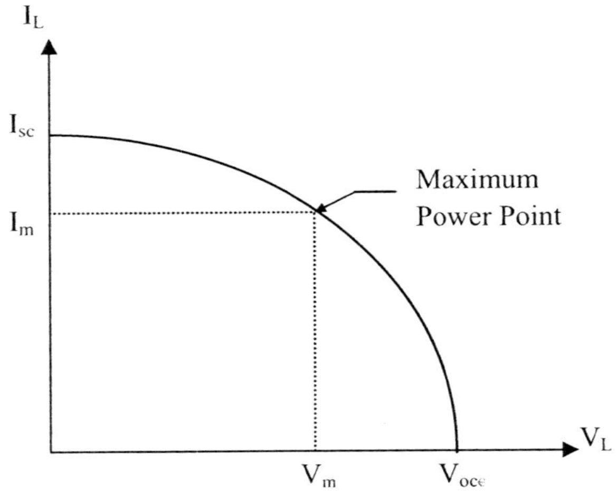

Figure 2: I-V characteristics of a PV cell

The load voltage at the maximum power point denoted by V_m (Figure 2) can be obtained by differentiating with respect to V_L the expression of power output i.e. the product of I_L (eqn.1) and V_L and then setting the derivative to zero. This results in a final expression for V_m as follows:

$$e^{K_1V_m}(1 + K_1V_m) = e^{K_1V_{oc}} \quad (2)$$

where,

$K_1 = q / (\gamma k T)$ (3)

Once V_m is obtained from eqn. 2 it can be substituted into eqn.1 to get the corresponding I_L i.e. I_m as follows.

$$I_m = I_{sc} - \left(\frac{I_{sc}}{e^{K_1 V_{oc}} - 1}\right)(e^{K_1 V_m} - 1) \quad (4)$$

It should be noted that the quantity V_{oc} termed open circuit voltage and used in eqns. 2 and 4 is expressed as follows.

$$V_{oc} = \frac{1}{K_1} \ln\left(\frac{I_{sc}}{I_o} + 1\right) \quad (5)$$

The above expression of V_{oc} can be found by substituting $I_L = 0$ into eqn.1 and denoting the corresponding V_L as V_{oc}.

Eqns. 2 and 4 show that at a given instant the maximum power which is the product of V_m and I_m can only be known if the corresponding instant's short circuit current I_{sc} and the open circuit voltage V_{oc} are available.

Proposed Methodology

The method proposed in this chapter for determining the MPP of a PV array consists of the following steps:

a) Transformation of load currents and load voltages measured in three intervals into corresponding instant's short circuit currents and open circuit voltages.
b) Forecasting the I_{sc} and V_{oc} for the required instant from those computed for the three immediate past intervals using a third order AR time series.
c) Identifying the MPP by finding V_m and I_m from forecasted I_{sc} and V_{oc}.

Transformation of Load Currents and Load Voltages

The load to be served by a stand-alone or a grid connected PV array is always variable and changes from instant to instant by a magnitude whatever small it may be. This aspect has been utilised in the proposed method to avoid measurement of I_{sc} and V_{oc} requiring disconnecting the PV array from load and instead, derive those in every interval in terms of two pairs of load currents and voltages which are very easy-to-measure. The values of I_{sc} and V_{oc} derived so in the last three intervals will be kept stored for use by a 3rd order AR model to forecast, well in advance, the I_{sc} and V_{oc} of the immediately succeeding interval.

If at a given instant two pairs of load currents and voltages respectively I_1, V_1 and I_2, V_2 are recorded in quick succession so that the same I_{sc} can be considered for both pairs, then it follows from eqns.1 and 3 that

$$I_{sc} = I_1 + I_0(e^{K_1V_1} - 1) = I_2 + I_0(e^{K_1V_2} - 1) \quad (6)$$

It follows from eqn.6 that

$$I_0 = \frac{(I_1 - I_2)}{(e^{K_1V_2} - e^{K_1V_1})} \quad (7)$$

Using eqn.7 and substituting $I_L = I_1$ or I_2 and $V_L = V_1$ or V_2 back into eqn.1 provides I_{sc} of an interval in terms of the corresponding two pairs of measured values of load currents and voltages, as expressed below.

$$I_{sc} = I_1 + \left(\frac{I_1 - I_2}{e^{K_1V_2} - e^{K_1V_1}}\right)(e^{K_1V_1} - 1) \quad (8)$$

Substituting I_{sc} from eqn.8 and I_0 from eqn.7 into eqn. 5 will provide the corresponding's instant V_{oc} in terms of the known pairs of load currents and voltages.

Forecasting the I_{sc} and V_{oc} Using AR Model

To forecast the short circuit current for any instant of time t from the short circuit currents (computed using eqn. 8) for the three immediately preceding and equally spaced instants (t-1), (t-2) and (t-3), a third order (p=3) AR model [7] can be applied as follows.

$$\hat{I}_{sc}(t) = \bar{I}_{sc} + \sum_{i=1}^{p=3}[\alpha_{p,i}\, y(t-i)] + w(t) \quad (9)$$

where,

$\bar{I}_{sc}$ is the average of the short circuit currents obtained for the three preceding intervals.

$\alpha_{p,i}$ are the AR model parameters determined for each forecast instant (t) using successively the eqns. 14 to 18 shown in the Appendix.

$$y(t-i) = \bar{I}_{sc} - I_{sc}(t-i) \quad (10)$$

w(t) is white noise represented by normally distributed random variables with zero mean and a variance σ^2_w.

Similarly the open circuit voltage for instant t is forecasted using the following equation from those computed for three preceding intervals.

$$\hat{V}_{oc}(t) = \overline{V}_{oc} + \sum_{i=1}^{p=3}[\alpha_{vp,i} y_v(t-i)] + w_v(t) \tag{11}$$

where α_v, y_v, w_v have significance respectively similar to those of α, y and w in eqn.9 and are determined similarly.

Identifying MPP

The open circuit voltage forecasted for the instant 't' using eqn.11 is substituted into eqn.2 and the natural logarithm is taken for both sides so that following equation results.

$$K_1 V_m + \ln(1 + K_1 V_m) - K_1 \hat{V}_{oc} = 0 \tag{12}$$

Eqn.12 is solved for V_m , the voltage at the MPP for the instant 't', using any numerical method e.g. Newton-Raphson method in the present work.

The open circuit voltage forecasted for the instant 't' using eqn.11, the short circuit current forecasted for instant 't' using eqn.9 and V_m obtained from eqn.12 are substituted into eqn.4 to find I_m , the current at the MPP for the instant 't'.

The maximum power at instant 't' is then

$$P_m = V_m I_m \tag{13}$$

RESULTS AND DISCUSSION

The various steps of the proposed method have been coded in FORTRAN 77 and run on a 225 MHz PC. The method has been tested using some readily available data sets on load currents and voltages which were recorded in different days from a stand-alone PV array. The PV array consisted of three panels in parallel each of which was designed for a maximum output of 43 watts subject to the availability of the correspondingly required insolation. Each panel again consisted of 30 blocks in series and each block comprises 53 cells in parallel so that the current through a single cell is the total PV array current divided by (3x53=) 159 and the voltage across a single cell is the array voltage divided by 30. The PV array was connected to a variable load.

Forecasted vs. Actual I_{sc} and V_{oc}

Table1 shows, for some typical cases, 10 minutes ahead forecasts on the short circuit currents and open circuit voltages for an interval of time from the recorded load currents and voltages of three 10 minutes spaced preceding intervals.

Table 1: Forecasted vs. actual short circuit current and open circuit voltages of the PV array

Days	Time of interval	I_1 amps	I_2 amps	V_1 volts	V_2 volts	Ambient temp. t_a°C	Forecasted $\hat{I}_{sc}$ amps	Actual I_{sc} amps	Forecasted $\hat{V}_{oc}$ Volts	Actual V_{oc} volts
	08.50	2.10	2.00	11.8	12.0	19.0	for 09.20: 3.16	at 09.20: 3.54	for 09.20: 14.16	at 09.20: 16.1
Day 1	09.00	1.88	1.85	12.6	12.7	20.5				
	09.10	1.78	1.70	12.8	13.0	21				
	09.40	1.44	1.40	13.9	14.0	22	for 10.10: 4.6	at 10.10: 4.3	for 10.10: 14.83	at 10.10: 16.0
Day 2	09.50	1.35	1.30	14.0	14.05	22.5				
	10.00	1.27	1.25	14.08	14.1	23				
	11.30	0.29	0.28	15.0	15.1	28.3	for 12.00: 1.75	at 12.00: 2.01	for 12.00: 13.31	at 12.00: 15.1
Day 3	11.40	5.25	4.91	5.0	9.0	28.2				
	11.50	1.14	1.04	13.7	14.0	28				

The forecasted values have been compared with the actual values of short circuit current and open circuit voltage at the corresponding instant. These actual values were measured (for the purpose of comparison) respectively by short circuiting and open circuiting the PV array just after every recording of the load currents and voltages.

It should be noted in Table 1 that the two pairs of load currents and voltages I_1, V_1 and I_2, V_2 shown in the same interval were recorded just at 1 minute's gap.

It is evident that the forecasted and actual short circuit current and open circuit voltages for the PV array are close to each other. Even the proposed method was able to accommodate interval to interval erratic variation in the input i.e. load currents and voltages as may be seen in the third one (day 3) of the typical cases reported in Table 1.

Maximum Power Using Forecasted and Actual Values of I_{sc} and V_{oc}

The load voltage and current at the MPP point, respectively V_m and I_m, were computed in the way mentioned in Section "Identifying MPP' using the forecasted as well as actual I_{sc} and V_{oc} for the same interval of time. Then the maximum power output of the PV array, P_m was computed using those V_m and I_m. Table 2 shows a comparison of the maximum power outputs corresponding to the typical cases reported in Table 1.

It should be noted that in the Newton-Raphson solution of eqn. 12 for each V_m , the starting value has been chosen as the corresponding open circuit voltage (forecasted or actual as the case may be) and for a tolerance margin of 0.001 volts the convergence was achieved in only 3 iterations for each case.

The ambient temperature (t_a°C) shown in Table 2 at an interval for which P_m is being calculated, was considered as equal to the last one among the temperatures recorded for the preceding three intervals. This is because the temperature did not vary significantly over the closely spaced (10 minutes) consecutive three intervals. However, during actual implementation of the algorithm, if needed, the temperature for the forecast interval can always be approximated using some simple extrapolation technique.

It is evident from Table 2 that the two values of P_m computed using forecasted and actual short circuit current and open circuit voltages for the PV array are close to each other.

Table 2: Maximum power using forecasted vs. actual short circuit current and open circuit voltages of the PV array

Days and interval of time	Ambient temp. t_a°C	P_m using forecasted I_{sc} and V_{oc}					P_m using actual I_{sc} and V_{oc}				
		Fore-casted $\hat{I}_{sc}$ amps	Fore-casted $\hat{V}_{oc}$ volts	V_m volts	I_m amps	Max. power P_m watts	Actual I_{sc} amps	Actual Voc volts	V_m volts	I_m amps	Max. power P_m watts
Day 1 09.20	21.0	3.16	14.16	10.67	2.70	28.83	3.54	16.1	12.37	3.09	38.22
Day 2 10.10	23.0	4.6	14.83	11.24	3.96	44.51	4.3	16.0	12.27	3.74	45.96
Day 3 12.00	28.0	1.75	13.31	9.89	1.48	14.59	2.01	15.1	11.49	1.73	19.81

Computer Time Requirement

In each of the cases investigated by the present work, the proposed algorithm required on average only about 0.5 seconds for all the three steps on a 225 MHz PC. The requirement of such a less time stems mainly from simple transformation equations and use of a short order (i.e. 3rd) AR model.

Scope for Further Improvement

The difference of the forecasts from the actual ones can further be reduced by adopting the following two means.

i) The results shown in Table 1 were achieved without using the random noise w(t) or $w_v(t)$ of the AR model (eqns. 9 and 11). As regards a PV system, this random noise is important because it takes into account a lot of uncertainties which affect insolation and may vary even minute to minute e.g. cloud coverage of the sky, moisture and dust content in the surrounding atmosphere. However, these two noise terms' standard deviations can be determined off-line from a long database on short circuit currents and open circuit voltages computed from recorded load current and voltages. Once the standard deviations are known a random number generator will add the noise terms on-line with the respective AR model (eqns.9 and 11) at every instant of forecast.

ii) The factor $K_l = q / (\gamma\ k\ T)$ used in various equations requires proper choice of the value for γ parameter of a PV cell. In the absence of manufacturer- supplied information, γ had to be determined by the present work. It was computed in the way shown in the Appendix for a few sets of measurements on short circuit current, load current and load voltage of the used PV array. From these the minimum one found to be 2.3807 was chosen for γ. Indeed, for accuracy γ is to be determined by the best curve fitting method from a large number of data sets on load currents and voltages. However, the method followed in the present work and also the method of curve fitting for γ, both need just off-line implementation.

The above two points though not investigated in the time frame set for the present investigation will be addressed in future investigations.

CONCLUSION

The widely used time series is still in its infancy so far as determining the MPP of a photovoltaic array is concerned. In exploring the prospects of the time series, a third order autoregressive model based algorithm has been proposed. This algorithm finds the MPP at an instant from the directly measurable load currents and voltages of the PV array in shortly spaced three preceding intervals of time.

The proposed algorithm has been presented in details. To have a preliminary assessment, it has been tested on several available sets of real data for a live PV system. The obtained results have been compared with the actually measured values and found very much promising.

It should be noted that in actual implementation of the proposed algorithm, load currents and voltages will be recorded through on-line data acquisition system. Then the three steps viz. transformation, forecasting and MPP identification will be done on-line. However, the on-line time needed then would be much trivial as confirmed by the requirement of only 0.5 seconds in each case on a 225 MHz PC in the present investigation.

However, further time saving could have been made by avoiding use and transformation of load current and voltages after first few intervals and instead, forecasting the I_{sc} and V_{oc} for an interval from those forecasted for the three immediately preceding intervals. But this may cost the accuracy to some extent. This is why the present work has used measured load currents and voltages to make the forecasting as realistic as possible.

The proposed method does not lack in generality and can easily be applied for a stand-alone or grid-interfaced PV array in any location. The suggested improvements can form the basis of an elaborate research and further investigations.

APPENDICES

Equations Successively Used to Determine AR Model Parameters α (in eqn.9) and α_v (in eqn.11) at the Instant 't'

$$C_k = \frac{1}{p}\sum_{i=1}^{p-k}[y(t-i)-\bar{y}][y(t-i-k)-\bar{y}];\ p=3, k=0,1,..p \tag{14}$$

where y (t-i) or y(t-i-k) are as in eqn. 10 and $\bar{y}$ is the average of y(t-i), i=1,2....p.

$$r(j) = \frac{C_j}{C_0};\ j = 1,....p \tag{15}$$

$$\alpha_{1,1} = r(1) \tag{16}$$

$$\alpha_{l+1,l+1} = \frac{r(l+1) - \sum_{j=1}^{l} \alpha_{l,j} r(l+1-j)}{1 - \sum_{j=1}^{l} \alpha_{l,j} r(j)};\ l = 1,...p-1 \tag{17}$$

$$\alpha_{l+1,j} = \alpha_{l,j} - \alpha_{l+1,l+1} \times \alpha_{l,l+1-j};\ l = 1,...p-1, j = 1,...l \tag{18}$$

The AR model parameters α_v for open circuit voltage forecast are determined using a set of equations similar to eqns. 14 through 18 and using y_v instead of y.

Determination of γ in the Present Work

From eqn.6 it can be written that

$$I_0 = \frac{(I_{sc} - I_1)}{(e^{\frac{qV_1}{\gamma kT}} - 1)} = \frac{(I_{sc} - I_2)}{(e^{\frac{qV_2}{\gamma kT}} - 1)} \tag{19}$$

It can be rearranged as

$$(I_{sc} - I_1)e^{\frac{qV_2}{\gamma kT}} - (I_{sc} - I_2)e^{\frac{qV_1}{\gamma kT}} = I_2 - I_1 \tag{20}$$

Replacing (I_{sc}-I_1) by A, (I_{sc}-I_2) by B, (I_2-I_1) by C and (q/kT) by K_0 it can be written that

$$Ae^{\frac{K_0V_2}{\gamma}} - Be^{\frac{K_0V_1}{\gamma}} - C = 0 \tag{21}$$

Using a measured value of I_{sc} and two pairs of load currents and voltages (I_1,V_1) and (I_2,V_2) recorded in quick succession, and downscaling them for a single cell, eqn.21 can be solved for γ through Newton-Raphson technique.

REFERENCES

[1] Benner JP, Kazmerski L. Photovoltaics gaining greater visibility. *IEEE Spectrum* 1999; 36(9): 34-42.

[2] Ilic M, Galiana F, Fink L. *Power systems restructuring: engineering and economics.* MA, USA: Kluwer Academic Publishers, 1998.

[3] Hu C, White RM. *Solar cells from basic to advanced systems.* New York: McGraw-Hill Book Company, 1983.

[4] Rahman S, Bouzguenda M. A model to determine the degree of penetration and energy cost of large scale utility interactive photovoltaic systems. *IEEE Trans. Energy Conv.*1994; 9(2): 224-230.

[5] Hiyama T, Kouzuma S, Imakubo T. Identification of optimal operating point of PV modules using neural network for real-time maximum power tracking control. *IEEE Trans. Energy Conv.*1995; 10 (2): 360-367.

[6] Al-amoudi A, Zhang L. Application of radial basis function networks for solar- array modelling and maximum power- point prediction. *IEE Proc., Gener. Transm. Distrib.* 2000; 147(5): 310-316.

[7] Box GEP, Jenkins GM. *Time series analysis - forecasting and control.* San Francisco: Holden Day, 1976.

[8] Galiana FD, Handschin E, Fiechter AR. Identification of stochastic electric load models from physical data. *IEEE Trans. Automatic Control* 1974; 19(6): 887-893.

In: Solar Energy: New Research
Editor: Tom P. Hough, pp.321-355

ISBN 1-59454-630-4

Chapter 11

CHARACTERIZATION OF SOLAR CELLS AND MATERIALS FOR THE CELLS BY SURFACE PHOTOVOLTAGE METHOD

***J. Toušková*[1], *J. Toušek*[1] *and D. Kindl*[2]**

[1]Faculty of Mathematics and Physics, Charles University Prague, Prague, Czech Republic

[2] Institute of Physics, Academy of Sciences of the Czech Republic, Prague, Czech Republic

ABSTRACT

The surface photovoltage method is described and its use for evaluation of some parameters of semiconductor materials and diodes is shown. Theory of the photovoltaic effect is presented first in a commonly used form suitable for thick samples. At present, advanced technology produces thin films and thin wafers, which is the reason for our generalization of the theory for arbitrary thick samples. Relations for the photocurrent and the open - circuit voltage are derived including the effect of reflections from the sample surfaces, surface recombination and thickness of the samples.

The surface photovoltage technique is modified for contactless measurements as well as for measurements with an auxiliary contact. The advantage of the latter method of measurement is that only a small number of parameters influence the photovoltage spectra. This allows reliable extraction of parameters by comparison of the experimental data with the theory. The main purpose is evaluation of diffusion length of minority carriers and surface recombination velocity in materials designated for solar cells and in the bulk of the absorber in the photocells. We have employed the method for this aim in silicon wafers, c - Si cells, microcrystalline Si:H layers, thin film CdS/CdTe cells and GaSb diodes.

The spectra are much more complex if radiative recombination plays an important role. A simple theory of photon recycling is derived and applied to the calculation of the photovoltage spectra of GaSb Schottky contacts and p-n junctions giving possibility to obtain parameters of the diodes.

The diffusion length may be influenced by effect of the sample thickness and by the properties of its surface. The diffusion length becomes an effective quantity L_{eff}. Analytical form of L_{eff} versus the sample thickness is presented.

INTRODUCTION

The surface photovoltage (SPV) method is a well-established and wide-spread technique for characterization of semiconductors providing information on various materials and structures. During the second half of last century the SPV method was continuously developed and more sophisticated methods for data analysis also appeared. A good survey of the methods can be found e.g. in the comprehensive work of Kronik and Shapira [1]. In 1961 Goodman developed an algorithm for easy extraction of the minority carrier diffusion length from the SPV data [2]. SPV was later used to study a great variety of semiconductor surfaces [3].

Scanning techniques, like scanning tunneling microscopy or atomic force microscopy, were employed for obtaining SPV measurements with high resolution. Lagowski et.al. [4] have scanned the SPV across Si wafers to trace heavy metal contamination and surface defects. Metallic contamination introduced by Si wafer handling has been mapped by means of minority carrier diffusion length measurement using the SPV [5]. A great number of works based on the SPV was directed to silicon, because of undying interest in this material. Nevertheless, the potential of this method has also been used for studying Ge, GaAs, $Cu(In,Ga)Se_2$, CdTe, or GaSb materials and structures as heterojunctions and solar cells.

The surface photovoltage method is often called "contactless method" because only capacitive probe "contacts" are used in the measurement. This is held as an advantage over other methods for measuring diffusion length, which require conducting contacts to the sample [2]. We have used the "contactless" (CPV) method for characterization of semiconducting materials and solar cells in some cases, but original electrode was employed if the measured sample was provided with a more complex contact system. Even a new extra electrode was deposited on the bulk of the sample to exclude the contribution of surface region to the photovoltage signal in special cases as this will be shown in Section 3.1.

The SPV signal is always generated in bulk and in one or more space charge regions (SCRs) of the sample, which means that its spectral dependence is a very complex function of many parameters. Illuminating of solar cells from the back-side and using auxiliary electrodes to its bulk we have practically excluded contribution from the top SCR, which greatly simplifies evaluation of the diffusion length. Similar arrangement has also been applied to semiconductor wafers. Such modification is very important in the case of thin samples where interference of both surfaces can be removed by this method.

If the SPV signal is taken from the bulk and both SCRs, the photovoltage is a sum of contributions from all these regions. The contribution from any SCR depends on photocarrier recombination, which strongly influences the measured spectra. This recombination is here introduced by a "gain factor" characterizing contributions of each photocarrier to the external current.

We have used the SPV method mainly for the minority carrier diffusion length and, in some cases, also for the surface recombination velocity evaluation. The SPV technique is introduced in Section 1. In Section 2, calculations of the photovoltage for various types of solar cells and various direction of illumination are carried out. In the case that radiative recombination plays an important role, the effect of photon recycling should be included. Simple theory leading to calculation of the photocurrent in such case is presented. Application of the theoretical calculations to experimental results is shown in Section 3. We have concentrated our attention to silicon wafers and solar cells, CdS/CdTe thin film heterojunctions and GaSb diodes.

The brief conclusion summarizes the theoretical and experimental results.

Surface Photovoltage Technique as a Method for Diffusion Length Evaluation

The surface photovoltage method is a nondestructive steady-state technique, the principles of which can be found e.g. in [2, 6, 7].

The sample to be measured is usually in the form of a wafer, but more complex structures as the solar cells can also be studied. Surface of the sample is chemically treated to induce a surface space-charge region, which is a consequence of charges trapped at the surface. Let us assume the width of the SCR at illuminated top surface to be w_1 and that at the bottom surface w_2. Chopped monochromatic light of energy is higher than the band-gap is used. During the measurement the wavelength is varied and absorbed photons generate electron-hole (e-h) pairs. The sample is denoted to be "thick" if no charge carriers are generated at the back surface or reach this surface. In thick sample photovoltage is created at the top surface only and results from the separation of electron-hole pairs generated in the SCR and those diffusing to SCR-bulk boundary from the bulk. For the case that absorption coefficient α is sufficiently low for $\alpha w_1 << 1$ and recombination in the top SCR is negligible the following relation for the photovoltage V_{ph} was derived [2]:

$$V_{ph} = C_1 \, I \alpha L/(1 + \alpha L) \tag{1}$$

where L is the minority carrier diffusion length, I is the entering photon flux density. C_1 is a constant under assumption that the reflectivity from the top surface is constant over a restricted wavelength range, front surface recombination velocity, diffusion length and diffusion coefficient can be considered constant as well.

The photon flux density is adjusted for each wavelength to hold V_{ph} constant. Then Eq. (1) can be written as

$$I = C_2 \, (L + 1/\alpha) \tag{2}$$

where C_2 is a constant.

Plotting I against $1/\alpha$ a line is obtained which after extrapolation intercepts $1/\alpha$ axes in the point – L. The width of the SCR is not often known. In the case of thick SCR the evaluation of diffusion length according to relation (2) fails as is illustrated in Figure 1. The intercept gives little reliability to the value of L because only a small number of measured points fulfills the needed linear dependence.

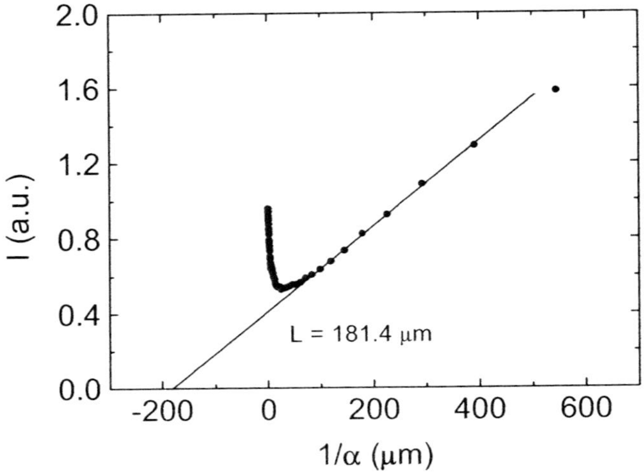

Figure 1. Illustration of simple evaluation of diffusion length L by SPV method.

The mentioned theory of SPV is not even well adaptable to samples with low thickness and wide SCR. Despite these limitations, the SPV is used as a test method and therefore it has been included for example in American Standard Test Methods [8]. An attempt to bring a more complete SPV theory has been made by Moore [6] who considered contribution from the SCR. Chiang and Wagner [9] presented an analysis for transport of minority carriers in a surface barrier under illumination and derived an analytical expression for the SPV. No one from the cited papers took into account the influence of a back SCR.

Our method calculates with two SCRs located at the top and the back surface. No constraint is imposed on the thickness of the bulk of the absorber or on the value of the generated photovoltage (small signal regime is not needed, but light intensity must not influence the minority carrier's lifetime). The theory applied to solar cells is derived in the following section. In direct semiconductors radiative recombination represents an important recombination mechanism. This fact leads to a photon recycling process which will be analyzed in section 2.5. Calculations of SPV in sections 2.1 to 2.4 do not include the photon recycling process.

Surface Photovoltage on Solar Cells

Contributions from the SCRs are Neglected

We have studied cells with p^+pn^+ structure. Such structure is usual in monocrystalline Si cells. After removing the ohmic electrode and the p^+ region, the sample was illuminated from the side of the absorber (Figure 2). This rear illumination of cells with only one SCR has been already used in Ref. [10]. Two space charge regions are assumed in the sample: the pn^+ junction of the cell and spontaneously created SCR on the free surface. Their thicknesses (w_2, w_1) and the width of the n^+ region (d_2) are negligibly small in comparison with the bulk of the absorber (d_1). Low absorption in w_1, w_2 and d_2 regions is assumed ($\alpha w_1 \ll 1$, $\alpha w_2 \ll 1$ and $\alpha d_2 \ll 1$). Electron-hole pairs generated in the bulk diffuse towards SCR1 and SCR2. At the boundaries they are separated by the field of the barriers establishing photovoltage.

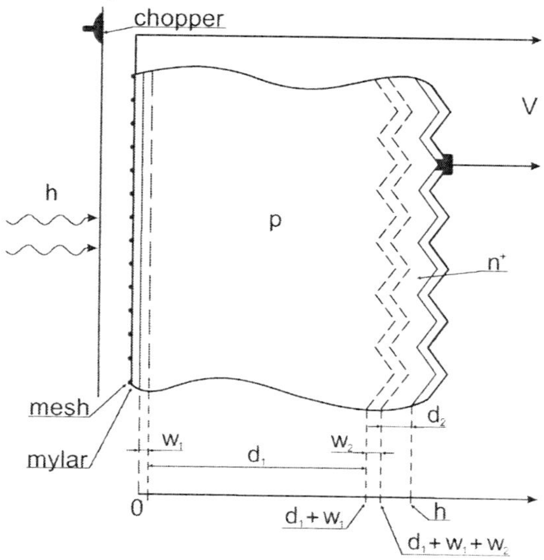

Figure 2. Structure of the sample and schematic arrangement of experiment, d_1 (d_2) is the thickness of bulk in p (n^+) region, w_1,w_2 are the thicknesses of the spase charge regions.

Its value at the SCR1 is:

$$V_{oc1} = (n_1 k_B T/e) \ln(1+J_{ph1}/J_{01}) \quad (3)$$

and at the SCR2 :

$$V_{oc2} = - (n_2 k_B T/e) \ln(1-J_{ph2}/J_{02}) \quad (4)$$

where symbols k_B,T and e have their usual meaning, n_1 and n_2 are the quality factors, J_{01} and J_{02} are the saturated current densities depending on the junction transport mechanism. Neglecting contributions from both SCRs as well as from the n^+ region the photogenerated current densities J_{ph1} and J_{ph2} can be calculated by solving the diffusion equation

$$Dd^2\Delta n(x)/dx^2 - \Delta n(x)/\tau = -g(x). \quad (5)$$

The $\Delta n(x)$ is the excess electron concentration at a depth x under the illuminated surface, D is the electron diffusion constant, τ is the lifetime of electrons. In the case of multiple reflection, the photogeneration g(x) may be expressed as

$$g(x) = \alpha(1-R_1)I_0\{\exp(-\alpha x) + R_2\exp[-\alpha (2h-x)]\}/[1-R_1R_2 \exp(-2\alpha h)], \quad (6)$$

where I_0 is the incident photon flux density, h represents the whole thickness of the cell, R_1, R_2 are the reflectances on the top and the bottom surfaces, respectively.

To find the photogenerated current two boundary conditions are required:

$$\Delta n(w_1) = 0 \quad (7)$$

$$\Delta n(w_1+d_1)=0 \quad (8)$$

The diffusion current density from the bulk at the boundary with the SCR1 is

$$J_{ph1} = eDd\Delta n/dx\,|_{x=w1} =$$

$$a_1 \exp(-\alpha w_1)\,\alpha L/(1-\alpha^2L^2)\,\{[\cosh(d_1/L)-\exp(-\alpha d_1)]/\sinh(d_1/L)-\alpha L\}$$

$$-a_2 \exp[-\alpha(w_2+d_2)]\alpha L/(1-\alpha^2L^2)\,\{[1-\cosh(d_1/L)\exp(-\alpha d_1)]/\sinh(d_1/L)-\alpha L\exp(-\alpha d_1)\} \quad (9)$$

where $a_1 \equiv eI_0\,(1-R_1)/(1-R_1R_2\exp(-2\alpha h))$

and $a_2 \equiv eI_0\,(1-R_1)R_2\exp(-\alpha h)\,/(1-R_1R_2\exp(-2\alpha h))$.

The diffusion current density at the SCR2 boundary can be calculated in the same way. Then

$$J_{ph2} = eDd\Delta n/dx\,|_{x=(w1+d1)} =$$

$$= a_1\exp(-\alpha w_1)\alpha L/(1-\alpha^2L^2)\,\{[1-\cosh(d_1/L)\exp(-\alpha d_1)]/\sinh(d_1/L) - \alpha L\exp(-\alpha d_1)\}$$

$$- a_2\exp[-\alpha\,(w_2+d_2)]\,\alpha L/(1-\alpha^2L^2)\,\{[\cosh(d_1/L)-\exp(-\alpha d_1)]/\sinh(d_1/L)-\alpha L\}. \quad (10)$$

The resulting voltage $V_{oc} = V_{oc1}-V_{oc2}$ can be written in the form

$$V_{oc} = (k_BT/e)\{(n_1\ln(1+\gamma_1 eI_0/J_{01}) - n_2\ln(1-\gamma_2 eI_0/J_{02})\} \quad (11)$$

where $\gamma_1 = J_{ph1}/eI_0$ and $\gamma_2 = J_{ph2}/eI_0$

According to the above mentioned assumptions $w_1 \approx 0$, $w_2 \approx 0$ and $d_2 \approx 0$ we can substitute zero for these quantities in Eqs. (9) and (10). Figure 3 and Figure 4 show a set of theoretical curves according to Eq. (11) normalized to their maximum for different diffusion lengths with other parameters fitted as in the captions. Wavelength – absorption coefficient dependence in the range 0.7-1.1μm was taken from ASTM [8], the other values of absorption coefficient originated from Ref. [11]. The zero point of the curves divides each curve into two parts differing by the sign of the signal. The longer wavelength part represents contribution from pn^+ junction of the cell, the shorter wavelength one is generated at the SCR on the illuminated surface. Influence of the diffusion length on zero-point location and the shape of the curves is shown in Figure 3. Sensitivity to the change in L decreases with increasing ratio L/d_1 and also depends strongly on the relation between the ratios eI_0/J_{01} and eI_0/J_{02} which should be different ($eI_0/J_{01} << eI_0/J_{02}$ or $eI_0/J_{01} >> eI_0/J_{02}$). Only in such cases we are able to extract L by fitting the theoretical curve to experimental points. Change in relative contributions of top and bottom junctions with ratio $(eI_0/J_{01})/(eI_0/J_{02})$ is shown in Figure 4. Despite the fact that the bulk is more than four times thicker than the diffusion length L the signal from the back junction is essential. Also remarkable is the impact of eI_0/J_{01} ,eI_0/J_{02} values on the FWHM of both curves, which is a consequence of nonlinear light intensity dependence of the signal.

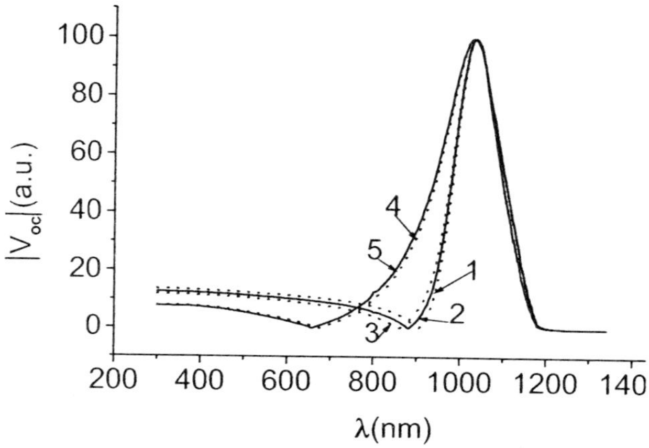

Figure 3. Theoretical dependence of $|V_{oc}|$ versus λ for several values of diffusion length L according to Eq.(11) (1) L = 70 μm, (2) L = 80 μm, (3) L = 90 μm (L«d_1 in these cases), (4) L = 365 μm and (5) L = 1000 μm (L»d_1), the other parameters: n_1 = 1.4, n_2 = 1.5, d_1 = 365μm, eI_0/J_{01} = 0.2, eI_0/J_{02} = 18, R_1 = 0.316 (the mean value of reflectance of free silicon surface in the region of 500-1200 nm), R_2 = 1.

The theoretical calculations and their results also hold for the case of wafers.

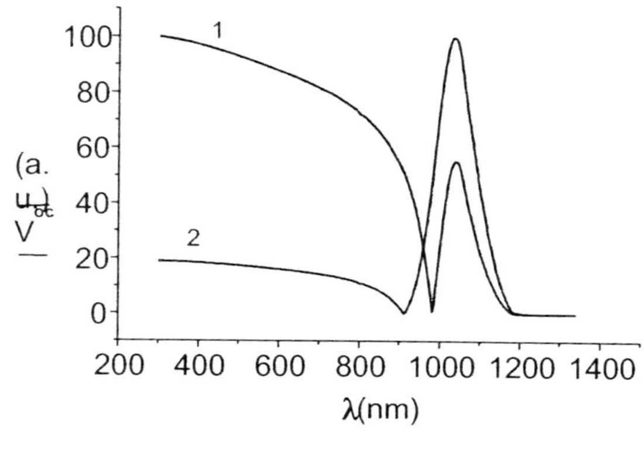

Figure 4. Influence of the ratio (eI_0/J_{01} / eI_0/J_{02}) on the signal $|V_{oc}|$ (1) eI_0/J_{02} = 1 and (2) eI_0/J_{02} = 10. The value eI_0/J_{01} is 0.2 in both cases. The diffusion length is 80μm and the other parameters are as in Figure 3.

Contributions from the SCRs are Assumed

In the case that generation of the carriers in the space charge regions cannot be neglected so that $\alpha w_1 << 1$ and $\alpha w_2 << 1$ do not hold, photocurrents from these regions must be included. The charge carriers generation rate g(x) is given by relation (6).

Let us assume recombination in the SCRs. Then every charge e generated by a photon in SCR1 acts in the photocurrent by a value eG_1 where $G_1 \leq 1$ is the gain factor. The generated photocurrent density is

$$J_{s1} = eG_1 \int_0^{w1} g(x)dx \tag{12}$$

The photocurrent density from the SCR2 is

$$J_{s2} = -eG_2 \int_{w1+d1}^{w1+d1+w2} g(x)dx \tag{13}$$

where the gain factor G_2 (≤ 1) accounts for recombination in the SCR2.

To calculate the total voltage V_{oc} (see Eq.(11)) the photocurrent densities J_{s1} and J_{s2} must be added to the photocurrents J_{ph1} and J_{ph2} from the bulk, respectively. At the first sight, the independence of J_b and J_s seems to be surprising. It means that recombination in the SCRs does not influence the value of J_b. Nevertheless, it is correct: bulk minority carriers enter into the SCRs and do not have an opposite sign partner for recombination.

Arrangement with one SCR on non-illuminated side of the Solar Cell Absorber

In the case that one surface region is enriched with majority carriers, no SCR is formed there. Influence of this surface is characterized by the surface recombination velocity. In the following calculation only SCR of the p-n junction is considered. The situation is illustrated in Figure 5 and corresponds to a heterojunction illuminated from the side of low band-gap p-type semiconductor-1 with quite large thickness. The main role of the SCR is to separate electron-hole pairs generated in the bulk. No contribution from the wide band-gap semiconductor-2 is assumed. The photovoltage is given by usual relation

$$V_{oc} = (nk_BT /e) \ln (1+J_{ph}/J_0), \tag{14}$$

In the following we will not consider multiple reflection so that the generation rate will be given by a more simple relation

$$g(x) = \alpha I_0(1-R)\exp(-\alpha x) \tag{15}$$

The photogenerated current density in the diode is derived as a sum of contributions from the bulk and from the SCR. The former is obtained from a diffusion equation (5) with the generation rate (15) and with boundary condition for the illuminated surface

$$D d\Delta n/dx |_{x=0} = s\Delta n(0). \quad (16)$$

Boundary condition between the bulk and the SCR is

$$\Delta n(d) = 0. \quad (17)$$

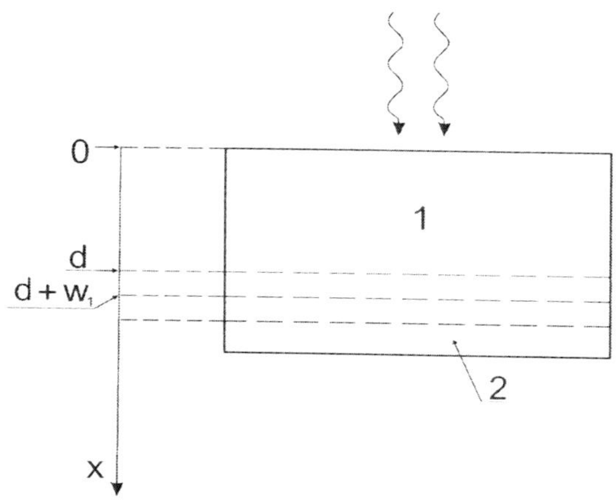

Figure 5. Scheme of the junction illuminated from the side of low band-gap semiconductor d is the thickness of its bulk, w_1 is the width of its SCR.

D is the minority carrier diffusion coefficient, s is the surface recombination velocity on the illuminated surface and d is the width of the bulk of the low band-gap semiconductor. The diffusion contribution from the bulk is

$$J_b = eD\, d\Delta n/dx |_{x=d}. \quad (18)$$

After solving of the equation (5) and substituting for Δn into (18), the current density J_b is derived as

$$J_b = eI\alpha L /(1-\alpha^2L^2) \{A \exp(-d/L) - B \exp(d/L) - \alpha L \exp(-\alpha d)\}, \quad (19)$$

where $A = \{(1-S) \exp(-\alpha d) + (S+\alpha L) \exp(d/L)\}/ \{(1+S) \exp(d/L) + (1-S) \exp(-d/L)\}$,

$B = \{(1+S) \exp(-\alpha d) - (S+\alpha L) \exp(-d/L)\}/ \{(1+S) \exp(d/L) + (1-S) \exp(-d/L)\}$ and

$S \equiv sL/D$.

The photocarriers generated in the SCR with the rate of generation g(x) produce the current density

$$J_s = -eG \int_{d}^{d+w1} g(x)dx = eGI_0(1\text{-}R)\exp(-\alpha d)[\exp(-\alpha w_1)\text{-}1], \quad (20)$$

The total photocurrent density J_{ph} in its absolute value is the sum

$$J_{ph} = | J_b + J_s | , \quad (21)$$

In the case of $J_{ph}/J_0 <<1$ $V_{oc} \sim J_{ph}$.

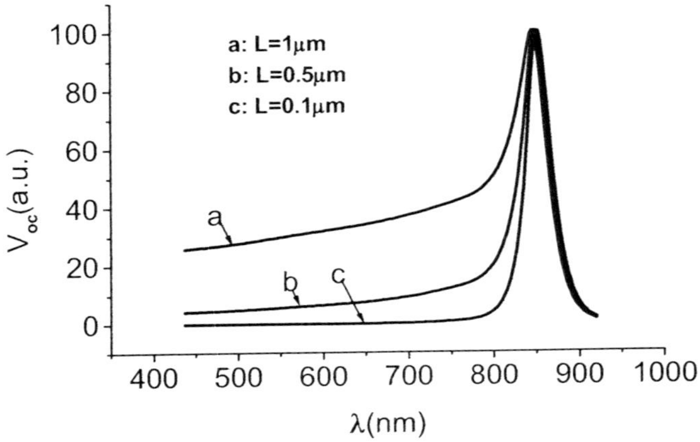

Figure 6. Photovoltage spectra demonstrating influence of diffusion length L on the theoretical dependence V_{oc} versus λ according to Eq. (14) for $w_1 = 0.5$ μm, S = 2, G = 0.7, and d = 2.5 μm in the case of low light intensity ($eI_0/J_0 = 0.01$).

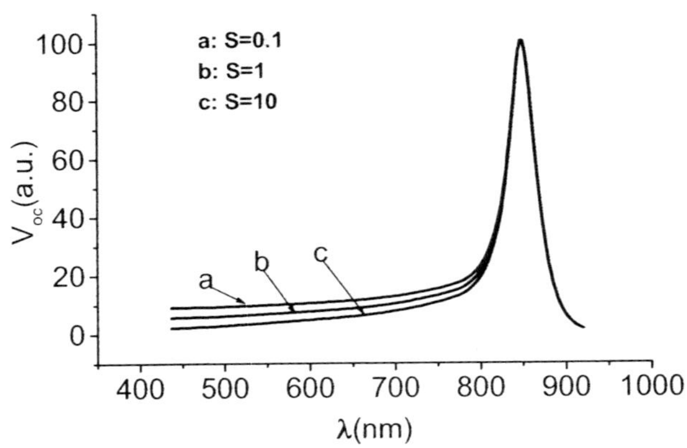

Figure 7. Effect of surface recombination velocity characterized by value S = sL/D on the theoretical V_{oc} spectra calculated for $eI_0/J_0 = 0.01$, L = 0.5 μm, and the other parameters as in Figure 6.

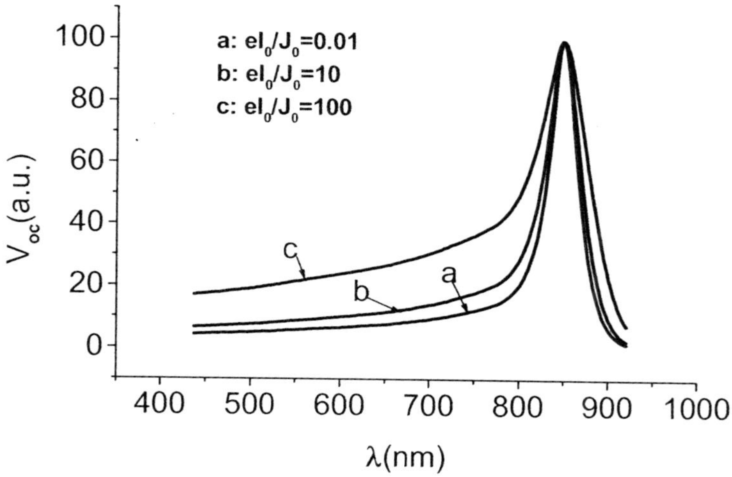

Figure 8. Effect of different levels of illumination expressed in ratio eI_0/J_0 on the theoretical spectra of V_{oc} calculated for L = 0.5 μm and the other parameters as in Figure 6.

The expression (14) does not lead to linear dependence of the I versus α^{-1} which is supposed in the older SPV methods [2, 8] mentioned in the section 1. Figures 6,7 and 8 show V_{oc} calculated according to Eqs. (14), (19)-(21) (at T = 300 K and with the quality factor n = 1) as a function of wavelength λ.The spectra are normalized to the values in maximum. In all cases, it holds that increasing radiation energy (decreasing wavelength) leads to an increase in the light absorbed by the cell and the signal consequently grows. After reaching maximum the signal strongly falls as photons with higher energy generate electron-hole pairs in longer distance from the junction. This region is predominantly characterized by minority carrier diffusion length (see Figure 6). The following slowly decreasing part is mainly influenced by surface recombination on the illuminated surface, which is manifested in Figure 7. The set of the curves in Figure 8 shows that increasing eI_0/J_0 ratio widens the curves without changing position of the maximum. The calculations show that the light intensity used in the measurement should be low enough for $eI_0/J_0 < 0.1$. In the opposite case, this ratio is another parameter of the V_{oc} spectral dependence.

Arrangement with One SCR on Illuminated Side of the Solar Cell Absorber

This case is shown in Figure 9 where the heterojunction is illuminated from the side of the wide band-gap n-type semiconductor-2. No contribution to the photocurrent from this material is assumed and multiple reflection is not considered. Origin of coordinates is put to interface of the semiconductors. Diffusion equation Eq. 5 with the rate of generation

$$g(x) = I_0(1-R)\exp(-\alpha w_1)\exp(-\alpha x) \quad (22)$$

is solved under following boundary conditions:

$$\Delta n(w_1) = 0 \tag{23}$$

$$D d\Delta n/dx \,|_{x=d+w1} = -s\Delta n\,(d + w_1) \tag{24}$$

The parameter s is the surface recombination velocity on the non illuminated surface. Substituting for Δn into the relation

$$J_b = eDd\Delta n/dx \,|_{x=w1} \tag{25}$$

the diffusion contribution from the bulk is derived as

$$J_b = eI\exp(-\alpha w_1)\frac{\alpha L}{1-\alpha^2 L^2}\,\frac{(1+S)\exp(d/L)-(1-S)\exp(-d/L)-\alpha L+2(\alpha L-S)\exp(-\alpha d)}{(1+S)\exp(d/L)+(1-S)\exp(-d/L)} \tag{26}$$

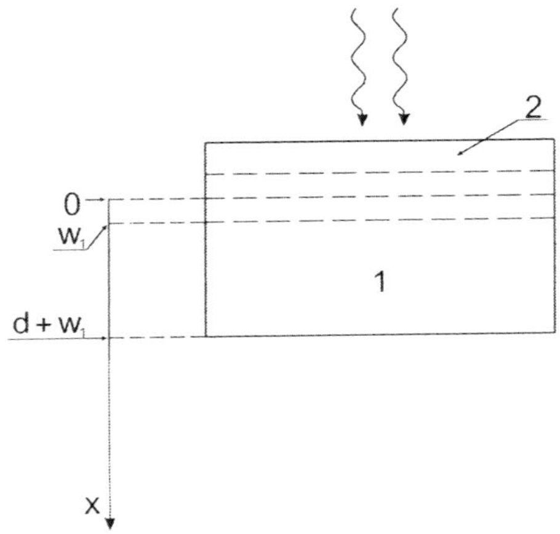

Figure 9. Scheme of the junction illuminated from the side of the wide band-gap semiconductor, d is the thickness of the bulk of the low band-gap material, w_1 is the width of its SCR

The photocarriers generated in the SCR produce the current density

$$J_s = eG\int_0^{w1} g(x)dx = eGI_0(1-R)(1-\exp(-\alpha w_1)). \tag{27}$$

The total photocurrent to be inserted into Eq.(14) is $J_{ph} = J_b + J_s$ (28)

If G = 1 the photovoltage is saturated in the region of high absorption while for G < 1 carriers generated in bulk near the SCR/bulk boundary produce higher current than those generated in the SCR, which leads to the formation of a maximum followed by a decreasing signal towards the low wavelength as it is shown in Figure 10.

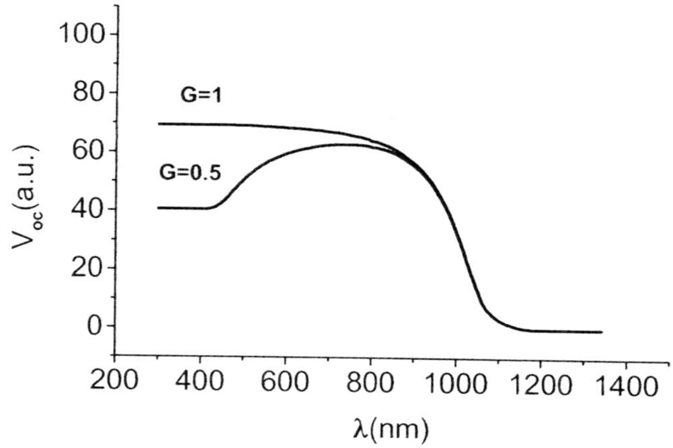

Figure 10. Theoretical spectra of V_{oc} calculated for two different values of G-factor (d = 300 μm, S = 1, L = 100 μm, eI_0/J_0 = 1).

Photovoltage Influenced by Photon Recycling

Photon recycling (PR) occurs in direct semiconductors, where radiative recombination is significant. As the photon energy of the luminescent light exceeds the energy gap, it is reabsorbed and secondary e-h pairs are created which can further radiatively recombine. The process is repeated many times. Numerous studies on PR have been published because of its essential effect on the properties of solar and thermophotovoltaic cells, semiconductor luminescent and light-detecting devices.

Studies of this phenomenon were reported by Moss, Landsberg and Dumke in 1957 [12, 13, 14]. Its impact on the diffusion length L and the internal quantum efficiency in GaAs was studied by Kuriyama et al. [15], the influence on luminescence characteristics of GaAs:Si and $Ga_{1-x}Al_xAs$:Si LEDs was published by Roedel and Keramidas [16]. Badescu and Landsberg [17] demonstrated that PR led to increased efficiency of GaAs solar cells and Parrott [18] showed a similar effect for InP solar cells. Influence of PR upon GaSb photovoltaic cells was included into the comprehensive paper by Stollwerck et al [19]. Deep theoretical analysis of these processes can also be found in publications of Badescu and Landsberg [20] and Enders [21]. The published results are too complicated to be used for the calculation of spectral photovoltage response. We have presented a simple theory of PR which can be directly applied to experimental photovoltaic spectra of diodes [22].

Theoretical Background

Photovoltage produced by p-n junctions and Schottky diodes schematically shown in Figure 11 and Figure 12 is analyzed taking into account photon recycling. The photocurrent is calculated under the following assumptions:

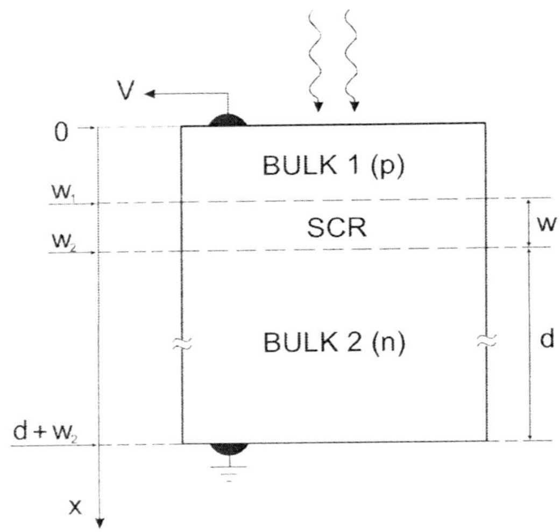

Figure 11. Schematic view of front illuminated p-n junction diode.

1) The recombination of photogenerated carriers is characterized by the lifetime τ given by the relation $1/\tau = 1/\tau_r + 1/\tau_n$, where τ_r and τ_n are the lifetimes of radiative and non-radiative processes, respectively.
2) The mechanism of photogeneration can be chronologically divided into two steps:
   a) Primary process – external light generation of photocarriers followed by their diffusion which establishes a distribution of the primary non-equilibrium photocarriers.
   b) Secondary process – generation and movement of secondary photons and secondary e-h pairs.

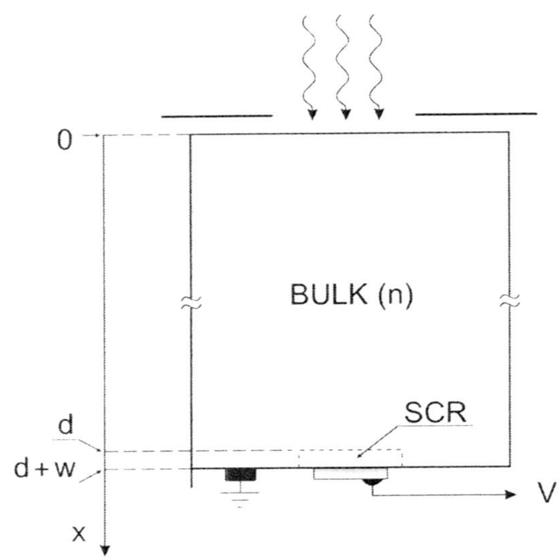

Figure 12. Schematic view of back-illuminated Schottky diode.

Every primary e-h pair which radiatively recombines after diffusion creates a secondary photon. The secondary photon is reabsorbed at a different place and generates a secondary e-h pair. This process may be repeated many times, which, in fact, results in a large effective penetration depth of the light beam. Consequently, behavior of the photons created in the sample (secondary photons) can be characterized by an effective value of the absorption length $1/\alpha_{eff}$ being much larger than in the case without PR. According to assumption 1) losses of photons in the sequence of succeeding absorption of photons, generation and diffusion of e-h pairs and their recombination are caused by nonradiative events. It can be characterized by a loss coefficient γ which, in addition, includes the mean value of losses caused by the escape of photons from the sample. If $\gamma << \alpha$ where α is the absorption coefficient of the primary radiation, the secondary e-h pairs are generated much more homogeneously compared to the primary e-h pairs. The nearly uniform distribution of the secondary e-h pairs can also be understood as a consequence of a diffusion process with a diffusion length much longer than the diffusion length L not influenced by PR. After absorption of the secondary photons in the SCR or when the pairs from the bulks reach the SCR by diffusion the photon recycling is terminated. Both e-h pairs created in the SCR by the secondary photons and the pairs having reached this region by diffusion are separated and contribute to the photocurrent.

Calculation of the Photocurrent

Both primary and secondary processes contribute to the total photocurrent.

1) The primary current is created by the external radiation photogenerated e-h pairs reaching the bulk-SCR boundary without recombination (this is the usual mechanism existing in materials where PR is not present). Calculation of the current needs evaluation of photocarrier distribution as a result of absorption of impinging radiation with photon flux I_0 and diffusion of the created pairs. Distribution of the primary photogenerated carriers $\Delta n(x)$ can be found by solving of usual diffusion equation with appropriate boundary conditions. The dimensions of the samples justify application of one dimensional treatment of the problem.
2) The secondary current results from radiative recombination of the primary e-h pairs of concentration $\Delta n(x)$ generating secondary photons with the rate equal to $\Delta n(x)/\tau_r$.

Calculation of the Primary Photocurrent Using the Diffusion Equation

In the following photocarrier distribution and photocurrent of an illuminated photodiode (Figure 11) will be calculated. The photogenerated current density in the diode is a sum of diffusion contributions from bulk 1 and bulk 2 and drift contribution from the space charge region. Assuming Beer's law distribution of photogenerated carriers in bulk 1 can be found by solving the diffusion equation

$$d^2\Delta n/dx^2 - \Delta n/L_1^{\,2} = -(\alpha I_0(1-R)/D_1)\exp(-\alpha x) \tag{29}$$

under the boundary conditions:

$$D_1\, d\Delta n / dx_{x=0} = s_1 \Delta n(0) \quad (30)$$

and $\Delta n(w_1) = 0$ (31)

L_1 is the electron diffusion length, R is the reflectance from the front surface, D_1 is the electron diffusion coefficient, s_1 is the surface recombination velocity on the surface of bulk 1. Replacing L_1, D_1, in Eq.(29) by equivalent parameters L_2, D_2, a similar equation for excess hole concentration Δp generated in bulk 2 is solved under the following boundary conditions:

$$\Delta p(w_2) = 0 \quad (32)$$

and $D_2\, d\Delta p / dx_{x=d+w2} = -s_2 \Delta p(d + w_2)$ (33)

(s_2 characterizes the recombination velocity on the surface of bulk 2)

The analytic expressions for the minority carrier concentrations $\Delta n(x)$ and $\Delta p(x)$ for bulk 1 and bulk 2 are:

$$\Delta n = \frac{c_1 \tau_1}{N_1}\{[(\alpha L_1 + S_1)\exp(w_1 / L_1) + \exp(-\alpha w_1)(1 - S_1)]\exp(-x / L_1)$$
$$+[(\alpha L_1 + S_1)\exp(-w_1 / L_1) - \exp(-\alpha w_1)(1 + S_1)]\exp(x / L_1)\} + c_1 \tau_1 \exp(-\alpha x) \quad (34)$$

and

$$\Delta p = \frac{c_2 \tau_2}{N_2}\{[-\exp(d / L_2)(1 + S_2) + \exp(-\alpha d)(S_2 - \alpha L_2)]\exp(-x / L_2)$$
$$-[\exp(-d / L_2)(1 - S_2) + \exp(-\alpha d)(S_2 - \alpha L_2)]\exp(x / L_2)\} + c_2 \tau_2 \exp(-\alpha x) \quad (35)$$

The diffusion current from bulk 1 is $J_{b1} = -\left| eD_1 d\Delta n / dx_{x=w1} \right|$ and that from bulk 2 is $J_{b2} = -\left| -eD_2\, d\Delta p / dx_{x=w2} \right|$. (1 cm^2 area of the electrodes is assumed). Final expressions for both currents are

$$J_{b1} = \frac{ec_1 L_1}{N_1}\{2(\alpha L_1 + S_1) + (1 - S_1)\exp(-\alpha w_1 - w_1 / L_1) - (1 + S_1)\exp(-\alpha w_1 + w_1 / L_1)$$
$$-\alpha L_1 N_1 \exp(-\alpha w_1)\} \quad (36)$$

where

$$J_{b2} = -\frac{ec_2L_2}{N_2}\{ 2(\alpha L_2 - S_2)\exp(-\alpha d) + (1+S_2)\exp d/L_2 - (1-S_2)\exp(-d/L_2) - \alpha L_2 N_2\} \tag{37}$$

$$N_1 = [(1-S_1)\exp(-w_1/L_1) + (1+S_1)\exp(w_1/L_1)],$$

$$N_2 = [(1-S_2)\exp(-d/L_2) + (1+S_2)\exp(d/L_2)]$$

$$c_1 = \frac{\alpha I_0(1-R)}{1-\alpha^2 L_1^{\ 2}}, \quad c_2 = \frac{\alpha I_0(1-R)\exp(-\alpha w_2)}{(1-\alpha^2 L_2^{\ 2})}.$$

τ_1, τ_2, are the total lifetimes of the photocarriers in bulk 1 and bulk 2, respectively, $S_1 \equiv s_1L_1/D_1$, $S_2 \equiv s_2L_2/D_2$.

The photocarriers generated in the SCR with the generation rate g(x) produce a current

$$J_{SCR} = eG\int_{w1}^{w2} g(x)dx = eI_0(1-R)\,G\,(\exp(-\alpha w_2) - \exp(-\alpha w_1)) \tag{38}$$

where G characterizes non radiative recombination of photocarriers in the SCR. The total short-circuit current J_{ph} is the sum of the individual contributions

$$J_{ph} = J_{b1} + J_{b2} + J_{SCR} \tag{39}$$

Inserting (39) into the usual relation for the photovoltage e.g., into Eq. (14), V_{oc} can be obtained as a function of L_1 ,L_2, G, eI_o/J_o, S_1, S_2, w_1, and thickness of the SCR $w = (w_2 - w_1)$. By fitting V_{oc} to experimental points the unknown parameters can be evaluated. An unexpectedly large value of diffusion length L_2 is practically found as it will be shown in Section 3.4. To explain this effect photon recycling will be included.

Currents Generated by Secondary Photons

Contribution from Bulk 1

As it has been already mentioned above the spreading of secondary photons will be described by means of an effective loss coefficient γ. Photon recycling will be terminated in the SCR where the electric field separates the generated e-h pairs preventing their radiative recombination. In the following the problem will be analyzed and the photocurrent from bulk 1 will be calculated.

The secondary radiation photon density generation rate $\varphi_1(x) = \Delta n(x)/\tau_{1r}$ at a point x decreases to the value $\varphi_{01}(x)$ at the SCR-bulk 1 boundary. It holds

$$\varphi_{01}(x) = \varphi_1(x)\exp(-\gamma_1(w_1 - x)) \tag{40}$$

$\Delta n(x)$ is the concentration of minority carriers generated by the primary radiation, τ_{1r} is their radiative recombination lifetime in bulk 1, γ_1 is the loss coefficient in bulk 1. Simplifying the approach we assume that the losses take simple exponential form. The magnitude of photocurrent from bulk 1 is proportional to the total number of photons reaching the SCR or its vicinity and absorbed there. The dimensionless coefficient of proportionality is K_1. In the presence of non-radiative recombination the rate of photon generation is reduced by the factor τ_1/τ_{1r} where τ_1 is the resulting lifetime in bulk 1. Consequently, the current J_{b1rec} created in bulk 1 is

$$J_{b1rec} = e(\tau_1/\tau_{1r})\,K_1 \int_0^{w1} \varphi_{01} dx = e(\tau_1/\tau_{1r})\,K_1 \int_0^{w1} (\Delta n(x)/\tau_1)\exp(-\gamma_1[w_1 - x])\,dx =$$

$$= a_1 \frac{L_1}{(1-\gamma_1 L_1)}(\exp[-\gamma_1 w_1] - \exp[-w_1/L_1]) + b_1 \frac{L_1}{(1+\gamma_1 L_1)}(\exp[w_1/L_1] - \exp[-\gamma_1 w_1])$$

$$+ c_1^* \frac{1}{\alpha - \gamma_1}(\exp[-\gamma_1 w_1] - \exp[-\alpha w_1]) \qquad (41)$$

where

$$a_1 = -\frac{c_1^*}{N_1}[(\alpha L_1 + S_1)\exp(w_1/L_1) + (1 - S_1)\exp(-\alpha w_1)],$$

$$b_1 = \frac{c_1^*}{N_1}[(\alpha L_1 + S_1)\exp(-w_1/L_1) - (1 + S_1)\exp(-\alpha w_1)],\quad c_1^* = \frac{\tau_1}{\tau_{1r}}\frac{eK_1\alpha I_0(1-R)}{1-\alpha^2 L_1^2}$$

and $N_1 = [(1 - S_1)\exp(-w_1/L_1) + (1 + S_1)\exp(w_1/L_1)]$.

Contribution from Bulk 2

The photocurrent $J_{b2\,rec}$ from the bottom bulk was calculated similarly just shifting the origin of coordinates to the SCR-bulk 2 boundary. The secondary radiation photon density $\varphi_2(x)$ produced per second at point x decreases at the SCR-bulk 2 boundary to value $\varphi_{02}(x)$ .It holds that

$$\varphi_{02}(x) = \varphi_2(x)\exp(-\gamma_2 x) \qquad (42)$$

where γ_2 is the loss coefficient in bulk 2. The photocurrent from bulk 2 is:

$$J_{b2\,rec} = e(\tau_2/\tau_{2r})\,K_2\int_0^d \varphi_{02}dx = e(\tau_2/\tau_{2r})\,K_2\int_0^d (\Delta p(x)/\tau_2)\exp(-\gamma_2 x)dx =$$

$$= a_2\,\frac{L_2}{\gamma_2 L_2+1}(1-\exp(-[\gamma_2+1/L_2]\,d)) - b_2\,\frac{L_2}{\gamma_2 L_2-1}(1-\exp(-[\gamma_2-1/L_2]\,d))$$

$$+ c_2^*\,\frac{1}{\gamma_2+\alpha}(1-\exp(-[\gamma_2+\alpha]\,d)), \qquad (43)$$

where

$$a_2 = \frac{c_2^*}{N_2}[(S_2-\alpha L_2)\exp(-\alpha d)-(1+S_2)\exp(d/L_2)],$$

$$b_2 = \frac{c_2^*}{N_2}[(S_2-\alpha L_2)\exp(-\alpha d)+(1-S_2)\exp(-d/L_2)],$$

$$c_2^* = \frac{\tau_2}{\tau_{2r}}\,\frac{eK_2\,\alpha I_0(1-R)\exp(-\alpha w_2)}{(1-\alpha^2 L_2^{\;2})} \quad \text{and}$$

$$N_2 = [(1-S_2)\exp(-d/L_2)+(1+S_2)\exp(d/L_2)]$$

We assume K_1 and K_2 to be spectrally independent constants.

Contribution from the SCR

We suppose that under our quasiopen circuit conditions the pairs photogenerated in the SCR by the primary radiation are immediately separated by the internal electric field so that the photon recycling need not be considered there. Hence the photocurrent is given by Eq. (38).

Resulting Photocurrent from the Diode

The photocurrent J_{phrec} that respects photon recycling is the sum:

$$J_{ph\,rec} = J_{b1rec} + J_{b2rec} \qquad (44)$$

It must be emphasized that in the bulk of diodes superposition of the photon recycling process and the usual diffusion process (without secondary photon generation) generally exists because some of the e-h pairs reach the bulk-SCR boundary without recombination. The diffusion mechanism cannot be neglected especially in layers with thicknesses comparable with the diffusion length. On the other hand, the effect of PR is essential for sufficiently thick samples [15]. In general, the resulting current J is a sum of the current J_{phrec} caused by PR and that generated by the usual mechanism J_{ph} :

$$J = J_{b1rec} + J_{b2rec} + J_{b1} + J_{b2} + J_{SCR} \tag{45}$$

In the case of small injection ($J_{ph} / J_0 \ll 1$), the photovoltage is proportional to the current J. In the calculations of J reflection R was considered as independent of λ.

Let us write $J_{b1rec,b2rec} = eI_0\,(1\text{-}R)\;\xi_{1,2}\,J'_{b1rec,b2rec}$, $J_{b1,b2} = eI_0\,(1\text{-}R)\,J'_{b1,b2}$,

$J_{SCR} = eI_0\,(1\text{-}R)\,J'_{SCR}$ and $J = eI_0\,(1\text{-}R)\,J'$ where $\xi_{1,2} = (\tau_{1,2}/\tau_{1r,2r})\,K_{1,2}$. Then from (45) it follows:

$$J' = \xi_1 J'_{b1rec} + \xi_2 J'_{b2rec} + J'_{b1} + J'_{b2} + J'_{SCR} \tag{45α}$$

which will be used hereafter. The coefficients $\xi_{1,2} = (\tau_{1,2}/\tau_{1r,2r})K_{1,2}$ are not dependent on wavelength of the impinging radiation. Their values can be found experimentally. For theoretical curves we take $\xi_{1,2} = 1$. This value is near to that found in our experiment (see Section 3.4). Now, application of the relation (45a) will be shown for a more simple case of a Schottky diode (Figure 12). As an example GaSb Schottky diode illuminated from the side of the bulk will be analyzed using absorption coefficients from Figure 13. In the region of λ<930 nm data of Aspnes, Studna [23] were used, for region 930 nm <λ<1710 nm Adachi's [24] data were applied and for the wavelengths above 1710 nm absorption coefficient of our GaSb: Te were determined by reflection and transmission measurements. The photocurrent J' has two parts only, namely that from the bulk and that from the SCR. Usually, $d \gg w$ holds, which in our case means $J'_{SCR} \ll J'$. Thus, the contribution from the SCR can be neglected. Reflection from the electrodes is neglected too and the photocurrent is obtained similarly as that from bulk 1 of the p-n junction diode but calculated for n-type material.

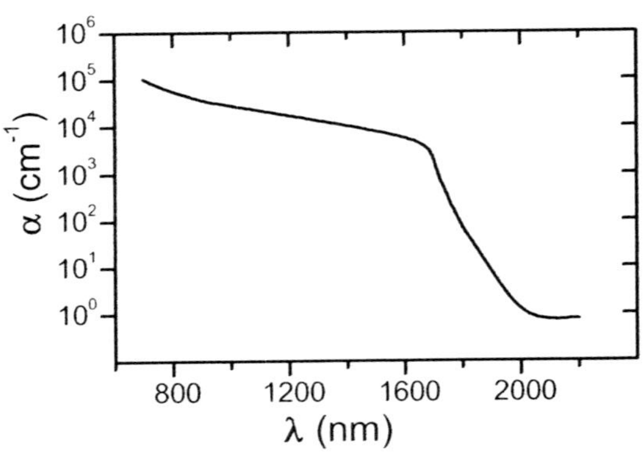

Figure 13. Absorption coefficient versus wavelength dependence of GaSb:Te.

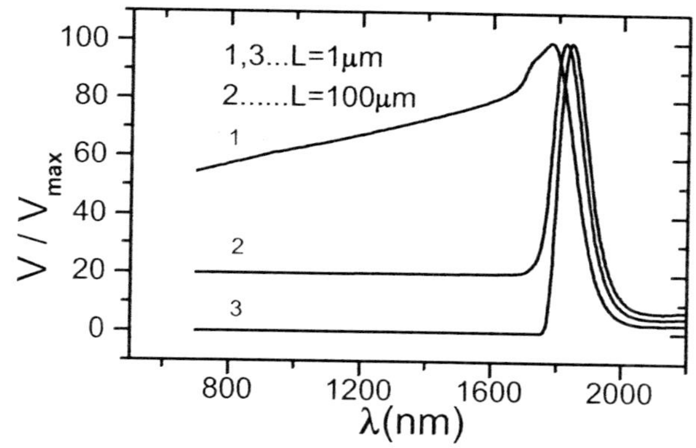

Figure 14. Theoretical photovoltage response of Schottky diode including photon recycling (1) and without this process (2,3) normalized to the values in maximum.
1: L = 1μm, $\gamma = 7\ cm^{-1}$, d = 365 μm, S = 0.9
2: L = 100 μm, d = 365 μm, S = 0.9
3: L = 1μm, d = 365 μm, S = 0.9

The large difference between the shape of theoretical curves accounting and not accounting for PR is shown in Figure 14. It is interesting that the diode with relatively large diffusion length L = 100 μm where photon recycling is not considered generates much lower signal than the diode with L = 1 μm where this process is accounted for. In fact, the value of the photovoltage at the maximum of curves 2 and 3 is lower than that of curve 1 by a factor of 6 and 800, respectively. Due to PR the radiatively recombining carriers are not lost and contribute to the current in dependence on the loss factor γ.

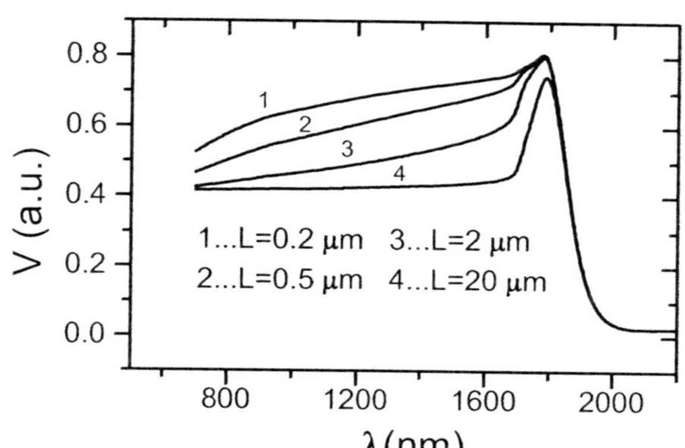

Figure 15. Influence of diffusion length on the theoretical photovoltage spectra of a Schottky diode - photon recycling included ($\gamma = 7\ cm^{-1}$, d = 365 μm, S = 0.9). It is shown that the signal increases with decreasing diffusion length.

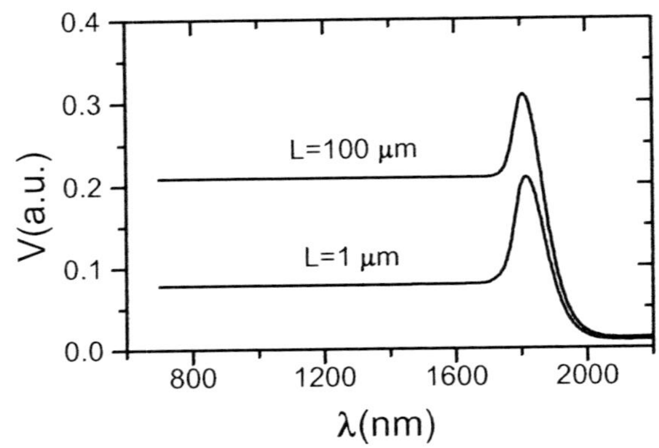

Figure 16. Theoretical photovoltage response of a Schottky diode including photon recycling ($\gamma = 70$ cm^{-1}, $d = 365$ μm, $S = 0$). In the case of zero surface recombination velocity the photovoltage spectra are influenced by diffusion length in the usual way.

Photon recycling leads to a rather surprising effect shown in Figure 15. At constant value of γ the signal increases with decreasing diffusion length L. Such phenomenon is a consequence of the non-zero surface recombination velocity, the influence of which increases with L. Larger L means a higher number of carriers recombining non radiatively at the front illuminated surface. If $S = 0$ the signal changes with L in the usual way as is shown in Figure 16, where $\gamma = 70$ cm^{-1} was chosen to emphasize the dependence of photovoltage on parameter L. For $S = 0$ and small value of γ the contribution caused by photon recycling depends on the parameter L only weakly and for $S = 0$, $\gamma = 0$ the dependence on L entirely disappears.

Application of the Theoretical Calculations to Results of Measurements

Application to Silicon Solar Cells

The experimental arrangement which we have used [25] is illustrated in Fig 2. The cell was illuminated from the side of the absorber through a metallic mesh separated by a thin (6μm) sheet of mylar from the specimen. The mesh served as one electrode while the original grid of the cell was used as the second one. Using illumination of the free back surface of the cell our method removes difficulties with spectral dependence of electrode transparency and scattering processes on the textured surface of the emitter. The light penetrating the specimen was chopped by a mechanical chopper. Pulses of voltage induced on the electrodes were further amplified. The SPV as a function of wavelength is measured by keeping photon flux density I_0 constant during the experiment. Typical spectral response of photovoltage together with fitted theoretical dependence of SPV according to Eq. (11) is presented in Figure 17. Let us note that in Section 3 the curves in all photovoltage spectra are normalized to their maximum. Two junctions give rise to two signals of opposite sign. The pn^+ junction generates higher voltage despite of the fact that intensity of the light absorbed in its vicinity is weaker.

It is a consequence of much lower saturated current J_{02} in this junction than is that in the spontaneously created one. The signal is a nonlinear function of light intensity andconsequently the plot of photon flux density I_0 against $1/\alpha$ according to Ref. [2] evidently cannot lead to correct value of the diffusion length. Goodman's assumptions mentioned in

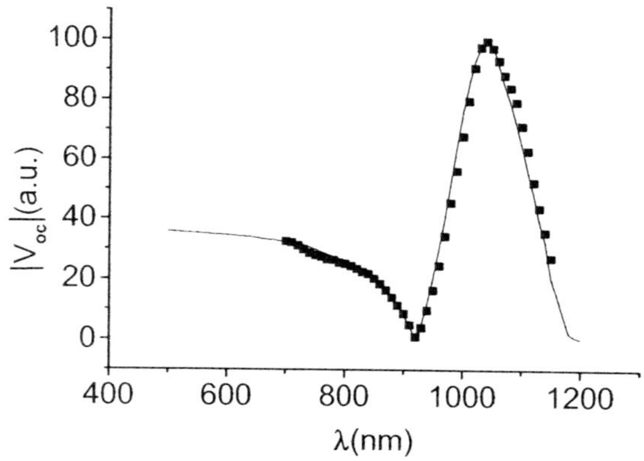

Figure 17. Theoretical (full line) and experimental (points) photovoltage spectrum for the case of two SCRs. Two branches separated by zero point represent opposite signs of the SPV signal. The theoretical curve is plotted for L = 88 μm, d_l = 350 μm,
$n_1 = 1.1$, $n_2 = 1.0$, $eI_0/J_{01} = 1$, $eI_0/J_{02} = 35$, $R_1 = 0.316$,and $R_2 = 1$.

Section 1 cannot be fulfilled in most solar cells also due to their small thickness in comparison with L. From that point of view the fitting of the theoretical dependence (11) to experimental points provides a better possibility for the estimation of diffusion length in solar cells.

This method can be substantially improved by excluding the contribution V_{oc1} from the surface region to the whole signal [26]. In the arrangement according to Figure 18 the photovoltage V_{oc2} is taken between an original grid electrode and an ohmic electrode to the absorber. The relative spectral response of V_{oc} is now characterized by four parameters namely L, eI_0/J_{02}, R_2 , d_l , but only the first three must be fitted. This fact simplifies the evaluation as in Eq (11) the term relating to the surface junction can be omitted and, consequently it holds

$$V_{oc} = V_{oc2} = -(n_2 k_B T/e) \ln(1 - J_{ph2}/J_{02}) \quad (46)$$

where J_{ph2} is given by Eq.(10).

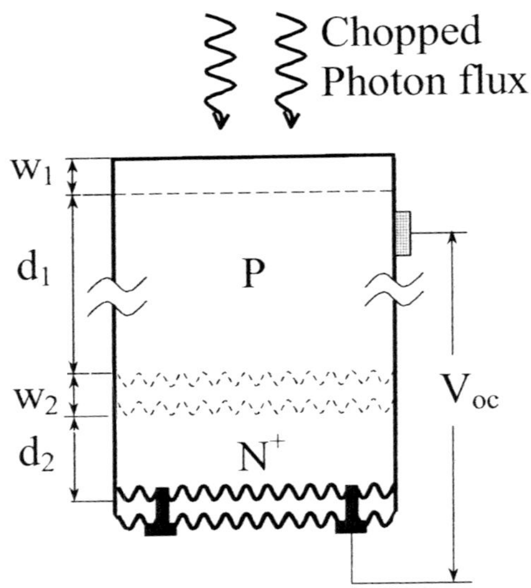

Figure 18. Experimental arrangement recording signal from the bottom junction only.

The measured V_{oc2} spectrum and theoretical curve where all four parameters were fitted are plotted in Figure 19. The evaluated thickness of the bulk agrees well with its real value and L = 71 μm. The typical value of L for the Si cells measured was 70 - 90 μm. Analysis of the signal V_{oc} from the whole structure (measured as on Figure 2) gives the same diffusion length but for successful measurement a fresh surface prepared by etching is needed.

In conclusion, both of the last discussed techniques can diagnose the quality of solar cells with regard to diffusion length of minority carriers in samples of arbitrary thickness and without limitations to light intensity. An advantage of the improved method is that only three parameters need to be fitted and the measurement is not influenced by the quality and aging of the illuminated surface.

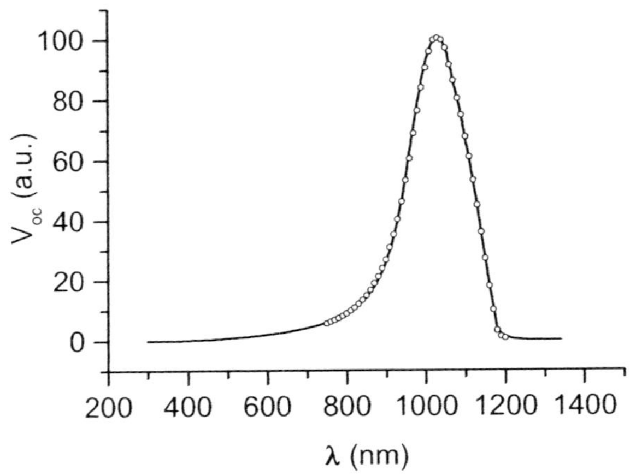

Figure 19. Experimental photovoltage spectrum of the p-n$^+$ junction (points). Theoretical curve according to Eq. (46) (full line) is plotted for d_1 = 285 μm, L = 71 μm, R_1 = 0.8, R2 = 0.9, and eI_0/J_{02} = 35.

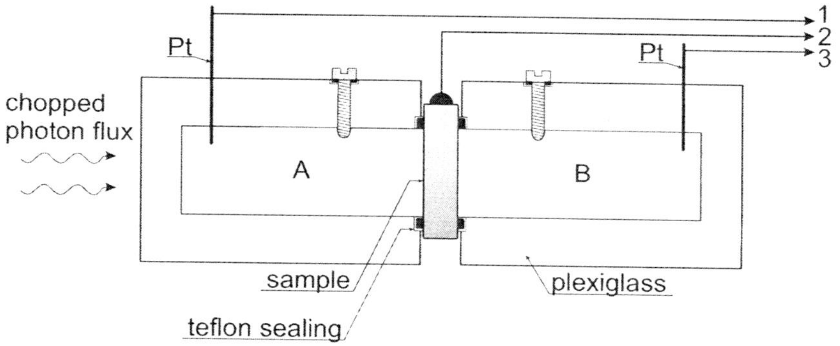

Figure 20. Cross-sectional view of the electrochemical cell

Application to Silicon Wafers

Silicon wafers for solar cells were examined in an electrochemical cell and signal from the junctions on the top and on the bottom surfaces was analyzed using the SPV technique. The signal from each junction can be measured separately using special arrangement of the experiment. The principle of this method is shown in Figure 20 and described e.g. in [27]. Silicon wafers were inserted into an electrochemical cell and surrounded by diluted (1%) HF. Platinum wire electrodes (1,3) were used as a contact to the electrolyte in chambers A, B, the wafer edge was contacted through soldering of In (2). The SCRs created on both silicon surfaces, which are covered with the electrolyte, generate photovoltage after illumination. Transmittance spectral dependence of plexiglass and electrolyte was determined. The thickness of the wafer was high (1.3 mm). It allows measurement of diffusion length according to [2]. The voltage between points 1 and 2 (chamber A) for each wavelength was maintained on a constant value V_{const} by adjusting the photon flux density I and the flux versus reciprocal value of absorption coefficient [2] in the material was plotted. Linear part of the curve fulfills relation (2) : $I = const(L+1/\alpha)$ which can be used for evaluation of L. In Figure 21 there are three curves for decreasing V_{const} as a parameter.

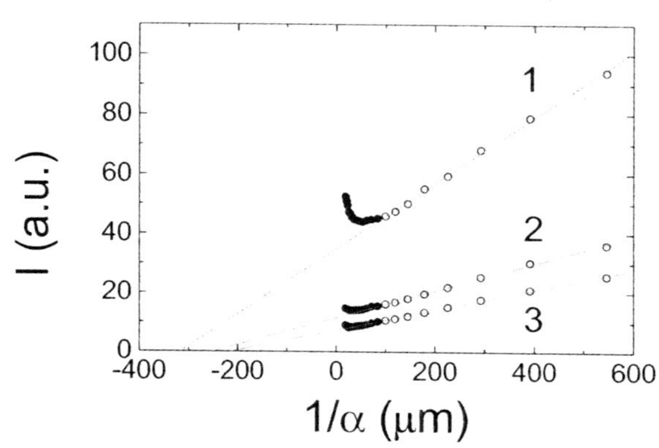

Figure 21. Evaluation of the diffusion length using chambre A of the electrochemical cell. The thickness of measured Si wafer was 1.3 mm.

It is noteworthy, that the lower V_{const} the more experimental points fulfill the linear dependence (2) and the evaluated values of L decrease, showing tendency to saturation for small V_{const}. The lowest value of diffusion length L = 200 μm is supposed to be the most reliable result as it is non- sensitive to further decrease of V_{const}. The spectrum taken between points 2 and 3 (the chamber B) at constant number of photons entering the same sample is in Figure 22.The value of the diffusion length L = 225 μm evaluated from the fitting according to relation (46) is in good agreement with the minimum value of L obtained by the previously described technique. Samples of different thickness, namely d = 2mm, 1.3 mm and 0.5 mm were examined.

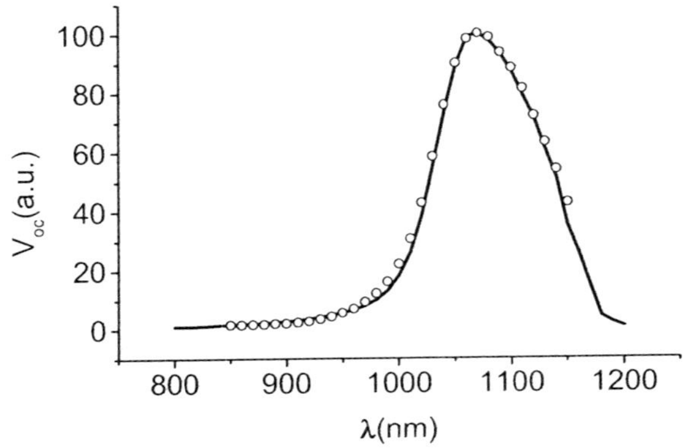

Figure 22. Photovoltage spectrum of the bottom junction of 1.3 mm thick Si wafer taken in B chamber of the electrochemical cell (points). L = 225 μm was obtained by fitting relation (46) (full line).

There was no difference in L evaluated by our method in the case of 2 mm and 1.3 mm thick samples. This method applied to 0.5 mm thick sample gives L = 150 μm only. In this case, influence of the surface of the sample modifies diffusion length L to an effective diffusion length L_{eff}(see Section 3.5). It means that the material is strongly influenced by the surfaces in the last case.

From the above results we can conclude that using the experimental arrangement as in the case of chambre A, influence of the SCR at the surface manifests whereas the photovoltage generated by the junction located farther from the illuminated surface does not show sensitivity to the properties of the SCRs and is noted for a very good reproducibility. Consequently, this method was used for evaluation of the minority carrier diffusion length by fitting of the theory to experimental points. "Classical" ASTM method was used here for Si wafers for comparison with our technique. Good agreement was found in the case that the "classical" method was applied at low light level.

We have also measured better quality Si wafers with thickness 0.35 mm. The samples were etched in CP4 (1 HF : 3 HNO_3 : 7 CH_3COOH) etchant to lower the surface recombination velocity essentially and to enable surface space charge regions formation. Typical SPV spectra of the samples, where 25 μm thick layer was removed from both upper and bottom sides, are shown in Figure 23. Fitting of Eq. (46) to the experiment provides value of the diffusion length L = 280 - 300 μm.

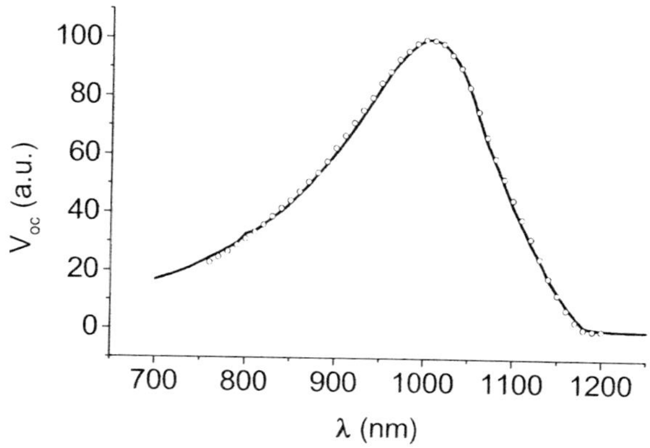

Figure 23. Photovoltage spectrum of thin (0.3 mm) Si wafer. Experiment-points, theoretical curve according to Eq.(46) - full line (L = 300 μm).

The carrier diffusion length LSO was also studied on microcrystalline silicon (μc-Si) [28], which is a very promising material for photovoltaic applications, mainly because it does not suffer from light induced degradation (in contrast to amorphous Si). Because of the inherent anisotropy of the μc-Si:H the surface photovoltage method is a very suitable one for evaluation of the diffusion length perpendicular to the substrate. There is often need to study thin samples, usually with the thickness of less than 2μm. Additional complication is due to the light scattering at the inhomogeneities within the volume of the sample, which affects the concentration and spatial distribution of the photogenerated carriers. The μc-Si:H layers were grown on glass substrates covered with ITO film which served as an ohmic back electrode. For evaluation of L we have to use our model which takes into account both the current from the bulk and the current from the space-charge region and recombination on the back surface. Correct optical data, especially precise knowledge of the absorption coefficient spectra is important. In μc-Si:H samples with a considerable surface roughness the surface light scattering could affect the results of measurements of $\alpha(h\nu)$. For elimination of scattering from absorption coefficient a multiple scattering theory and the process described in [28] were used. Then calculation according to our model was applied and diffusion length extracted as fitting parameter was 0.7 μm for 2 μm thick sample.

Application to CdS/CdTe Solar Cells – Diffusion Length in CdTe by Measurement of Photovoltage Spectra

CdS/CdTe solar cells are one of the prospective candidates for widespread commercial success in solar energy conversion. They can be produced at a low price with good efficiency, stability and short payback time [29]. The efficiency depends strongly on properties of the CdTe absorber, namely on the diffusion length of minority carriers and on the recombination rate of fotogenerated carriers in the SCR and on the thickness of the SCR. We have studied the cells with thin CdS layer deposited by the chemical bath technique and CdTe layer which was electroplated [30, 31].

The SPV method was applied to glass/indium-tin-oxide (ITO) /CdS/CdTe heterostructures differing from the complete cells only by the absence of the contact to CdTe [32]. This free surface was used to illuminate the absorber (Figure 24). We assume that the electric field is only in the SCR of the CdS/CdTe junction. The thickness of this region is (w_1 $+w_2$). The free surface of CdTe has quite a high p-type conductivity because of enriching the surface with tellurium during its bromine -in-methanol etching [30] and consequently no SCR is supposed to be there. If a SCR at the surface was spontaneously created there would be compensation of the photovoltage signals from both junctions and the spectral response would show a zero value at certain wavelength. Such form of the spectra was not measured in our case. The ITO layer acts as a ohmic contact to n-CdS film, hence no SCR is formed there as well.

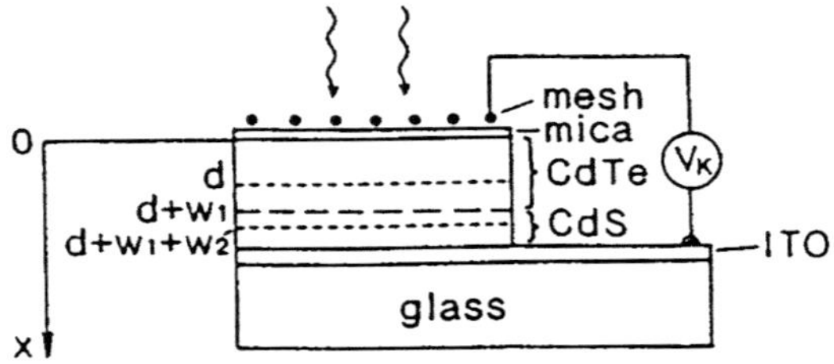

Figure 24. Schematic arrangement of photovoltage measurement (d is the thickness of the bulk in CdTe, w_1, w_2 is width of the space charge region in CdTe and CdS, respectively.

The experimental arrangement used does not allow light to penetrate into the wide band-gap CdS because of high absorption of short wavelengths in lower band-gap CdTe consequently no photovoltage is supposed to come from CdS.

Calculation of the theoretical photovoltage spectra performed according to Eqs. (14-21) requires knowledge of the absorption coefficient in CdTe material. The spectral dependence of the absorption coefficient in the subband-gap region was measured and evaluated by the constant photocurrent method (CPM) [33]. The results were completed by the data obtained from transmission measurements [32] and by the data compiled by Palik et al. [34]. This α versus λ dependence is shown in Figure 25.

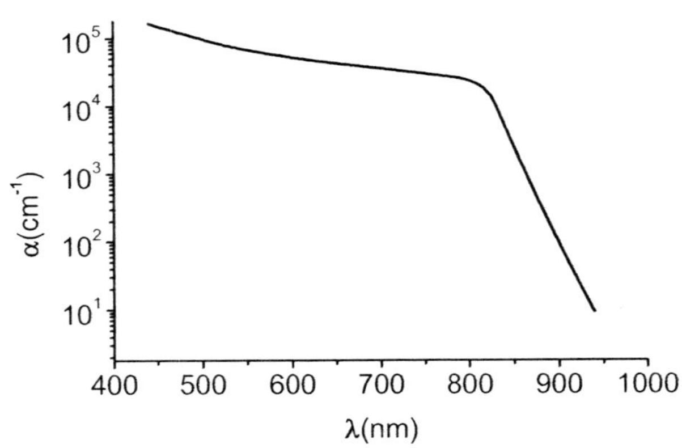

Figure 25. Spectral dependence of the absorption coefficient of CdTe.

Fitting of the theory to experimental points (Figure 26) gives the following results: L = 0.43 μm, S = 0.9 and d = 1.54 μm. The parameters of the SCR do not essentially influence the values of V_{oc} and consequently cannot be determined from this experiment.

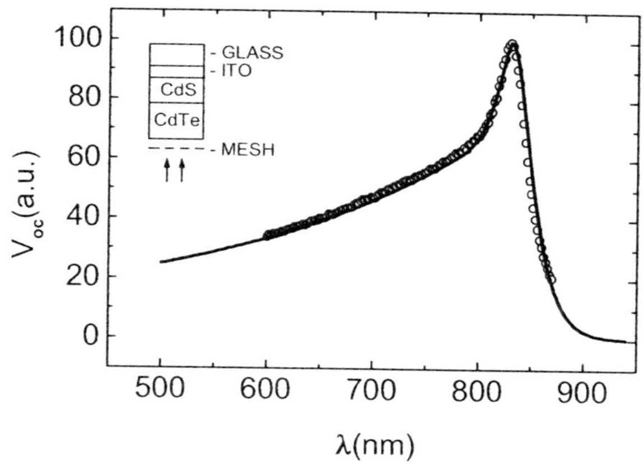

Figure 26. Comparison of the theoretical photovoltage spectrum (full line) with measured CdS/CdTe spectrum (points). Parameters of the fitting are: L = 0.43 μm, d = 1.54 μm, and S = 0.9.

The SPV experiment was completed with measurement of the spectral response of the front-illuminated cells. In this case light penetrates through the CdS layer. The total photocurrent is given by Eq.(28). In contrast to the preceding case, front illumination leads to results which are strongly influenced by the SCR, which can be seen in Figure 27. As L and d are already known and V_{oc} is not sensitive to s, only the thickness and the gain factor G of the SCR are evaluated by fitting of the theory to measured photovoltage spectra (Figure 27). The deviation for λ lower that 650 nm is probably due to effect of the CdS layer which was not taken into consideration. Very low gain factor (G = 0.003) means high recombination in the SCR, which may be explained by the lattice mismatch of CdS and CdTe materials.

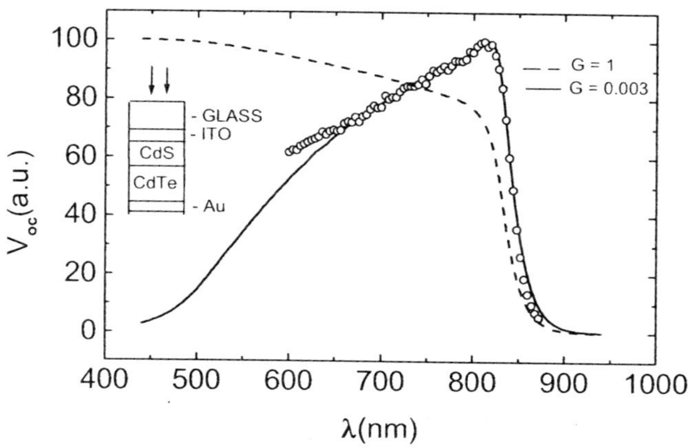

Figure 27. Photovoltage spectrum of CdS/CdTe cell (points) as compared with theoretical dependence for L = 0.43 μm, d = 1.54 μm, w_1 = 0.33 μm, and G = 0.003 (full line). The dashed line is theoretical dependence for G = 1 and the other parameters as for the full line curve.

Diffusion Length of Minority Carriers in GaSb Diodes - Influence of Photon Recycling on Photovoltage Spectra

GaSb is a semiconductor with low direct energy band-gap (0.7eV at 300K). Its use in the field of the photovoltaics is mainly in thermophotovoltaic cells. Radiative recombination is important recombination mechanism in this material so that photon recycling process must be accounted in calculation of the photocurrent.

Our simple theory of photon recycling was applied for evaluation of parameters of GaSb diodes. Two types of samples were investigated : Schottky diodes and p-n junction diodes. Rectifying gold contact was evaporated on Te doped n-type GaSb substrate to obtain Schottky diode. The same wafers were used for preparation of p-n junction structures. GaSb p and n epitaxial layers were deposited on the substrate by the metalorganic vapor phase epitaxy process. The layer preparation was carried out with two different growth rates. The lower growth rate was 20 nm/min., the higher growth rate was 40 nm/min. The technological process is described in detail in [35].

Experimental arrangement is in Figure 11 and 12. To study the photovoltage spectra the wavelength dependence of the absorption coefficient was taken from the data obtained as it is described in Section 2.5.2 (Figure 13). These coefficients were used for the substrate and for both epitaxial layers.

The photovoltage spectrum at constant photon flux density obtained from the back-illuminated Schottky diode is plotted in Figure 28 (points). In spite of back-illumination of the thick bulk the signal in the short wavelength region is only slowly decreasing from its value at maximum. The spectrum cannot be explained not considering PR. The theoretical dependence (full line) was fitted to the experiment with L = 1 μm, S = 0.9, $\gamma = 7\ cm^{-1}$ and ξ = 1 corresponding to the situation in Figure 14, curve 1.

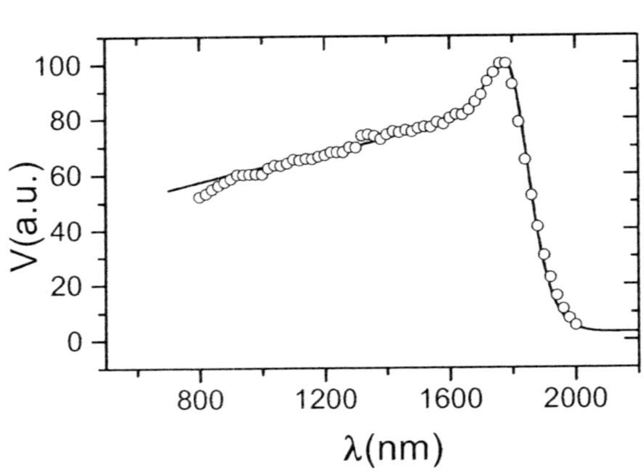

Figure 28. Experimental photovoltage spectrum of the back-illuminated Schottky diode (points) and the theory including photon recycling (full line) for the parameters: L = 1μm, $\gamma = 7\ cm^{-1}$, d = 365 μm, S = 0.9, ξ = 1.

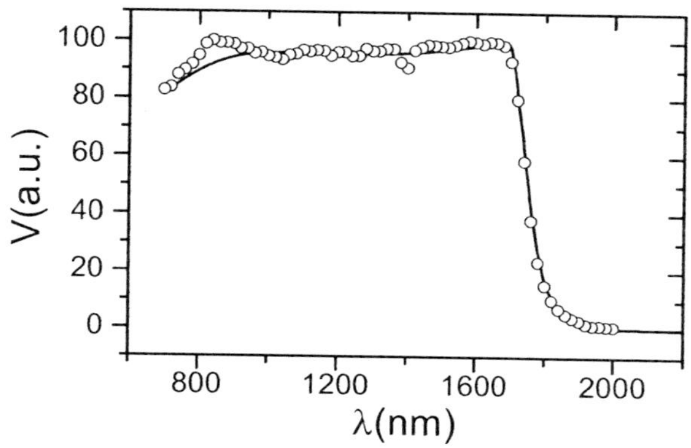

Figure 29. Experimental photovoltage spectrum of the p-n junction of slowly grown diode (points) compared with the theory including photon recycling (full line) for following parameters: $L_1 = 0.55$ μm, $L_2 = 0.6$ μm, $w_1 = 0.5$ μm, d = 365 μm, w = 0.4 μm, $S_1 = 0.4$, G = 0.9, $\gamma_1 = \gamma_2 = 500\ cm^{-1}$, $\xi_1 = 0.6$, $\xi_2 = 1$. Photon recycling enhances the signal at short wavelengths.

Figure 29 and Figure 30 show the photovoltage spectra of p-n junction diodes illuminated from the front side at constant photon flux density. Diffusion length L_2 in their n-bulk as large as 20-30 μm is obtained if relation (39) is fitted to the experiment. Similar values have been published for n-GaSb in [36]. On the other hand, theoretical curves (full lines) according to (45a) yield $L_2 = 0.6$ μm with other fitting parameters listed in the captions. It was found that the influence of surface recombination S_2 on the photovoltage spectra was negligible because of large thickness of the samples.

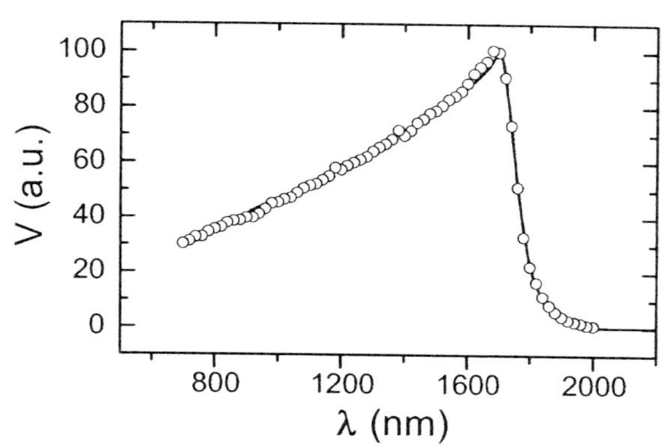

Figure 30. Experimental photovoltage spectrum of the p-n junction of faster grown diode (points) compared with the theory including photon recycling (full line) for following parameters: $L_1 = 0.4$ μm, $L_2 = 0.6$μm, $w_1 = 0.9$ μm, d = 365μm, w = 0.05 μm, $S_1 = 0.3$, G = 0, $\gamma_1 = \gamma_2 = 700\ cm^{-1}$, $\xi_1 = 0$, $\xi_2 = 1$. The suppression of PR in p-emitter results in a decrease of the signal at short wavelengths.

Calculation of the diffusion length of holes in n-GaSb controlled by radiative recombination requires the knowledge of radiative lifetime and minority carrier mobility. The latter was reported [37] to be $\leq 1000\ cm^2V^{-1}s^{-1}$, in general. For $3x10^{17}cm^{-3}$ doping concentration the hole mobility is 700 $cm^2V^{-1}s^{-1}$ according to [36]. The radiative recombination lifetime at this concentration is 38 ns (coefficient of the radiative recombination being $8.7x10^{-11}cm^3s^{-1}$ [36]). Consequently, for the diffusion length influenced by radiative recombination only, it follows that the diffusion length is 8.3 μm for this doping concentration. This is lower than that obtained from evaluation which does not consider PR (20-30 μm). On the other hand, it exceeds the values found from our measurements because in our samples the lifetime is probably reduced due to additional effect of non radiative recombination processes. Regarding the low level of doping in n-GaSb base we consider the Auger recombination is not significant. On the other hand, Shockley-Read-Hall recombination is expected to result in diffusion lengths as low as 0.6 μm.

The value of the L_2 should correspond to L (1 μm) obtained from the Schottky diode because, in fact, it characterizes the substrate as well. The reduction of the diffusion length in the n-type bulk of the p-n junction is probably the result of thermal treatment during the p-n junction preparation. Similar values of diffusion length (1-1.3 μm) on Au/n-GaSb Schottky diode obtained from EBIC and photoresponse spectra were reported [38].

In the p-type epitaxial layers $L_1 = 0.55$ μm and 0.40 μm for slowly and fast grown epitaxial layers were found, respectively. There is a wide range of almost spectrally independent response in the first case caused by PR. On the other hand, the spectra of the fast deposited photodiodes demonstrate a sharp decrease of photoresponse towards the short wavelengths. The evaluation shows that PR is practically missing. Higher rate of growth leads to higher recombination center concentration, which has been shown by dark I-V characteristic measurements [35]. Such recombination centers probably cause that non-radiative recombination is the main recombination mechanism, thus PR is suppressed.

Effect of the Thickness of Sample and Properties of its Surface

We would like to note that the diffusion length which was derived and/or experimentally determined in the previous sections may be influenced by effect of sample thickness and properties of its surface. Evaluation of L from the SPV for thin wafers always represents a problem because it reflects not only properties of the material but also the state of its surface and, consequently, becomes an effective quantity L_{eff} depending on the thickness d of the sample.

An attempt to find an analytical form of L_{eff} versus d dependence has been presented in [39] assuming homogeneous samples and homogeneous photogeneration rate. The surface represents a sink for the photocarriers and causes their redistribution which can be found by solving the diffusion equation for the photogenerated carriers Δn given by

$$\frac{d^2}{dx^2}\Delta n - \frac{\Delta n}{L^2} = -\frac{g}{D} \qquad (47)$$

with the boundary conditions

$$D \frac{d}{dx} \Delta n \Big|_{x=0} = s \ \Delta n(0) \tag{48}$$

$$D \frac{d}{dx} \Delta n \Big|_{x=d} = -s \ \Delta n(d) \tag{49}$$

D is the ambipolar diffusion coefficient. The solution of (47) with respect to (48) and (49) gives distribution $\Delta n(x)$. The current density flowing to the surface ($x = 0$) is

$$J_n(0) = eD \frac{d}{dx} \Delta n \Big|_{x=0} =$$

$$= egLS \frac{2S + (1-S)\exp(-d/L) - (1+S)\exp(d/L)}{(1-S)^2 \exp(-d/L) - (1+S)^2 \exp(d/L)} \tag{50}$$

For a thick sample ($d/L >> 1$) it holds that

$$J_n(0) = egL\, S/(1+S) \tag{51}$$

Comparing (50) and (51) we see that for a sample with arbitrary d/L ratio the diffusion length is represented by L_{eff}, where

$$L_{eff} = L(S+1) \frac{2S + (1-S)\exp(-d/L) - (1+S)\exp(d/L)}{(1-S)^2 \exp(-d/L) - (1+S)^2 \exp(d/L)} \tag{52}$$

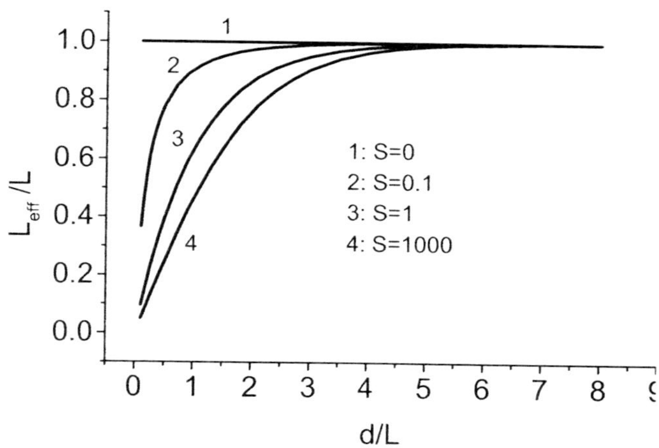

Figure 31. Ratio of the effective diffusion length (L_{eff}) and the true diffusion length (L) calculated according to Eq. (52) plotted against the ratio of d/L (d is the thickness of the sample, the parameter $S = sL/D$).

For $S = 0$ L_{eff} is equal to L independently of the thickness of the sample in agreement with our conception. The L_{eff} is the quantity that is obtained experimentally. In Figure 31 the ratio L_{eff}/L is plotted against d/L with S as a parameter. For $S \neq 0$ the ratio L_{eff}/L increases with increasing d/L and for large d/L approaches unity. If $d \geq 4L$, the difference between L_{eff} and L is less than 5% for arbitrary S. This substantiates the assumption concerning the thickness of samples in ASTM Standard [8]. Progress in technology leads to longer diffusion lengths and the mentioned assumption is often not fulfilled. For this reason a table was added to the Standard that enables one to prescribe the true values of L (corresponding to thick samples) for the measured values L_{eff}.

CONCLUSION

In this chapter photocurrents and photovoltage are calculated for illumination of the sample from the side of the SCR or from the bulk. In some cases, the photogeneration of the charge carriers includes multiple incoherent reflection of the light. For the direct band-gap materials a simple theory of the spectral dependence of photovoltage considering photon recycling is derived. In this case, total photocurrent is a sum of primary current and the secondary one resulting from the primary e-h pair recombination.

The SPV method used here proved to be convenient for minority carrier diffusion length evaluation even for thin samples. Using arrangement with an auxiliary electrode to the bulk we have reduced the number of parameters in the relation describing the measured structures. Application of illumination from back side of the cells means that the most of photocarriers are generated in the absorber. In this case diffusion current is prevailing component of the generated photocurrent with diffusion length as a main parameter. The same arrangement excluding contribution from upper spontaneously created SCR was applied to wafers. The diffusion length in the photocells and the wafers could be easily obtained by this way. Good agreement of experiments with presented theory was attained.

The diffusion length was evaluated for the following materials and structures: silicon wafers, silicon solar cells, microcrystalline μ-Si:H layers, CdS/CdTe heterojunctions on the basis of thin layers, GaSb Schottky diodes and homojunctions. The relations containing the multiple reflection were used in the case of bulk silicon because of low light absorption of this material and the typical value of L for the Si cells was 70-90 μm. Some of the silicon wafers were studied in electrochemical cell, which enabled us to compare the diffusion lengths determined using dependence of photon flux density versus reciprocal absorption coefficient with the result of our calculations. It was shown that the "classical" ASTM method can be used at low light levels only. The diffusion length in various types of the wafers was in the range from 225 to 300 μm. The SPV technique to advantage was employed for thin (2 μm) μ-Si:H layers. Fitting the calculation according to our model to the experiment we have obtained $L = 0.7$ μm.

The thin film CdS/CdTe heterojunctions were illuminated from the side of CdTe absorber or from the side of wide-gap CdS. Diffusion length in CdTe was 0.43 μm and surface recombination velocity written as $S = sL/D$ was 0.9. Illumination of the structure from the side of CdS yielded values of gain factor and thickness of the SCR of the junction. Low value of the gain factor shows a substantial recombination in the SCR probably originating in difference of the lattice parameters of the materials.

Our simple theory of photon recycling was applied for evaluation of the parameters of GaSb Schottky diodes and p-n junction diodes. Very small decrease of the photovoltage signal at short wavelengths for Schottky diodes illuminated through their bulk cannot be explained not considering the photon recycling process. Diffusion length in n - bulk of the p-n junction diodes as large as 20-30 μm was obtained fitting the theory disregarding this process whereas L = 0.6 μm was evaluated using our theory. Value of L = 8.3 μm calculated by means of the data published in the literature is higher than our result. Shockley-Hall-Read non-radiative recombination is expected to cause this reduction. The diffusion length in p-type part of the junction was about 0.5 μm. Other parameters as the surface recombination velocity S and the gain factor were also determined.

In general, diffusion length may be influenced by effect of the sample thickness and its surface so that L in thin samples becomes an effective quantity L_{eff}. The function describing dependence of L_{eff} on the sample thickness is presented which enables the finding of correct diffusion length from measurements of two samples with different thickness.

Acknowledgements

Complete text of "Influence of photon recycling on photovoltage spectra of GaSb diodes", figures and equations concerning of diffusion length in CdTe and properties of μ-Si:H were reprinted from our articles in JAP (cited in the reference list as [22, 28. 32]) with permission from Office of the Publisher, Journal and Technical Publications.

Figures and equations from our articles listed as [25, 26, 39] were reprinted with kind permission from Elsevier.

References

[1] Kronik, L; Shapira,Y. *Surf. Sci. Rep.* 1999, 37,1-206.

[2] Goodman, A.M. *J.Appl. Phys.*1961, 32, 2550-2552.

[3] Brillson,L.J. *Surf. Sci. Rep.*1982, 2,123-326.

[4] Lagowski,J.; Edelman,P.; Dexter,M.; Henley,W. *Semicond. Sci. Technol.* 1992,7, A185-192.

[5] Beaudoin,F.; Simand-Normandin,M; Meunier,M. In *Recombination Lifetime Measurements in Silicon;* Gupta,D.C.; Bacher,F.R.; Hughes,W.M.; Ed.; ASTM Stock STP1340; ASTM West Conshohocken, PA USA,1998; 219-225.

[6] Moore,A.R. *J.Appl. Phys.*1983, 54, 222-228.

[7] Schroder,D.K.; *Semiconductor material and device characterization*; John Wiley & Sons, Inc. New York, NY USA,1990; 359-447.

[8] *Annual book of ASTM Standards* ; ASTM F 391-96; American Society for Testing Materials, Pholadelphia USA, 1990;1-9.

[9] Chiang,Ch.L.; Wagner S. *IEEE Trans. El. Devices* 1985, ED-32,1722-1726.

[10] Gee, J. M. In *Proceedings of the 25th IEEE PVSEC Conf.*; WA, USA,1996, 437-440.

[11] Aspnes,D.E. In *Properties of Silicon*; EMIS Datareview series No 4, Inspec London, GB,1988; 72-79.

[12] Moss,T.S. *Proc. Phys. Soc.* Sect.B; London, GB, 1957, 70, 247-250.
[13] Landsberg,P.T. *Proc Phys. Soc.* Sect.B; London, GB, 1957, 70, 1175-1176.
[14] Dumke, W.P. *Phys.Rev.* 1957, 105, 139-144.
[15] Kuriyama,T.; Kamiya, T.; Yanai,H. Jpn. *J. Appl. Phys.* 1977, 16, 465- 477.
[16] Roedel, J.; Keramidas, V.G. *J. Appl. Phys.* 1979, 50, 6353-6362.
[17] Badescu, V.; Landsberg , P.T. *Semicond. Sci. Technol.* 1997, 12, 1491-1497.
[18] Parrott, J.E. *Sol. Energy Mater. Sol. Cells* 1993, 30, 221- 231.
[19] [Stollwerck, G.; Sulima, O.V.; Bett, W.A. *IEEE Trans. Electron Devices* 2000, 47, 448- 457.
[20] Badescu,V.; Landsberg,P.T. *Semicond. Sci. Technol.* 1993, 8,1267- 1276.
[21] Enders,P. *Phys. Status Solidi* 1986, B 137, 353- 360.
[22] Toušek,J.; Toušková,J.; Hulicius,E.; Šimeček,T.; Pangrác,J.; Melichar,K.; Výborný,Z.; Jurka,V. *J. Appl. Phys.* 2004, 95, 5104-5110.
[23] Aspnes, D.A.; Studna, A.A. *Phys. Rev.B*, 1983, 27, 985-1009.
[24] Adachi, S. *J. Appl. Phys.* 1989, 66, 6030-6040.
[25] Toušek,J.; Kindl,D.; Toušková,J. *Sol. Energy Mater. Sol. Cells* 2000, 64, 29-35.
[26] Toušek,J.; Kindl,D.; Toušková,J.; Dolhov,S. *Sol. Energy Mater. Sol. Cells* 2001,69, 297-302.
[27] Toušek,J.; Kindl,D.; Toušková,J.; Dolhov,S. In *Proc. 17th PVSEC*; McNelis,B.; Palz,W.; Ossenbrink,H.A.; Helm,P.; Ed.; WIP-Renewable Energies: Munich, Germany, 2001; Vol. 1, 143-145.
[28] Švrček, V.; Pelant, I.; Kočka, J.; Fojtík, P.; Rezek, B.; Stuchlíková, H.; Fejfar, A.; Stuchlík, J.; Poruba, A.; Toušek, J. *J. Appl. Phys.* 2001, 89, 1800-1805.
[29] Bloss, W. H.; Pfisterer, F.; Schubert, M.; Walter, T. *Progress in Photovoltaics* 1995, 3, 3-24.
[30] Toušková, J.; Kindl,D.; Toušek,J. *phys.stat.sol. (a)* 1994, 142, 539-549
[31] Toušková, J.; Kindl, D.; Toušek, J. *Thin Solid Films* 1997, 293, 272-276.
[32] Toušek, J.; Kindl, D.; Toušková J.; Dolhov,S.; Poruba,A. *J. Appl. Phys.* 2001, 89, 460-465.
[33] Vaněček, M.; Kočka, J.; Stuchlík, J.; Tříska, A. *Solid State Commun.* 1981, 39, 1199-1202.
[34] Amirtharaj, P. M. In *Handbook of Optical Constants of Solids II;* Palik, E. D.;Ed.; Academic New York, USA, 1991;670-675.
[35] Toušková, J.; Kindl, D.; Blahitka, B.; Toušek, J.; Hulicius, E.; Pangrác, J.; Samochin, E.; Šimeček, T.; Výborný, Z. *Solid - State Electr.* 2003, 47, 1471-1478.
[36] Stollwerck, G.; Sulima. O.; Bett, A. W. *IEEE Trans. Electron Devices* 2000, 47, 448-457.
[37] Vul, A.Y. In *Handbook Series on Semiconductor Parameters* ; Levinshtein, M.; Rumyantsev, S,; Shur, M.; Ed.; World Scientific, Singapore, 1996;1, 125-146.
[38] Polyakov, A. Y.; Stam, M.; Milnes, A. G.; Schlesinger, T. E. *Mater. Sci. Eng. B* 1992, 12, 337- 343.
[39] Toušek, J.; Dolhov, S.; Toušková, J. *Sol. Energy Mater. Sol. Cells* 2003, 76, 205- 210.

INDEX

A

B

C

D

E

N

O

P

Q

R

U

V

W

X

Z